# Synchronization of EEG Activity in Epilepsies

A Symposium Organized by the
Austrian Academy of Sciences, Vienna, Austria
September 12–13, 1971

Edited by
H. Petsche and Mary A. B. Brazier

Springer-Verlag
Wien New York 1972

Prof. Dr. H. Petsche
Neurological Institute of the University of Vienna
and Brain Research Institute of the Austrian Academy of Sciences

Prof. Dr. Mary A. B. Brazier
Brain Research Institute, University of California, Los Angeles, U.S.A.

© 1972 by Springer-Verlag/Wien
Softcover reprint of the hardcover 1st edition 1972
Library of Congress Catalog Card Number 72-75730

With 157 partly colored figures

ISBN-13:978-3-7091-8308-3     e-ISBN-13:978-3-7091-8306-9
DOI: 10.1007/978-3-7091-8306-9

# Foreword

The onset of an epileptic seizure has become a matter of prime interest in the last two decades. For successful therapy it is of the greatest importance to understand how the different inhibitory mechanisms involved in the normally functioning brain are carried out, and how the devastating avalanche of the seizure activity may overrun wide areas of grey matter. This problem of how an attack may arise has been dealt with in several monographs, but from different points of view: on the one hand chiefly from the viewpoint of the clinical EEG (GASTAUT et al. 1969), on the other hand from the viewpoint of experimental epilepsy, starting with the activities observed at the level of the nerve cell (JASPER et al. 1969).

This volume has quite a different perspective. It contains the papers and discussions presented at a Symposium organized by the Austrian Academy of Sciences, the aim of which was to arrive at a better understanding of some of the electrical phenomena accompanying the epileptic seizure.

A centre of interest was the problem of synchronization. This term—familiar enough to electroencephalographers and clinicians—was originally coined to describe the basic electrical mechanisms which cause large regions of the brain to be brought into apparently uniform activity, i.e., to a kind of unison of cellular activity. But even in the gross EEG, the tendency towards acting uniformly may be seen in the similarity in shape of the component waveforms over large regions. A more detailed study of these phenomena, however, has shown that the basic events underlying this apparently uniform activity are extraordinarily complex. Therefore, no simple concept of "synchronization" is any longer sufficient to represent exactly enough the mechanisms of apparently stereotyped brain function.

It has become clear that in order to study these mechanisms in more detail, not only physiological data are needed but a comprehensive knowledge of the cortical circuitry is also indispensable. Moreover, a close collaboration with mathematicians and physicists must be sought if these processes are to be understood. The programme was therefore designed to include these facets of the main problem.

The phenomenon of "synchronization"—the appropriate nomen-
clature for which met with difficulties even among the panel members
themselves—is not only of interest in the framework of the epilepsies.
Seizures offer the optimal condition for both displaying and studying
this phenomenon but synchronization is also a basic phenomenon of
the EEG itself. It would not be possible at all to record an EEG
with the usual scalp electrodes were there no tendency of bioelectrical
unitary events to group roughly simultaneously over extended brain
areas. The EEG, as a kind of summated activity, would not be
recordable if the electrical units discharged at random. Thus, this
symposium deals not only with the origin and spread of seizure itself;
it is just as much concerned with the electrical signs that constitute
the EEG.

H. Petsche
Mary A. B. Brazier

# Contents

# List of Participants

ANGELERI, F., Università di Perugia, Italia.

BANCAUD, J., Service de Neurochirurgie fonctionnelle, Paris, France.

BRAITENBERG, V., Max-Planck-Institut für biologische Kybernetik, Tübingen, BRD.

BRAZIER, MARY A. B., University of California, Los Angeles, U.S.A.

CALVET, J., C.H.U. Pitié-Salpêtrière, Paris, France.

CASPERS, H., Universität Münster, BRD.

ELUL, R., University of California, Los Angeles, U.S.A.

FLEISCHHAUER, K., Universität Bonn, BRD.

FREY, Zs., Postgraduate Medical School, Budapest, Hungary.

GERIN, P., Institut National de la Santé et de la Recherche Medicale, Lyon, France.

GLOOR, P., Neurological Institute, Montreal, Canada.

HERZ, A., Max-Planck-Institut für Psychiatrie, München, BRD.

KUHNT, U., Max-Planck-Institut für biophysikalische Chemie, Göttingen, BRD.

LEHMANN, D., Universität Zürich, Schweiz.

LUX, H. D., Max-Planck-Institut für Psychiatrie, München, BRD.

MARCHESI, G. F., Università di Perugia, Italia.

NAQUET, R., Institut de Neurophysiologie et de Psychophysiologie, C.N.R.S., Marseille, France.

PERONNET, F., Institut National de la Santé et de la Recherche Médicale, Lyon, France.

PETSCHE, H., Neurologisches Institut der Universität und Hirnforschungsinstitut der Österreichischen Akademie der Wissenschaften, Wien, Österreich.

PFURTSCHELLER, G., Technische Hochschule, Graz, Österreich.

RAABE, W., Max-Planck-Institut für Psychiatrie, München, BRD.

RAPPELSBERGER, P., Hirnforschungsinstitut der Österreichischen Akademie der Wissenschaften, Wien, Österreich.

RIEGER, H., Universität Mainz, BRD.

SCHERRER, P. J., C.H.U. Pitié-Salpêtrière, Paris, France.

SERVIT, ZD., Czechoslovak Academy of Sciences, Praha, ČSSR.

SHAW, J. C., MRC Clinical Psychiatry Research Unit, Chichester, England.

SINDOU, M., Institut National de la Santé et de la Recherche Médicale, Lyon, France.

SPECKMANN, E. J., Universität Münster, BRD.

SZENTÁGOTHAI, J., University of Budapest, Hungary.

TÖMBÖL, T., University of Budapest, Hungary.

TRAPPL, R., Universität Wien, Österreich.

VERZEANO, M., University of California, Irvine, U.S.A.

VITAL-DURAND, F., Institut National de la Santé et de la Recherche Médicale, Lyon, France.

VOLANSCHI, D., Romanian Academy of Sciences, București, Romania.

ZIEGLGÄNSBERGER, W., Max-Planck-Institut für Psychiatrie, München, BRD.


## Sponsors

Österreichische Akademie der Wissenschaften
Bundesministerium für Wissenschaft und Forschung
Comesa, Wien
Creditanstalt-Bankverein
Erste Österreichische Spar-Casse
CIBA, Wien
Geigy, Wien
Hoffmann-La Roche, Wien
Wander, Wien
Beckmann-Instruments, Wien
Siemens-Reiniger-Werke, Wien

# Comparison of Different Cortices as a Basis for Speculations on Their Function

V. BRAITENBERG

Max-Planck-Institut für biologische Kybernetik, Tübingen, Germany

For several years I have been interested in the visual ganglia of the fly and in their role as an automatic pilot during the flight of this humble animal. Very generously the organizers of this meeting, gave me an opportunity to participate even if I could not bring any evidence of seizures in flies. I feel it is only fair, then, that I should repay their magnanimous view of the field of neurology with some remarks sufficiently broad to embrace both flies and men. I shall use the concept of cortex as a convenient bridge between such distant objects of research.

*The Griseum.* Cortices are made of grey matter. A cortex is, therefore, a delimitable piece of grey matter of the brain or, borrowing a term from O. VOGT, a cortex is a *griseum* (we shall disregard for the moment the possibility that a cortex, such as the cerebral cortex of man, may be composed of various areas each representing a separate griseum). Before we discuss cortices, it may be opportune that we ask ourselves whether this concept of the griseum has any meaning in a functional, rather than classificatory neuroanatomy.

In order that we may recognize it as something unitary, it is necessary that a griseum have some structural traits that are recognizable throughout its extent. The density of neurons, the average orientation of their processes, the proportion of myelinated vs. unmyelinated fibers are examples of structural characteristics that are used to define the physiognomy of a griseum in descriptive neuroanatomy. These structural traits have another aspect: they imply that throughout the griseum the interactions of the neurons are of a similar kind. Reasoning in terms of patterns of excitation, this means that a constellation of active neurons in the griseum, representing a message, is subjected to a transformation which is the same for all parts of the pattern of excitation, being locally defined by the connectivity of the neurons in the griseum. Stated more simply, the structural homogeneity implies a homogeneity of the function performed.

Another aspect of the griseum may be relevant especially in connection with the main theme of this symposium, synchronization and epilepsy. The most trivial delimitation of a griseum is given by the white substance which surrounds it. Within the griseum communication is mediated by the synaptic network between one griseum and the next, on the other hand the message is transmitted generally through myelinated fibers, *i.e.*, the message is coded into spatio-temporal patterns of spikes in bundles of fibers. In terms of electro-physiology, I think everybody agrees that graded depolarizations, field effects and volume conduction play a very important role within the synaptic network of a griseum but only a very limited role in the white substance. To use an anthropomorphic expression for lack of a more precise definition, neurons within a griseum are much more intimately connected to each other than neurons of different grisea. The state of a neuron within a griseum depends on the states of other neurons in the same griseum in a way defined perhaps by "collective modes" of oscillation, to borrow a term from physics, *i.e.*, in a manner of reciprocal dependance conceptually quite different from the *projection* of one set of neurons onto another, which may be the more appropriate description for the dependence of one griseum on another.

Finally, in many instances we find the coordinates of some sensory space (not necessarily spatial coordinates) mapped onto the anatomical coordinates of a griseum. The griseum is thus an image of the sensory space, and its delimitation and internal consistency reflects the limitation and internal consistency of the sensory space represented.

*Cortices.* Most of the grey substance of the human brain, most of the grey substance of the frog's brain as well as of the fly's brain appears in the form of cortices. These are uniform sheets of synaptic tissue, mostly clearly layered, which is to say, a certain pattern of synaptic relationships between neurons, reflected by the layering, is repeated over the entire extent of the sheet. Input and output are, as a rule, connected to different layers of the sheet, and the input derived from an orderly arrangement of elements somewhere else (*e.g.*, in a sense organ) is projected onto the cortex in a manner which reflects the order of the original arrangement. The homogeneous arrangement of fibers over the sheet indicates that the interaction of inputs to give an output depends only on their relative position, not on where it occurs in the sheet. Thus the input is subjected to a transformation which is locally defined, at least to a first approximation. To what extent collective modes of activity, boundary effects and other effects not immediately related to the homogeneous fine structure play a role in the function of cortices, is not known, as far as I know, in any particular case.

Cortices are found in the brains of many very different animals, connected with various sense organs and sometimes not clearly connected with any one sense in particular. Molluscs, insects, crustaceans, and vertebrates have nervous tissue organized as cortices sufficiently similar in their general layout, to make it difficult for a non-neuroanatomist to distinguish microphotographs taken from the brains of such widely different species as flies and frogs. We select a few examples of cortices in order to exemplify possible criteria for their classification. The examples are: the cerebral cortex of man, the cerebellum of man, the tectum of the frog, the lamina ganglionaris (the first visual ganglion) of the fly, the medulla (the second visual ganglion) of the fly and the lobuli (the third visual ganglion) of the fly.

*Type of Symmetry.* Although in the plane of the sheet of any cortex a certain constant pattern of connections is repeated, there are differences describable in terms of the symmetry of the structure. A convenient pragmatic definition of symmetry can be obtained by considering the orientations and localizations of histological sections which can or cannot be distinguished by a histologist. Thus, neglecting the architectural areas, in the cerebral cortex of mammals I can tell the angle at which a histological section has been cut with respect to the surface of the cortex (any angle between a "vertical" and a "tangential" or "horizontal" section), but I cannot tell apart two vertical sections cut at right angles to each other: to a first approximation, at least, the connections in the plane of the cortex are the same in all directions. The tectum of the frog, as far as is known, appears to be of the same type, which we may call *isotropic continuous*. On the contrary, the cerebellar cortex is *anisotropic continuous*: it is clearly possible to distinguish preparations cut in the direction of the folia from preparations cut at right angles to the folia. But we cannot tell from what point of the folium, or from which folium the section was taken, since the structure is continuous in both directions.

The three visual ganglia of the fly show a curious pattern of symmetry, unfamiliar to the vertebrate neuroanatomist. The whole structure, in each case, is basically built up of identical columns arranged in a hexagonal array which maps the hexagonal array of the lenses in the compound eye. The individual column, however, is clearly asymmetrical, so that the orientation of a section can be established even on a small fragment of tissue. The *finite translational symmetry* of this arrangement is further complicated by the mirror symmetry which is valid not only, as in other brains, between the right and left half of the brain, but also between the upper and lower half of each ganglion.

We may try to make functional sense out of these very general remarks on the symmetry of nerve nets. In the cerebral cortex the same basic operation is obviously performed everywhere, in a continuous fashion, and the interactions are the same in all directions. In the cerebellum, clearly, the two rectangular (antero-posterior and latero-lateral) coordinates have totally different functional meaning, but again the basic operation, judging from the homogeneity of the structure, is precisely the same everywhere. In the ganglia of the fly, the periodic arrangement of the synaptic network, together with its asymmetry, leads us to expect oriented small range computations, perhaps connected with the basic asymmetry of the most frequent visual perception of a flying fly, the backward streaming of the environment in the visual field.

*Gradients.* Irrespective of the basic repetitiveness of a structural scheme over the entire extent of a cortex, certain structural traits may show quantitative variation. Thus there is an increase of the density of myelinated fibers in the strias of Baillarger from the frontal pole to the central sulcus of the human brain, and in the fly's ganglia there are very evident gradients of the thickness of some components of the columns. When we succeed in putting these gradients in relation with some quantity which varies similarly in the sensory space represented, we may gain insight in the details of the basic mechanism represented by that cortex.

*Size of Different Cortices.* The surfaces of the cortices of our sample vary by more than a factor of one million. The following table gives the approximate maximum and minimum diameters and the approximate surface area.

| | | |
|---|---|---|
| Cerebral cortex, man | 30 cm × 30 cm | 1000 cm² (×2) |
| cerebellum, man | 17 cm × 120 cm | 2000 |
| tectum, frog | 0.4 cm × 0.3 cm | 0.12 cm² (×2) |
| lamina, fly | 0.1 cm × 0.04 cm | 0.004 cm² (×2) |
| medulla, fly | 0.07 cm × 0.03 cm | 0.002 cm² (×2) |
| lobuli, fly | 0.04 cm × 0.02 cm | 0.0008 cm² (×2) |

Does this enormous range of variation really imply that a fly can digest only about one millionth of the information which a man can handle? Or is the operation performed a much simpler one in the case of the fly? One would be tempted to relate the complexity of the operation performed with the thickness rather than with the surface area of a cortex. The thickness, in our sample, varies much less than the surface, only by one or two orders of magnitude, as can be seen from the following table. Lobulus and lobula of the fly are two parallel and intimately connected sheets which should be considered as one unit and infact in many insects are fused into one.

| | | |
|---|---|---|
| Cerebral cortex, man | | 2500 $\mu$m |
| cerebellar cortex, man | | 600 $\mu$m |
| tectum, frog | | 750 $\mu$m |
| lamina, fly | | 50 $\mu$m |
| medulla, fly | | 100 $\mu$m |
| (lobula, fly | 8 $\mu$m) | |
| (lobulus | 20 $\mu$m) | |
| lobuli, fly | | 100 $\mu$m |

Here the mammalian cerebral cortex holds the first place uncontested, but the differences of the thicknesses being on the whole not very impressive, one may conclude that nervous integration is quite complex wherever it occurs, be it applied to the vast body of human knowledge and experience or to the relatively modest problems of flight-navigation in the fly.

If, then, we consider thickness as representing the complexity of the operation, surface should be related to channel capacity in the sense of information theory. Is the amount of information which reaches a cortex proportionate to its size, or, in other words, is the density of information per unit surface area constant in large and small cortices? It is difficult and dangerous to estimate the information rate of a neural channel, but supposing the channel capacity of an axon to be roughly constant, the number of axons reaching the cortex may be a good measure for the amount of information which it receives. Here are three examples:

$$\text{medulla, fly} \quad \frac{15{,}000 \text{ fibers}}{0.2 \text{ mm}^2} = 75{,}000 \frac{\text{fibers}}{\text{mm}^2}$$

$$\text{tectum, frog} \quad \frac{400{,}000 \text{ fibers}}{12 \text{ mm}^2} = 33{,}000 \frac{\text{fibers}}{\text{mm}^2}$$

$$\text{area striata, man} \quad \frac{1{,}000{,}000 \text{ fibers}}{2600 \text{ mm}^2 \, [1]} \approx 400 \frac{\text{fibers}}{\text{mm}^2}$$

Two striking facts emerge from this comparison. The densities of fibers in the medulla of the fly and in the tectum of the frog are remarkably similar. On the other hand, the area striata seems to provide much more space per input fiber than the other two cortices. Obviously it subjects its input to much more elaborate transformations, or confronts it with a much more complex background information than is the case in the other two visual analyzers. We learn from this that not only the thickness of the cortex, but also its surface is related to the complexity of the operations performed.

But it is not simply the space provided that garantees the complexity of the information handling. What counts is the number of elements which are there to receive the input. Here the comparison of different cortices leads to surprising results. I select four examples for which relatively reliable cell counts as well as counts of input or output fibers are available.

cerebellum, man                  3,000 neurons/1 output fiber
area striata, man                 500 neurons/1 input fiber
tectum, frog                       1 neuron/1 input fiber
lamina, fly                        5 neurons/6 input fibers

(5 neurons/channel, since groups of 6 input fibers carry the same message).

The cell count in the striata is from Glezer (1960), in Blinkov and Glezer (1968). The cell count in the tectum is from Lázár and Székely (1967) and from Kemali and Braitenberg (1969). The figure for the fibers in the frog's optic nerve is from Maturana (1959).

It is difficult to make general sense out of the widely varying figures. Individually, the following comments can be made. The very large number of neurons in the cerebellar cortex is largely provided by the

---

1 Preobrazhenskaya and Filimonov (1949), from Blinkov and Glezer (1968).

granular cells. It is doubtful whether this large population of neurons, as numerous as all the neurons in the cerebral cortex, makes use of all the freedom which their large number would seem to provide, since certainly groups of them receive the same input and are connected to the same output cell. There may be good reasons for distributing messages in parallel onto a large number of identical lines, such as reliability or reduction of the signal to noise ratio. It would be difficult to reconcile the position of the cerebellum at the top of the list with the modest opinion which we have otherwise of the cerebellum, as compared to the cerebral cortex.

The figure of 500 neurons per input fiber in the striate area is not surprising in view of the large volume of grey substance per fiber which we found in this region. What needs explanation, on the other hand, is the very low ratio between interneurons and input fibers found both in the optic tectum and in the lamina of the fly. The latter may be little more than a relay station between the retina and the much more elaborate second visual ganglion. But the optic tectum is certainly the most impressive piece of grey substance of the frog's brain, dominating the scene in the amphibian as much as the cerebral cortex dominates the mammalian brain. Yet, the number of internal neurons in the tectum is not larger than the number of fibers in its principal input, the optic nerve. This is a remarkable fact in view of some theories of the function of brains, and in particular of perception, which operate carelessly with the number of available neurons. With impossibly large numbers of interneurons, too large to be housed not in a skull, but in the entire universe, every well defined task can be solved trivially. But the frog masters the task of perception with a remarkably small number of neurons, obviously using collective states, rather than individual neurons to map the states of the external world on. How this is done effectively, is a question which automata theorists will eventually have to answer.

*Conclusion.* Far removed as these remarks may be from the pressing questions of seizure activity in brains, I can only hope that they contribute to making three points more popular: (a) that any attempt at making our ignorance of the normal function of the brain less complete than it is now, may be a worth while approach to pathology; (b) that the contemplation of neuroanatomical structures may be a useful tool in neurophysiology and (c) that quantitative reasoning, even at this crude level, is always apt to introduce new insights into neuroanatomy.

## Summary

As an exercise in quantitative comparative neuroanatomy, the cerebral cortex of man, the cerebellar cortex of man, the tectum of the frog and the three visual ganglia of the fly are compared. The surfaces of these cortices vary by a factor of one million, their thickness much less, the number of input fibers per unit area is similar in the tectum and in the insect visual ganglia. The number of internal neurons is very large in the cerebellum, large in the cerebral cortex and equal to the number of input fibers in the tectum. Such comparisons would make good material for discussion in terms of automata theory.

## References

BLINKOV, S. M., and IL'YA I. GLEZER: The human brain in figures and tables. Basic Books, Inc. Publ., Plenum Press. 1968.

KEMALI, M., and V. BRAITENBERG: Atlas of the frog's brain. Berlin-Heidelberg-New York: Springer. 1969.

LÁZÁR, G., and G. SZÉKELY: Distribution of optic terminals in the different optic centers of the frog. Brain Res. *16*, 1—14 (1969).

MATURANA, H. R.: Number of fibers in the optic nerve and the number of ganglion cells in the retina of anurans. Nature *183*, 1406 (1959).

PREOBRAZHENSKAYA, N. S., and I. N. FILIMONOV: The occipital region. In: Cytoarchitectonics. Moscow: 1949.

# The Basic Neuronal Circuit of the Neocortex

J. SZENTÁGOTHAI

First Department of Anatomy, Semmelweis University Medical School,
Budapest, Hungary

## Introduction

Let it be assumed, for the time being, that something like a basic
neuronal circuit of the neocortex does indeed exist and that the well
known structural and connectivity differences between various
cortical regions do not render such a notion *a priori* meaningless.
Under basic neuronal circuit (or circuits) the fundamental neuronal
chain(s) is (are) meant that connect(s) the elements of input (afferents)
with the elements of output (efferents) of any grey matter under
consideration. Such a chain is rarely if ever a linear succession of
neurons: (i) The input channels break up often into several parallel
lines that finally converge again upon one or several types of output
elements; (ii) interneuronal connexions are often established between
the points of synaptic articulations of the parallel lines; (iii) these
connexions may be established between articulation points (vertices
of the network) of the same order (relative to the input elements), *i.e.*,
simply cross connexions or (iv) the connexions may be recurrent from
articulation points of higher order to those of lower order (often to
secure feedback couplings), else (v) they may feed forward from
lower order articulations of the chain to vertices of higher order. But
in spite of these complications even the most complex neuronal network can be reduced to a number of functional chains and sometimes
these separated chains can be made available for detailed physiological studies on the unit level. The cerebellar cortex might serve as an
example, where this mode of approach proved to be quite useful
(ECCLES *et al.* 1967). Admittedly the cerebellar cortex offers unique
advantages owing to the virtual uniformity of the structure in all of
its regions, almost geometric regularity in "wiring", and low number
(five) of types of elements (neurons); all strictly uniform in arbori-

zation pattern and in numerical, metrical, and topological parameters of connexions.

The neocortex, conversely, has at least 50 types of much less equalized elements, high regional diversities, and a much less regular—not to say virtually irregular—"wiring". And yet there is still a certain degree of principal similarity in the structural design: (1) a fairly uniform principle of lamination, (2) a relatively uniform main cell type: the pyramids, (3) certain characteristic types of interneurons or Golgi Type 2 cells, (4) an essential similarity in the organization of input channels: association afferents, commissural afferents, specific and non- (or less) specific subcortical afferents, and (5) an essential similarity in the organization of the output lines, mainly the axons of pyramid neurons. This gives us the confidence that in spite of obvious differences in detailed structure and more even in connexions with other regions of the CNS, certain "units" of neocortical tissue might be built on the basis of the same fundamental principle, i.e., they might be essentially similar as devices for processing neural information. Here we have arrived at a crucial point: what can be considered as a "tissue unit" in the CNS? The elementary unit of nervous tissue is obviously the neuron, however, in considering the grey matter as a device for information handling we have to look for larger units. Some years ago I suggested the term "higher integrative unit" (SZENTÁGOTHAI 1967) for the smallest assembly of neurons that might be able to perform the basic task in information processing that this piece of neural tissue normally is built for. Thus I have tried to specify the higher integrative unit in the spinal cord, in the specific thalamic nuclei, in the cerebellar, and cerebral cortex. In the spinal cord it is generally a group of specific segments—branchial segments for anterior limb movement, lumbosacral segments for posterior limb movement—which satisfies the term higher integrative unit. In the specific sensory nuclei of the thalamus or the geniculate bodies the higher integrative unit would be the assembly of cells involved in processing the information arising from a receptive field. In the cerebellum the cells activated by a beam of parallel fibers (SZENTÁGOTHAI 1963, 1965 a, ECCLES et al. 1967) might be considered as such a higher tissue unit. In the neocortex the integrative unit would correspond to the columns of the primary sensory cortices established by MOUNTCASTLE (1957) and HUBEL and WIESEL (1962).

It will be readily understood that the two concepts of the "basic neuronal circuit" and "higher integrative unit" are different in start-ing point and definition, but nevertheless they are strongly convergent and lead up eventually to the same mode of looking at a certain piece of neural tissue. It is in this sense that a look at the principal neuro-nal arrangement of the neocortex will be presented.

## The Development of the Concept of Neuronal Circuits

Many important details in synaptic arrangement of neurons in the neocortex were already correctly understood and interpreted by RAMÓN Y CAJAL (1911). Particularly the arrival of specific sensory impulses in middle layers of the cortex, and the possibility to have direct synaptic connexions with basal dendrites of pyramidal cells is indicated in some of his diagrams. RAMÓN Y CAJAL and most other contemporary authors, even those of the cytoarchitectonics school, speculated about the possible significance of the large numbers and diverse types of short interneurons. LORENTE DE NÓ (1938 a) was the first to offer realistic concepts in terms of neuronal circuit by emphasizing two principles of interneuron arrangement: (i) the multiplicity (parallel arrangement) and (ii) the reciprocity (recurrent nature) of connexions. This aspect of neural organization has been considered already in the first paragraph of the Introduction. It is quite understandable that LORENTE DE NÓ became concerned at this stage mainly with the idea of so called "reverberating circuits". These were (and perhaps still are) essential tools that could be envisaged to produce certain timings of events and synchronizations. Lacking, at that time in crucial information about the existence of specific in-hibitory interneurons, authors were led to speculate about the pos-sibility of inhibition being produced by impulses circulating in the interneuronal network, possibly by the Vvedensky mechanism (LO-RENTE DE NÓ 1938 b). Obviously, these general concepts have lost much of their appeal today, when a number of elementary neuronal mechanisms of recurrent, parallel, and feed-forward inhibitions have become known in most parts of the CNS, and when the great wealth of mutual subcortico-cortical connexions offers an almost infinite number of possible neuronal loops in which reverberation of impulses might be arranged at much larger scale. But this eventually renders the original concept of reverberating circuits so universal and com-pletely lacking in boundaries that it becomes almost useless for explaining any specific mechanism of information processing.

The concept of elementary neuronal circuits came into focus again when by the introduction of artificial neurons (so called neuromines) it became possible to simulate the function of elementary neuron networks containing few (2–5) neurons, connected mutually in simulating either excitatory or inhibitory interactions (Harmon 1964). A very elementary circuit model for walking movement patterns has been conceived on the basis of neuroembryological observations and tested with neuromine simulation by Székely (1968); later the theory has been extended by Kling and Székely (1968) and, finally, a general mathematical theory of such types of networks has been developed by Ádám (1968).

Transmission probability models for various types of geometric relations in CNS (mainly neocortical) neuropil have been proposed by Uttley (1955) on the basis of Sholl's (1956) quantitative dendroarchitectonics and arborization studies. However, these still suffered from the fundamental disadvantage of having to assume "quasi random" connectivity. Their fundamental importance lies in having called attention to the importance of quantitative approaches to neural tissue histology, and having brought the study of the cortex again into focus after decades of virtual stagnation.

Larger scale models of relatively extended pieces of specific neural tissue have been proposed for the cerebellar cortex (Szentágothai 1963, 1965 a) and for the specific sensory thalamic (and geniculate) nuclei (Szentágothai 1967) mainly on the basis of speculations starting from the geometrical (and topological) features of the networks and a trail-and-error procedure in successive steps of quasi arbitrarily made assumptions on the probable physiological qualities (whether excitatory or inhibitory) of the several neuron types. For the cerebellar cortex these speculations have been relatively well supported and extended considerably by unit level analysis (Eccles et al. 1967), at least for some time. In spite of the unique advantage offered by the cerebellar cortex with its geometric regularity of connexions, the anatomical assumptions about the network had to introduce considerable simplifications in randomness of certain connexions that do not seem now to be justified. A recent quantitative analysis (Palkovits et al. 1971 a, b, c) of the neuronal network of the cerebellar cortex suggests that each parallel fiber might contact synaptically only 5–6 of the roughly 250 Purkinje cells the dendritic trees of which it crosses. This means no less than that the possibility has to be envisaged, that instead of a quasi homogeneous population of

Purkinje cells embedded in a beam of parallel fibers, as assumed by the original concept (SZENTÁGOTHAI 1963, 1965 a, ECCLES *et al.* 1967) there might be at least five but probably more specifically different populations of Purkinje cells; different with respect to their input through mossy fibers. Although such a change does not completely invalidate the basic model of the "cerebellar neuronal machine", this development shows how careful we have to be assuming certain randomness or "non-specificity" in neural connexions.

A new vogue of studies of the neocortical structure has been introduced by the report of COLONNIER (1966) on the structural design of the cortex given at a conference of the Pontifical Academy in 1964. After earlier attempts to reduce dendritic and axonal ramification patterns to two fundamental cell types: pyramids and stellates, having few subtypes each, COLONNIER (1966) emphasizes the importance of numerous mainly Golgy Type 2 interneurons having extremely specific patterns of axonal and dendritic arborizations. Although these cell types have all been described previously by RAMÓN Y CAJAL (1911), they have more recently been neglected by the authors speculating about neuron circuitry. COLONNIER (1966) points out particularly the significance of bundles or of individual terminal axon branches having vertical orientations. It is obvious that such vertically oriented terminal branches would have ample opportunity to get into repeated "climbing-type" contacts with the apical dendrites of pyramidal cells. Such "climbing-type" or "rope ladder" contacts between vertical axons and the spines of apical dendrites have been described previously by Soviet authors (POLJAKOV 1955). By analogy it is reasonable to assume that such climbing type contacts would insure a very effective synaptic action of such vertical axons on the apical dendrites of single cells.

A more specific basic circuit model has been proposed later (SZENTÁGOTHAI 1967) on the basis of some new information gained from studies on isolated cortical slabs (SZENTÁGOTHAI 1962, 1965 b) particularly on the local intracortical origin of pericellular basket synapses surrounding pyramid cell bodies. A crucial bit of information has been added to this by the observation of COLONNIER (1968) that somatic synapses on pyramidal cells have generally flattened vesicles and, hence, according to the concept of UCHIZONO (1965, 1967) might be suspected as being of inhibitory nature (see also SZENTÁGOTHAI 1969). Eventually a host of most recent Golgi, EM level degeneration and EM neuropil architectonics studies (to be

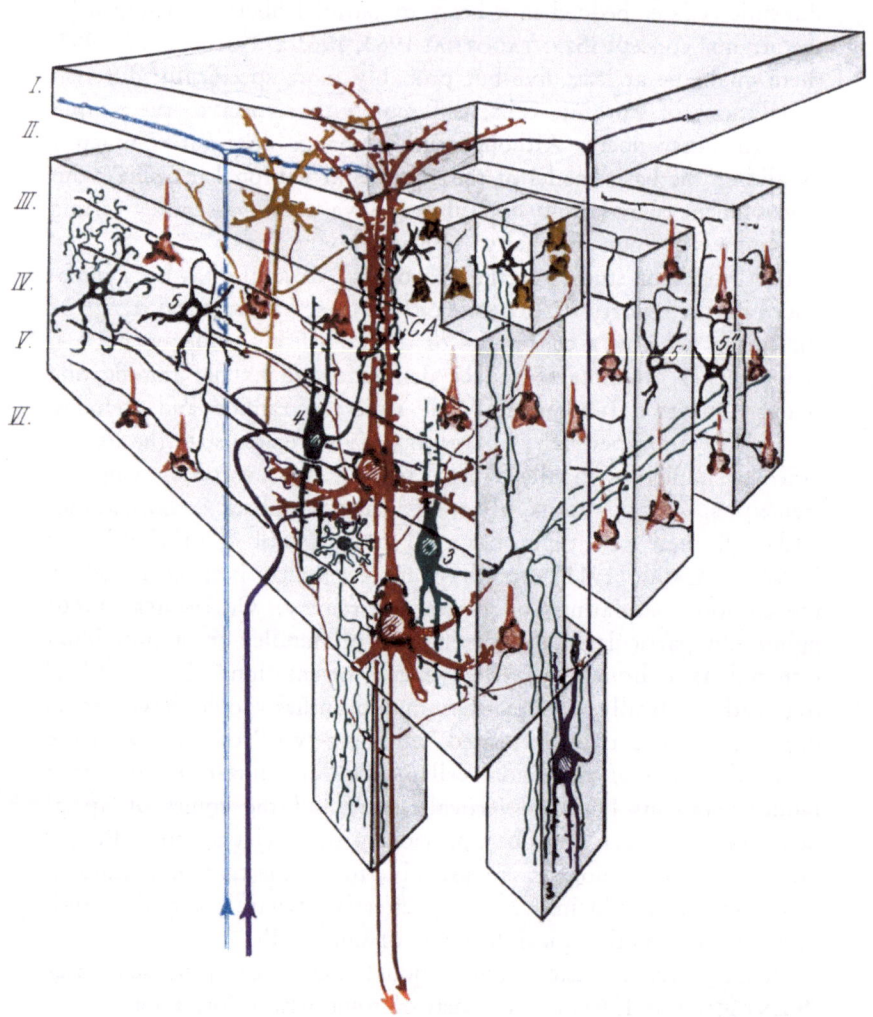

Fig. 1. Stereodiagram illustrating some of the general features of neocortical circuitry. The blocks of various shapes and sizes indicate definite spaces of cortical tissue (conventional layering is indicated at left margin) in which the cells (or their dendrites) share certain types of connexions. In lamina I apical dendrite branches of star-pyramid (yellow) and pyramid cells (red) have nothing but crossing-over contacts with the surface parallel axon system of this layer. In the vertical discs cutting through laminae III to V all pyramidal cells share common axosomatic input from the same basket cell(s) (in black 5, 5′, 5″). In the smaller cube-shape spaces of lamina II all star pyramids (yellow) share input from small basket cells (black). Vertical prisms of lamina VI correspond to spaces

discussed in the following Chapter) led over a number of stepwise modifications (SZENTÁGOTHAI 1969, 1970, 1971) to the still very hypothetical and preliminary circuit model to be presented.

## The Basic Circuit Model Proposed

The circuit indicated very diagrammatically in Fig. 1 is based on combining (1) classical and some more recent Golgi information on the general structure of the neocortex, particularly on arborization patterns of specific sensory and other afferents, (2) a host of light and EM degeneration observations on somatosensory (JONES and POWELL 1969), visual (COLONNIER and ROSSIGNOL 1969, GAREY and POWELL 1971), and on auditory (own unpublished) cortex sensory afferents; (3) Golgi and EM information gained from chronically isolated cortical slabs containing intact pieces of layers I + II, I + II + III, etc. up to I + II + III + IV + V + VI (SZENTÁGOTHAI 1962, 1965 b, 1972); (4) a new inventory of specific kinds of Golgi Type 2 neurons (SZENTÁGOTHAI 1972); (5) an attempt to analyse on the EM level neuropil geometry of lamina I and lamina III (SZENTÁGOTHAI 1969, 1971); and finally—(6) some crucial new observations of basket cell axon arborizations by MARIN-PADILLA (1970). The diagram is largely self-explanatory and needs little comment. Only a few basic features will be discussed here very briefly; they have been treated *in extenso* in some of our earlier papers indicated above.

a) *Afferents.* All EM degeneration studies (cited in the preceding paragraph) seem to indicate that the vast majority of afferents, both specific sensory and association as well as callosal, establish synapses with dendritic spines. This might be interpreted in favour of the assumption that they contact mainly pyramidal neurons. However, a recent analysis of many types of stellate cells shows that only

---

in which fusiform cells would receive powerful inputs from vertically oriented descending axon arborizations. Martinotti type cell is shown in violet colour. Several Golgi Type 2 cells are shown in various shades of green: 1 = general type, 2 = neurogliform type with descending axonal ramification, 3 = Golgi cell with wide field of action, 4 = cellule à double bouquet of RAMÓN Y CAJAL with ascending horse-tail-shape terminal ramification making climbing type multiple contacts with the dendrite shafts of two neighbouring pyramid cell. This synaptic system has been labelled as "cartridge synapse" = CA. Specific sensory afferent is indicated in dark purple, cortico-cortical association (and/or callosal) afferent in light-blue colour

relatively few are entirely devoid of spines, while many have quite numerous spines (SZENTÁGOTHAI 1972). In addition to this it is much easier to recognize axon spine synapses in degeneration due to strong and persisting connexions between the degenerated axon and the spine, which is generally shed by the dendrite when its axonal contact degenerates. Conversely, synapses between axons and dendrite surface or soma surface proper lose contact in early stage of degeneration, so that it becomes difficult to recognize the original site of contact. Thus the observed numerical difference in favour of degenerated axon-spine synapses may well be due partly to an error of sampling. Nevertheless, the occurrence of bundles of parallel terminal branches oriented obliquely, observed frequently in sensory afferents (SZENTÁGOTHAI 1969) may well support the notion that sensory afferents can have multiple contacts with mainly basal dendrites of pyramidal cells. As has been discussed earlier (SZENTÁGOTHAI 1969) it seems to be unlikely that specific sensory afferents have much access to apical dendrites of the pyramidal cells. —Both cortico-cortical association afferents and callosal afferents seem to have abundant terminations on spines in lamina I and II, however, it would need more systematic studies to find out, whether (i) this is general, and whether (ii) there are no other layers where the number of contacts is high also.

b) *Golgi Type 2 Interneurons.* The main types of interneurons, found especially in the primary sensory regions are included by one or two representative cells and are numbered consecutively. The *general type* of Golgi Type 2 interneurons (1) is local, its dendritic tree as well as its axonal arborization appears to have no specific orientation. The *neurogliform type* (2) has a spatially very restricted but otherwise extremely richly arborizing dendritic tree, its axonal arborization may be restricted to a small local space too, but it may be also long and sometimes oriented in vertically descending fashion. An important cell type also with respect to the obvious asymmetry of its axonal arborization is the *stellate cell with wide field of action* (3). Its axonal arborization crosses several layers and bridges distances up to 1 mm. It may have vertically ascending, descending, and several horizontally oriented groups of terminal branches (SZENTÁGOTHAI 1972). The classification of this cell type as well as of the following as Golgi 2nd type cells is certainly subject to argument. A particularly important interneuron is a cell described originally by RAMÓN Y CAJAL (1911) as *"cellule à double bouquet"* (4).

Although indicating clearly the predominantly vertically oriented axonal arborization, CAJAL did not fully appreciate the narrow horsetail shape strands of terminal fibers that these axons generally give rise to. These cells appear to be particularly abundant in sensory regions, but they have been found in almost every part of the neocortex (SZENTÁGOTHAI 1969, 1970, 1971). It is quite obvious that if such strands—as can often be observed directly in the Golgi preparations (SZENTÁGOTHAI 1969)—envelopes an apical dendrite of a pyramidal cell, this arrangement would enable the presynaptic neuron (cellule à double bouquet) to establish (very) many (in the order of hundreds) contacts with the same postsynaptic neuron (pyramid cell). Such an arrangement might secure an exceptionally effective contact between these two elements. Finally the *basket cells* (5) have a particular position as, so far, they are the only elements that can be suspected with reasonable confidence of being inhibitory. They will, therefore, be discussed in the following paragraph.

c) *Spatial Arrangement of Inhibition.* The diagram of Fig. 1 takes advantage of some important new observations by MARIN-PADILLA (1970) on the distribution of the axon arborization of individual basket neurons (5) within narrow discs cutting through laminae III–V and arranged in parallel with each other. If this is so one might envisage the spatial arrangement of inhibition in these layers as a succession of parallel discs in which most pyramidal cells might be either inhibited, or non-inhibited, according to whether the respective basket cell (or cells) of the several laminae were or were not activated. The span of the discs may be several hundreds of microns, however, their thickness appears to be small (probably few tens of microns). Another kind of basket cells in lamina II, indicated already by RAMÓN Y CAJAL (1911) and illustrated in more detail by SZENTÁGOTHAI (1969), have a much smaller span. If the distribution of their axons were as would appear from Figs. 2–3 D of my paper in 1969 a mosaic pattern of much smaller module would result by such cells being either activated or silent and in turn either block the star-pyramid cells of lamina II in their immediate neighbourhood situated within the cubes indicated in Fig. 1 or leave them free to be activated. This probably very simplistic view of the spatial distribution of inhibition results from virtual lack of information about other kinds of inhibitory interneurons. However, from the fact that both cell bodies and dendrites of assumed stellate cells and dendrites of pyramid cells are contacted partly by synaptic terminals containing smaller "flattened" synaptic vesicles (Colonnier 1968) one may infer that there exist other inhibitory interneurons, or even, that some of the distant afferents have inhibitory terminals on cortical neurons.

d) *Neuropil Geometry.* Synaptic connectivity appears partly to be very specific: *i.e.*, a given type of axon contacts one (or few) certain type(s) of cell(s). There is even a rather specific somatotopic arrangement within two sets of neurons connected, with respect to specific topographic position that each of the neurons in connexions holds within its own set. From this viewpoint it appears to be irrelevant, whether we consider connexions between distant, or, conversely, between relatively closely neighbouring neurons. We do not know, whether and if so how far this specificity in connexions is deterministic or only statistical. It would seem likely from many examples from different regions of the CNS, that it is deterministic with respect to the kind of neurons contacted, but statistical only with respect to which individual neurons are contacted from the whole set within the corresponding somatotopic region. Another aspect of synaptic connectivity, only secondarily related to the problem of specificity, is that of neuropil geometry. It is quite obvious that if the terminal axon branches cross the dendrites at close to right angles it is unlikely that more than one or two synaptic contacts are established between any axon and any dendrite. A "crossing-over" arrangement between terminal axons and dendrites is a rather usual type of synaptic connexion. Its clearest example occurs in the molecular layer of the cerebellar cortex between parallel fibers and Purkinje cells. In such cases the synapses are generally established between the fusiform thickenings of the axons passing by and the spines of dendrites (Hámori and Szentágothai 1964). Conversely, if the terminal axon branches are arranged in parallel with dendrites, repeated contacts may occur between the same terminal axon and dendrite leading to either a true "climbing-type" contact or to the so-called "rope ladder" phenomenon mentioned already. It is obvious that such a parallel arrangement of the neuropil would lead to a much higher probability of powerful synaptic connexions than in the case of crossing-over contacts. This side of neuropil geometry has been considered for the cerebral cortex (Szentágothai 1969, 1971) and it has been shown that lamina I can be considered largely as a "crossing-over" synaptic system.

The terminal branches of pyramidal cells are oriented at roughly 45° angles to the pial surface, hence the crossings with the surface parallel axonal neuropil would occur mainly at such angles. Preliminary studies on the synapses of this layer showed (SZENTÁGOTHAI 1971) that the far majority of synapses in this lamina are established with dendritic spines. As stellate (Golgi Type 2) neurons do occur in this layer, more detailed studies would be needed to appreciate, whether or not such cells are also contacted mainly by overcrossing or whether they have more specific synaptic connexions. As the axons of lamina I are either the terminal branches of ascending axons of cells situated in the deeper layers (V and IV; SZENTÁGOTHAI 1962, 1969, 1970) or mainly cortico-cortical (association or callosal) afferents one would have to assume that the connexions in this layer are less specific (or powerful)—or at least would need very considerable spatial summation in order to become effective.

Conversely in layers II, III, V, and VI much of the neuropil is vertically arranged. As mentioned already, little of this corresponds to the arborizations of afferents, which show little vertical arrangement of their terminal branches. The same applies to the initial collaterals of pyramidal neurons, the terminal branches of which tend to have rather oblique courses and do not seem to attach themselves in parallel to the dendrites of other cells. The vast majority of vertically oriented terminal axon branches belong to Golgi Type 2 cells as easily seen from the drawings of RAMÓN Y CAJAL (1911) and more recently from those of RUIZ-MARCOS and VALVERDE (1970). A specific cell in this respect is the "cellule à double bouquet" (4) especially those with narrow vertically oriented strands of parallel terminal branches. Such an arrangement as shown in Fig. 1 would secure an extremely strong drive by these cells of one or two neighbouring (or of those placed vertically below one another) pyramid cells. For an analysis of the neuropil immediately surrounding the apical dendrites of pyramid cells the impression is gained (SZENTÁGOTHAI 1969, 1971) that in addition to some crossing over synaptic contacts this is a specific synaptic system—termed "cartridge synapse"— in which vertically oriented axons mainly of stellate cells would play a crucial role.

## Comments and Conclusions

However preliminary, this attempt at outlining the skeleton of the neuronal machine of the neocortex may show that modern neuroanatomy is gradually achieving a position from which certain concepts about the functional organization of the neocortical neuronal network might be advanced. In spite of considerable difficulties due mainly to regional differences and the irregularity of architecture, on the gross scale, a combined approach—using the classical Golgi techniques, degeneration procedures both on the light and electron microscope level, an EM analysis of neuropil architecture, and taking advantage of long term isolation techniques rendered possible by the excellent blood supply of cortical structures from the pial surface— offers attractive perspectives for future progress along these lines. This approach is entirely qualitative, at present, and from the example of the cerebellar cortex (Palkovits et al. 1971 a, b, c) it became obvious that lacking precise quantitative information only a very limited understanding of any neuronal network can ever be reached. After the pioneering studies of Sholl (1956) and EM techniques developed recently by Molliver and Van der Loos (1969) there is no reason why exhaustive stereological analyses of well defined cortical regions could not be accomplished in the near future. One of the most serious obstacles to be overcome is the identification in the EM picture of both cell bodies, and particularly of synaptic terminals belonging to certain specific (local) neurons. But even this task should not offer today insurmountable difficulties with electrophoretic techniques available for injecting radioactive substances into physiologically defined neurons and by subsequently tracing their bodies and local synaptic connexions by EM autoradiography.

From the viewpoint of electrogenesis in cortical structures the following general anatomical features might be considered of potential significance:

1. Vertically oriented mass of apical dendrites of pyramidal cells— and their vertically descending main axons—creating a system of vertical dipoles.

2. A tangentially oriented terminal axonal system of about 5 mm maximal spread of individual elements (Szentágothai 1962, 1965 a) in lamina I (and partly also II) fed mainly by association (ipsilateral cortico-cortical) and callosal afferents, and by ascending axons of cells located mainly in lamina VI and V (Martinotti type cells). Synaptic connectivity in this plexus is of the "crossing over" type, probably requiring considerable convergence for effective transmission.

3. Another predominantly tangentially oriented axonal system at the

transition between laminae III and IV is present in the sensory cortical regions and is fed mainly by the preterminal branches of the specific sensory afferents. The terminal branches of this axonal system are not specifically oriented in any preferential direction—certainly not vertically or tangentially—although relatively numerous terminal branches take oblique directions or terminate in irregularly arranged clusters of terminal thickenings. The spread in tangential direction of individual specific afferent arborizations ranges probably between 200–600 $\mu$m.

4. All other horizontally oriented axonal elements are mainly of local nature, since they are found intact in chronically isolated cortical slabs. Their origin is heterogeneous: pyramidal cell axon collaterals, long axon stellate cell collaterals, wide range Golgi Type 2 cell (Golgi Type 2/3) axons, and basket cell (5) axons contributing mainly to this part of the neuropil. [A considerable portion of the horizontally oriented larger axons in the deeper layers (V and VI) are not collaterals but main axons projecting to neighbouring areas.]

Vertically oriented axons have to be divided into two major categories:

5. Efferent axons, predominantly of pyramidal neurons, are prone to join into bundles. This bundling may convey the impression of a vertical (pseudo-)columnar arrangement of cell bodies, however, it is probably largely a "packing" phenomenon and perhaps without major functional significance. Afferent fibers may contribute to the vertically oriented mass of axons only in laminae VI, in lamina V they assume their characteristic oblique courses.

6. The vast majority of the vertically oriented axonal neuropil is furnished by stellate axon arborizations with the following main constituents: (i) the strictly vertically oriented terminal axonal strands of the "cellules à double bouquet" (4), (ii) the axonal systems of numerous transitional cell types (not specifically mentioned in this paper), the terminal axonal arborizations of which are also vertically oriented but not arranged into compact thin strands, (iii) the vertically oriented part of terminal arborizations of the cells labelled Golgi Type 2 cells with wide field of action (3), (iv) vertically oriented neurogliform (2) Golgi Type 2 cell axonal arborizations, and (v) vertically oriented parts of pyramid cell axon collaterals. This last group may not be very significant if one looks at individual pyramidal cell arborizations, however, in consequence of the large

abundance of pyramidal cells in general this constituent of the vertical neuropil might be very significant.

## Summary

A general concept of the elementary neuronal circuit for the mammalian neocortex is developed on the basis of (1) Golgi architecture, (2) light and electron microscope level degeneration studies, (3) ultrastructural analysis of synaptic and neuropil architecture and on that of (4) light and electron microscope studies on chronically isolated cortical slabs.

## References

Ádám, A.: Simulation of rhythmic nervous activities, II. (Mathematical models for the function of networks with cyclic inhibition.) Kybernetik 5, 103—109 (1968).

Colonnier, M. L.: The structural design of the neocortex. In: Eccles, J. C. (ed.), Brain and Conscious Experience, pp. 1—18. Berlin-Heidelberg-New York: Springer. 1966.

Colonnier, M.: Synaptic pattern of different cell types in the different laminae of the cat visual cortex. An electron microscope study. Brain Research 9, 268—287 (1968).

— and S. Rossignol: Heterogenity of the cerebral cortex. In: Jasper, H. H., A. A. Ward, and A. Pope (eds.), Basic mechanisms of the epilepsies, pp. 29—40. Boston: Little, Brown & Co. 1969.

Eccles, J. C., M. Ito, and J. Szentágothai: The cerebellum as a neuronal machine. Berlin-Heidelberg-New York: Springer. 1967.

Garey, L. J., and T. P. S. Powell: An experimental study of the termination of the lateral geniculo-cortical pathways in the cat and monkey. Proc. R. Soc. (London) B. 179, 41—63 (1971).

Hámori, J., and J. Szentágothai: The "crossing-over" synapse: an electron microscope study of the molecular layer in the cerebellar cortex. Acta biol. Acad. Sci. hung. 15, 95—117 (1964).

Harmon, L. D.: Problems in neural modeling. In: Reiss, E. R. F. (ed.), Neural theory and modeling, pp. 2—30. Stanford: Stanford University Press. 1964.

Hubel, D. H., and T. N. Wiesel: Receptive fields, binocular interaction and functional architecture in the cat's visual cortex. J. Physiol. (London) 160, 106—154 (1962).

Jones, E. G., and T. P. S. Powell: An electron microscope study of the mode of termination of cortico-thalamic fibers within the sensory relay nuclei of the thalamus. Proc. R. Soc. (London) B 172, 173—185 (1969).

Kling, U., and G. Székely: Simulations of rhythmic nervous activities, I. (Function of networks with cyclic inhibitions.) Kybernetik 5, 98—103 (1968).

LORENTE DE NÓ: The cerebral cortex: Architecture, intracortical connections and motor projections. In: FULTON, J. F. (ed.), Physiology of the Nervous system, pp. 291—321. London-New York-Toronto: Oxford University Press. 1938 a.

— Analysis of activity chains of internuncial neurons. J. Neurophysiol. 1, 207—244 (1938 b).

MARIN-PADILLA, M.: Prenatal and early postnatal ontogenesis of the human motor cortex: a Golgi study, II. The basket-pyramidal system. Brain Research 23, 185—192 (1970).

MOLLIVER, M. E., and H. VAN DER LOOS: The synaptic strata of the somesthetic cortex in neonatal dog. Anat. Rec. 163, 317—318 (1969).

MOUNTCASTLE, V. B.: Modalities and topographic properties of single neurons of cat's sensory cortex. J. Neurophysiol. 20, 408—434 (1957).

PALKOVITS, M., P. MAGYAR, and J. SZENTÁGOTHAI: Quantitative histological analysis of the cerebellar cortex in the cat. I. Number and arrangement in space of the Purkinje cells. Brain Research 32, 1—13 (1971 a).

— — — Quantitative histological analysis of the cerebellar cortex in the cat. II. Cell numbers and densities in the granular layer. Brain Research 32, 15—30 (1971 b).

— — — Quantitative histological analysis of the cerebellar cortex in the cat. III. Structural organization of the molecular layer. Brain Research 34, 1—18 (1971 c).

POLJAKOV, G. I.: On structural mechanisms of interneuronal connections in human cerebral cortex. (In Russian.) Arh. Anat. Gistol. Embriol. 32, 15—19 (1955).

RAMÓN Y CAJAL, S.: Histologie du système nerveux de l'homme et des vertébrés, Vol. 2. Paris: Maloine. 1911.

RUIZ-MARCOS, A., and F. VALVERDE: Dynamic architecture of the visual cortex. Brain Research 19, 25—39 (1970).

SHOLL, D. A.: The organization of the cerebral cortex. London: Methuen. 1956.

SZÉKELY, G.: Development of limb movements: embryological, physiological and model studies. In: Growth of the Nervous System, pp. 77—95. (Ciba Foundation Symposiom on Growth of the Nervous System, 1968, London.) London: Churchill. 1968.

SZENTÁGOTHAI, J.: On the synaptology of the cerebral cortex. In: Structure and Function of the Nervous System, pp. 6—14. (Proc. Conf. Structure and Function of the Nervous System 10—14, Dec. 1960.) Moscow: Medgiz. 1962.

— The structure of the synapse in the lateral geniculate body. Acta Anat. (Basel) 55, 166—185 (1963).

— The use of degeneration methods in the investigation of short neuronal connections. In: SINGER, M., and J. P. SCHADÉ (eds.), Progress in Brain Research, Vol. 14, pp. 1—32. Amsterdam: Elsevier. 1965.

— The synapses of short local neurons in the cerebral cortex. In: Modern Trends in Neuromorphology, pp. 251—276. (Proc. Internat. Conf. on Neuromorphology, July 5—6, 1963, Budapest.) Budapest: Akadémiai Kiadó. 1965.

— The anatomy of complex integrative units in the nervous system. In: LISSÁK, K. (ed.), Recent Development of Neurobiology in Hungary, pp. 9—45. Budapest: Akadémiai Kiadó. 1967.

Szentágothai, J.: Architecture of the cerebral cortex. In: Penry, K. (ed.), Basic Mechanisms of the Epilepsies, pp. 13—40. New York: Little, Brown & Co. 1969.
— Les circuits neuronaux de l'écorce cérébrale. Bulletin de l'Académie Royale de Médecine de Belgique. VII. Série 10, 475—492 (1970).
— Some geometrical aspects of the neocortical neuropil. Acta biol. Acad. Sci. hung. 22, 107—124 (1971).
— Neuronal and synaptic architecture of the lateral geniculate body. In: Jung, R. (ed.), Handbook of Sensory Physiology 7/2. Berlin-Heidelberg-New York: Springer (in press).
Uchizono, Koji: Characteristics of excitatory and inhibitory synapses in the central nervous system of the cat. Nature 207, 642—643 (1965).
— Synaptic organization of the Purkinje cells in the cerebellum of the cat. Exp. Brain Res. 4, 97—113 (1967).
Uttley, A. M.: The probability of neural connexions. Proc. R. Soc. B 144, 229—240 (1955).

# A Golgi Analysis of the Sensory-Motor Cortex in the Rabbit

T. Tömböl

First Department of Anatomy, Semmelweis University Medical School, Budapest, Hungary

## Introduction

Renewed interest for Golgi architecture of the neocortex was prompted by the desire to explain electrogenesis, especially during generalized and synchronized activities of the cortex, in terms of neuronal structure and arrangement. Both functional and structural information appears to favour the view that neuronal connectivity is arranged predominantly in the vertical direction, while horizontal connectivity appears to be more limited to relatively short distances —apart from the most superficial and the deepest strata. It is especially the deeper strata where it becomes increasingly difficult to distinguish true intracortical horizontal connections from subcortical relay between closely neighbouring regions. Attempts at explaining various events of electrogenesis have to consider primarily structures that might ensure sufficient numbers of systematically arranged dipoles, like the dendritic shafts of the pyramid cells, bundles of entering or leaving axons, neuropil regions of determined geometric order, however, the systematics of connexions—i.e., what is connected with what else—cannot be left out from such considerations.

RAMÓN Y CAJAL's (1911) account of neocortical structure still holds after 60 years, and a recent "second look" by COLONNIER (1966) and by GLOBUS and SCHEIBEL (1967) on the "structural design of the neocortex" has shown that current oversimplifications of subdividing cortical cells into pyramids and stellates is more misleading than helpful, hence it might be advisable to try to continue RAMÓN Y CAJAL's approach by concentrating on specific regions in specific species. It seems a difficult undertaking to improve on RAMÓN

Y Cajal who has recognized and illustrated in most of their expansions around 50 different cell types in the whole neocortex. But what is needed is certainly neither an increase in the number of identifiable cell types nor explaining away essential differences by various artificial categorizations. It is important instead to try to narrow down—if possible—the inventory of cells in various specific regions and to understand their specific internal and afferent connectivities. The present paper gives an account of a Golgi study of the sensorimotor cortex in the rabbit. It should be kept in mind, of course, that Golgi architecture is only one, even if a particularly important, aspect of the relevant structural features of any piece of grey matter.

## Methods

These studies are based on celloidin embedded series of perfusion Golgi-Kopsch material, prepared from the brains of 3 months old rabbits. Dendrite arborizations of various kinds of neurons are generally well stained with this procedure, and so are axons as well as their intracortical arborizations. However, also this procedure shares the common characteristic feature of all Golgi type impregnations in being unpredictable in the random selectivity by which various kinds of elements are stained or not stained. Pyramidal neurons are generally well stained in most layers of the cortex, also Golgi type II cells have shown up quite well in our material. Unfortunately basket cells were left in most cases unstained. The terminal arborizations of afferent fibres of distant source were stained well in some of our material, however for a more detailed account on the afferent fibers axonal degeneration studies would be needed. Although dendritic and axonal arborizations can readily be traced in this kind of material, the synaptic relations between the two remain largely conjectural. Golgi architectonics, in general, gives information only about the probable connectivity, certain information about *de facto* synapses between any type of axon and cell body or dendrite needs to be confirmed by EM level degeneration observations.

## Results

### a) The Principal Cell Types

At the interior border of the first layer, about 150 $\mu$ from the surface, medium size multiangular cells may be observed in the rabbit cortex (Fig. 1 a). The cell body is 15–18 $\mu$ in diameter. The dendritic tree

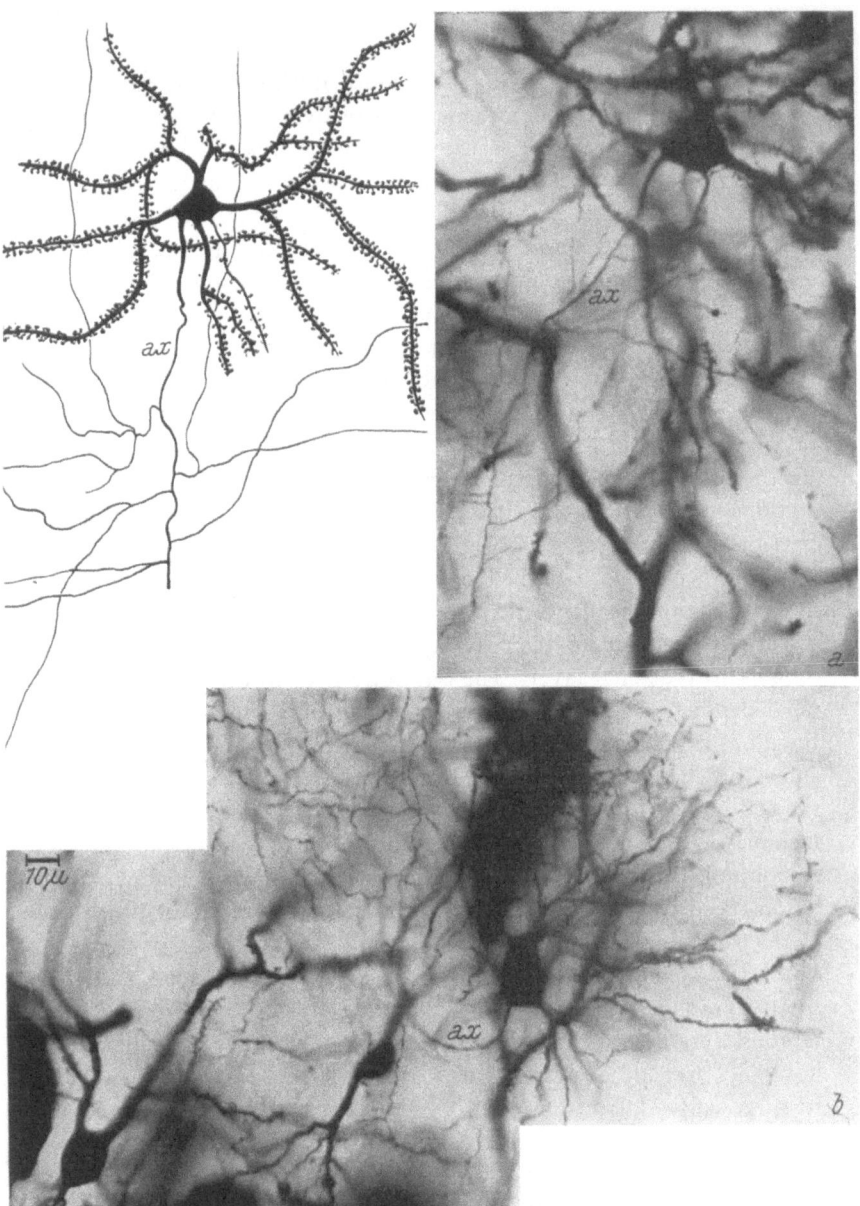

Fig. 1. *a*) drawing and photomicrograph of relatively large multiangular cells in lamina I having a descending axon and very numerous initial collaterals. *b*) Large neurogliform neuron in lamina I with axon recurving towards the zonal layer. The fine feltwork at left of this cell corresponds to the arborization of this axon

invades mainly the first layer, 5–6 main dendrites arise from the cell body branching mostly dichotomically already at short distances. The secondary dendrite branches are of wavy course and are densely covered with characteristic spines. The axon originates from the cell body and is directed vertically towards the depth of the cortex. Due to incomplete impregnation or because leaving the plane of sectioning, the axons could be followed only for about 1000 $\mu$, i.e., they reach the fourth layer without having terminated. In the second layer the axon gives off 5–6 or more fine collaterals. Some of them return obliquely into the first layer, others run horizontally at the border of the first and second layers. Finally they also enter into the first layer and both types contact the dendrites of adjacent cells or the apical dendrite bouquets of various pyramidal cells.

Another unusually large cell type can be observed in the inner-most part of the first layer, at 200 $\mu$ distance from the surface (Fig. 1 b). A similar cell was described by CAJAL in the second layer; since, however, both the dendritic and axonal trees of this cell type are essentially confined to the first layer, it is better to consider it belonging to the zonal layer. The cell body is piriform and 8–10 principal dendrites are found to originate from it. The arborization of these has a "tufted" character. The dendrites are smooth, lacking in spines but have often varicose swellings. The axon originates from the cell body and curving towards the surface breaks up in the usual fashion of the neurogliform cells into a dense arborization corresponding in size to that of the dendritic tree. Some contacts may be observed between this axon-arborization and the apical dendrites of the pyramidal cells.

From the second layer two observations deserve mentioning. The small pyramidal cells with long cortico-fugal axons and initial collaterals were described already by RAMÓN Y CAJAL. The collaterals, however, are not exclusively of horizontal direction, the first two to three collaterals can be seen to have an ascending course and to establish contacts with dendrites in the first layer. The other observation in the second layer concerns Golgi type II neurons sending their axons into the third layer (Fig. 2). This cell type is of small ovoid shape, its ascending dendrites are long and bear spines. The axons arising from the cell body branch after a certain distance; their arborizations reach the third layer and may join the apical dendrites of the pyramidal cells of the third layer.

In the third layer the small pyramidal neurons belong to the cells with short axons according to the present observation. Most of the axonal branches may be seen to terminate in the neighbourhood, some of them, however, enter into deeper layers. The local axon-branches terminate on the dendrites of large pyramidal cells of the third layer (Fig. 3). No evidence of branches leaving the cortex is found in these small pyramids, so that these cells are apparently short axon

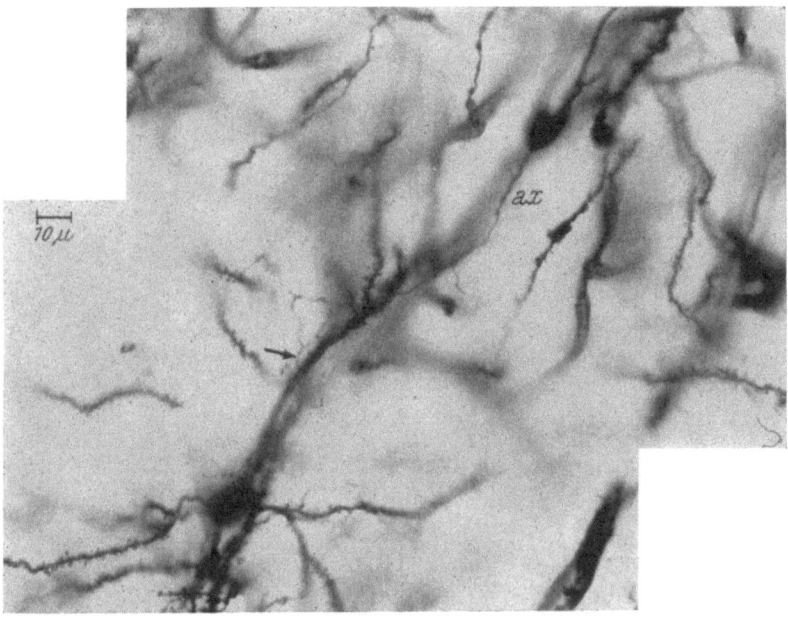

Fig. 2. Small Golgi type II cell in the second lamina and the close relations of its axonal branches (arrow) to apical dendrite of lamina III pyramidal neuron

neurons, which with respect to axonal arborization do not differ significantly from Golgi type II cells.

The stellate neurons of the *fourth layer* can be classified according to their patterns of axonal arborization into three types: (i) Large stellate neurons having a vertically descending corticofugal axon and collaterals issued in horizontal and ascending directions; (ii) small Golgi II type stellate neurons with axons ramifying in the immediate vicinity. Its axon arborization spreads scarcely beyond the space occupied by the dendritic tree; (iii) Another type of small stellate neuron has a large spiny dendritic tree with an axon that emits collaterals into the fifth and third layers, while the main trunk

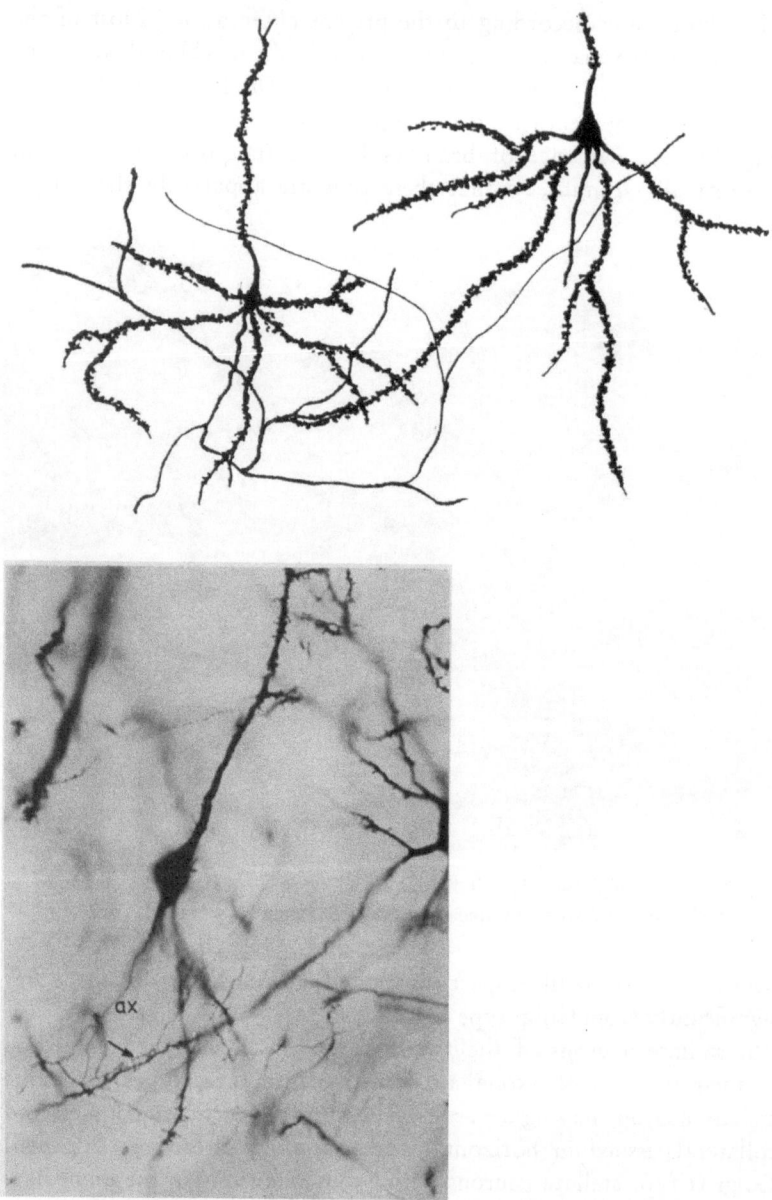

Fig. 3. Large and small pyramidal cell of lamina III. Axon of the small pyramid breaks up into Golgi type II arborization. Below: photomicrograph of such a small pyramidal cell showing close attachment of delicate axon branches to basal dendrite of large pyramidal cell

extends horizontally at larger distances within the fourth layer. Some
of the initial axonal branches can be traced back to the vicinity of the
cell body, other vertical terminal branches descend into the fifth layer

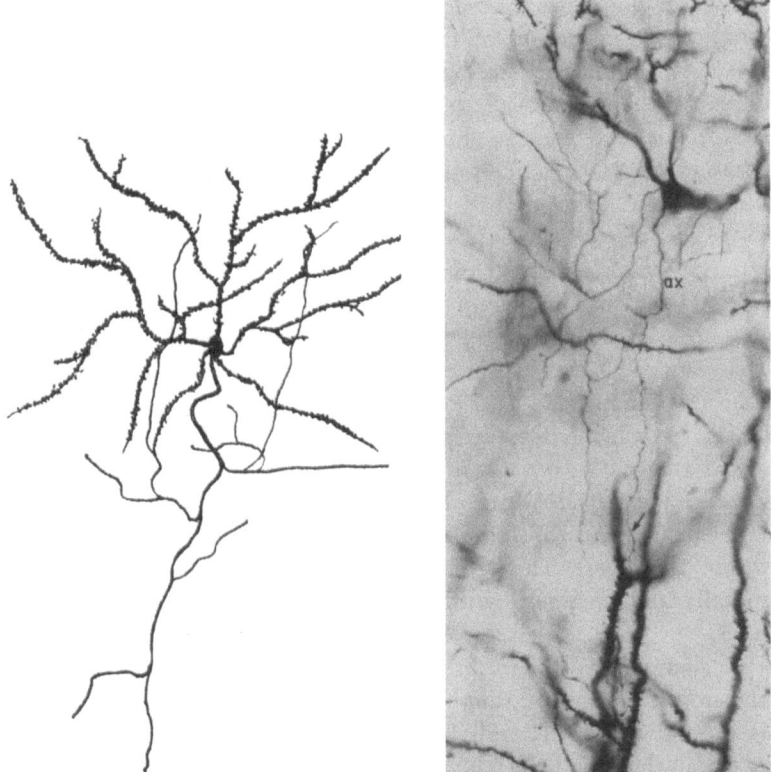

Fig. 4. Small stellate cell of lamina IV. At right: photomicrograph showing
close relation of an axon branch to apical dendrite of large pyramidal cell of
lamina V (arrow)

and attach themselves to the apical dendrites of large pyramidal cells
(Fig. 4). Occasionally these branches contact several apical dendrites.
Especially in the fourth, but also in the third and fifth layer, a very
characteristic neuron type may be observed, which is generally only
incompletely impregnated. The cell body of these neurons is variable
in size: medium, large, and even giant cells being found. The 5–8
dendrites arising from the cell body spread in all directions, they
are thick, varicose, and sparsely ramifying. The dendrites are about

300–400 $\mu$ long. They are supplied with very few spines, the dendritic tree can extend over two or three cortical layers. The axon originates from the cell body and is directed towards the surface. At a short distance the axon gives off side branches. These curve backwards and appear to terminate on adjacent cells. Due to incomplete impregnation of the axon, it could not be ascertained, whether these neurons are basket cells but this is probable also with regard to the thickness of the axon. Rather coarse horizontal axons can be observed issueing short side branches towards cell bodies of pyramidal cells. It is likely that these axons are those of incompletely stained basket cells.

There are large pyramidal cells in the fifth layer, the axons of which, after giving off a number of collaterals, turn in horizontal direction. The collaterals partly ramify in the fifth layer, partly ascend for a certain distance. Another interesting observation in the fifth layer is, that axon collaterals of pyramidal cells establish contacts with neighbouring pyramids.

In the sixth layer inverse medium size pyramidal cells may be observed frequently, often with the apical dendrite oriented obliquely towards the depth. Their dendrites are covered with spines. The axons of these neurons arise from the cell body and run horizontally within the layer. The collaterals of the axon spread partly horizontally and terminate within the sixth, others ascend into the fifth layer.

The neurons in the sixth layer having corticofugal axons issue numerous horizontally running collaterals. The horizontally running axons and axon-collaterals form a well developed horizontal fiber-system in the sixth layer.

## b) Fibers

Few of my own observations concern the afferent fibers of the sensori-motor cortex. The cortico-cortical fibers (commissural and association fibers) are thin, running obliquely through the inner layers and turn in vertical direction in the third layer. They may be followed to the surface of the cortex. The fibers have short terminal side branches along their course in the cortex. Their terminal parts are fine fibers with occasional varicosities.

In the fourth layer medium thick fibers may be found, that break up into repeated branchings (Fig. 5). Branches of the third and

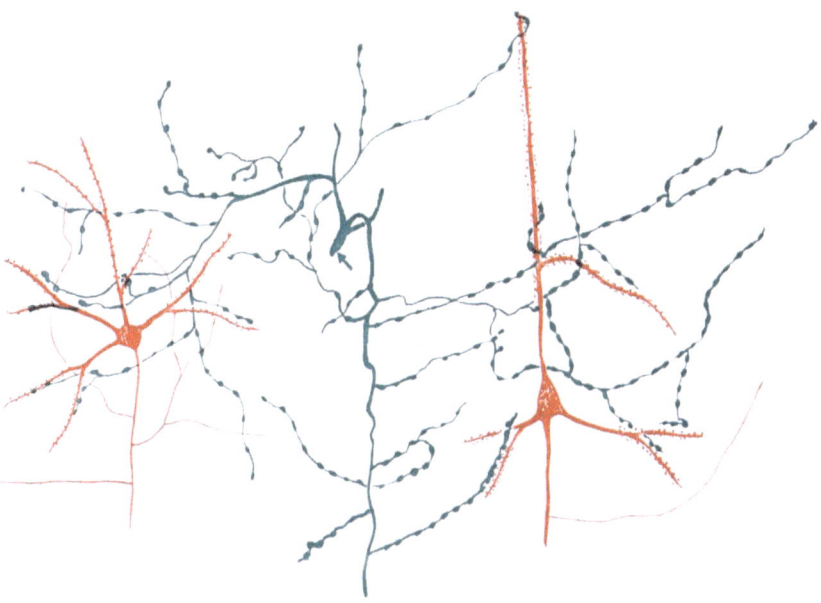

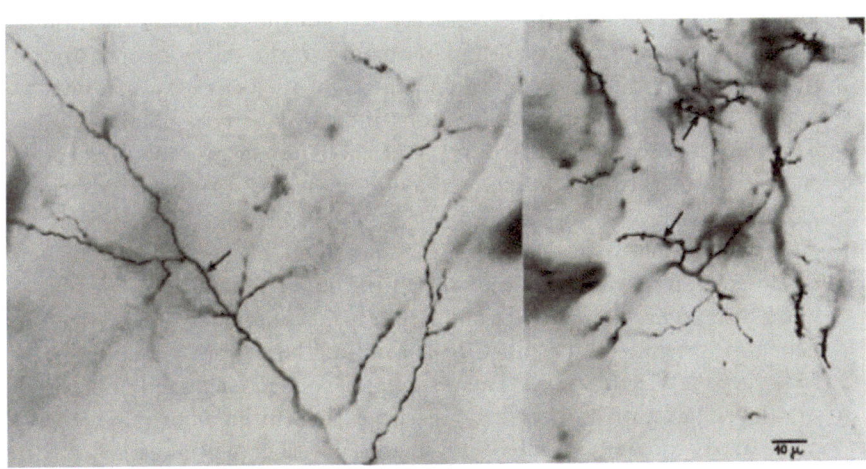

Fig. 5. Arborization of afferent fibre (probably sensory afferent) in lamina IV, showing relation of terminal branches to dendrites of pyramidal and stellate neuron. Arrow indicates cut trunk of afferent fiber. Photomicrographs below show details of arborization

further order are thin and varicous. Little can be said about the extent of the arborization of these fibers. The largest fiber arborization that could be measured had a total span of 350–450 $\mu$ in diameter, the thickness of the slide was 120 $\mu$. The varicose terminal branches of the fibers join the dendrites of small and large pyramidal and stellate neurons in the fourth layer. This type of arborization would suggest them to be probably the specific sensory afferents. A similar type of fiber may be found also in the second layer, it has, however, a poorer arborization. Here these fibers contact the small pyramidal cells of the second layer.

## Discussion

The all-over structural arrangement that emerges from these preliminary studies is shown in Fig. 6. The efferent neurons, mainly pyramidal but also some of stellate shape are shown in red. Apart from two horizontal plexus formed by axon collaterals in the first, second and in the sixth layer there is nothing new in their arrangement. The afferents are shown in green colour; it can be seen that while the specific afferents terminate predominantly in the fourth and to lesser extent in the second layer, the callosal and association (cortico-cortical) afferents terminate mainly in the third, second, and in the first layer. The vertical organization of the cortex is warranted mainly by the Golgi type II neurons indicated here in blue colour. The Golgi type II cells are very variable, some are pyramid or stellated, others have the conventional non-spiny dendritic arborizations.

A special, hitherto neglected architectural and connectivity feature disclosed by our studies is shown here in black. There appear to be quite strong mutual connections between a superficial and a deep neuronal system. The deep neurons have been known since long as the Martinotti (1890) cells. Their synaptic connection mainly in the first layer has already been assumed by earlier authors but it became particularly obvious in our own material. The multiangular cells described now in the first layer have axons that descend into the middle of the cortex, however, it seems reasonable to assume that their main target is the cells of the sixth layer.

As a morphologist, I ought perhaps not to speculate too much about the significance of this hitherto unknown structural principle. But still, one might venture to assume that such an abundant mutual

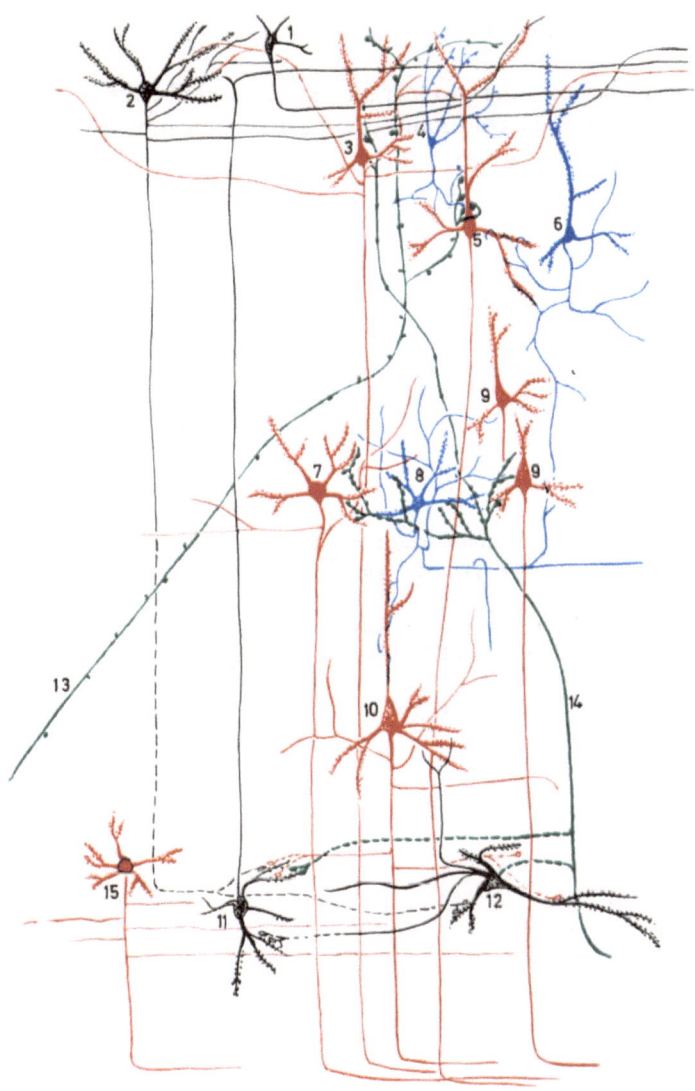

Fig. 6. Diagram showing the main cell types and connexions in sensori-motor cortex. Efferent elements are shown in red, Golgi type II cells in blue, cells of lamina I and VI in black; afferents in green colour. Conjectural connexions are indicated by dashed lines. 1. Horizontal cell, 2. large multiangular cell, 3. small pyramidal cell, 4. Golgi type II neuron, 5. large pyramidal cell, 6. small pyramidal cell, 7. large stellate neuron, 8. small stellate neuron, 9. large pyramidal cell, 10. large pyramidal cell, 11. Martinotti cell, 12. medium size pyramidal cell in lamina VI, 13. callosal fiber, 14. afferent fiber, 15. medium size cell in lamina VI with corticofugal axon

interlocking between the two marginal layers might be the anatomical basis of some general synchronization in the function of the neuronal network through the entire depth of the cortex.

## Summary

A report of Golgi studies in rabbit sensory-motor cortex. Some new hitherto unknown cell types are described. Connexions established between layers II and III, III and IV, and IV and V by means of Golgi type II neurons are considered in detail. The horizontal fiber system in the first and sixth layers and their connexions are discussed.

## References

Colonnier, M. L.: The structural design of the neocortex. In: Eccles, J. C. (ed.), Brain and Conscious Experience, pp. 1—23. Berlin-Heidelberg-New York: Springer. 1966.

Globus, A., and A. B. Scheibel: Pattern and fields in cortical structure: The rabbit. J. comp. Neurol. *131*, 155—172 (1967).

Martinotti, C.: Beitrag zum Studium der Hirnrinde und dem Centralursprung der Nerven. Int. Monatsschrift Anat. Physiol. 7, 69—90 (1890).

Ramón y Cajal: Histologie du système nerveux de l'homme et des vertébrés II. Paris: Maloine. 1911.

# Invited Discussion

K. FLEISCHHAUER

Anatomisches Institut der Universität Bonn, Germany

Dr. PETSCHE has asked me for a few remarks from the anatomical point of view, and I should like to use this opportunity of pointing out that apart from the neuronal circuitry there are some other morphological aspects which we must bear in mind when trying to explain the electrophysiological findings. There are actually two points I should like to make:

1. When seeing the sustained laminar discharges observed by Professor PETSCHE, I immediately thought of the anatomical fact that the various laminae of the cerebral cortex are not only different with respect to the form, number, and connexions of their neurones, but also with respect to the content of oxidative and other enzymes as was shown by means of histochemical as well as microchemical methods. These laminar differences are quite pronounced and easily detectable, whereas the occurrence of chemical differences between adjacent cytoarchitectonic fields has long been questioned and is still subject to discussion as far as the neocortex is concerned. The laminar differences, on the other hand, have been shown to be present in a great number not only of mammalian but also of submammalian species (for literature see SCHARRER and SINDEN 1949, FLEISCHHAUER 1959, FRIEDE 1966).

In this connexion it is also worthwhile to point out that within the cortex various systems of neurones seem to exist which make use of different transmitter substances. Thus some years ago KRNJEVIC and SILVER (1965) have shown that in cats the neurones which give rise to the U-fibres contain great amounts of acetyl-cholinesterase whereas the nerve cells of the upper layers are almost devoid of this enzyme.

Another fact which is easily forgotten and may yet be of importance is the number and arrangement of the blood vessels and the accompanying perivascular structures. The differences occurring in this context are illustrated by projecting a figure from the classical book of PFEIFER (1940) depicting a section through the postcentral gyrus of the macaque after injecting the blood vessels with Indian ink. The following facts are recalled:

a) There are pronounced laminar differences in the number of blood vessels.

b) There are marked quantitative and qualitative changes corresponding to the borders of cytoarchitectonic areas and

c) the larger blood vessels are orientated perpendicularly to the surface of the cerebral cortex.

This latter observation may be of importance for the explanation of the so-called columns in the cortex, since there may be a relation between the arrangement of large perpendicular blood vessels, the vertical striations brought about by bundles of myelinated fibres and the distribution of nerve cells. The question of whether or not there is such a relation between the vertical striations formed by the bundling of myelinated fibres—as illustrated in all myeloarchitectonic maps of the brain—and the arrangement of nerve cells, neuroglia, and blood vessels is being investigated in my laboratory at the present time.

2. The second point I should like to make concerns the geometry of current distribution in the cerebral cortex. It is actually a question which puzzles me a great deal but this may only be due to my ignorance in electrophysiology. So I should like to ask the electroencephalographers present for an explanation:

In the papers of Dr. ELUL and Dr. LUX and also in some papers today we have seen a number of drawings and heard a lot about dipoles of varying descriptions. It was said that excitation of a pyramidal cell creates a field of current flow and that the shape and gradient of this field can be detected by having a second electrode nearby.

On this and other occasions one got the impression as if a nerve cell can be considered as a complicated electrical machine situated in a homogeneous and empty outside, say in the sea. But the anatomist sees that a great percentage of the surface of the nerve cell is taken up by synapses and the rest by glial cells or processes; that in the immediate neighbourhood of the nerve cells there are axons or bundles of axons and dendrites and of course blood vessels; and as we have just seen in the beautiful electron micrographs of Prof. SZENTÁGOTHAI, the electron-microscopist sees all this as a lot of "profiles" separated by an exceedingly small amount of extracellular space.

Now my question is: Do we know whether or not, and if so, how, the spread of current and the shape of the electrical fields originating from the firing of a cell are influenced and distorted by the membrane properties and by the manifold electrical events taking place in the "profiles" forming the environment of the nerve cell under consideration? Am I wrong in the assumption that the activity in these dendrites influences the discharge from a single cell to such a degree that its nature as a dipole would be completely obscured?

### References

FLEISCHHAUER, K.: Zur Chemoarchitektonik der Ammonsformation. Nervenarzt 30, 305—309 (1959).

FRIEDE, R. L.: Topographic brain chemistry. New York-London: Academic Press. 1966.

KRNJEVIĆ, K., and A. SILVER: A histochemical study of cholinergic fibres in the cerebral cortex. J. Anat. 99, 711—759 (1965).

PFEIFER, R. A.: Die angioarchitektonische areale Gliederung der Großhirnrinde. Leipzig: Thieme. 1940.

SCHARRER, E., and J. SINDEN: A contribution to the "chemoarchitectonics" of the optic tectum of the brain of the pigeon. J. comp. Neurol. 91, 331—336 (1949).

# Discussion to the Papers
## of Braitenberg, Szentágothai, and Tömböl

PETSCHE: What do you think about the possibility of distinguishing, from the morphological point of view, between excitatory and inhibitory synapses?

SZENTÁGOTHAI: I cannot make up my mind about UCHIZONO's concept on spherical and flattened vesicles. In the cerebellum, where seven different types of synapses can be identified—four of which are inhibitory and three excitatory—the hypothesis appears to hold. Also in the spinal cord there is some quite good evidence for spherical vesicles in excitatory and flattened ones in inhibitory terminals. But this is certainly not universal. Monoaminergic terminals may be inhibitory but they do not contain flattened vesicles.

GLOOR: Our results indicated that inhibition in the hippocampus may also come about by axodendritic synapses located on the long apical dendrites.

LUX: I should like to ask Prof. SZENTÁGOTHAI to point out the arrangement and role of recurrent collaterals of pyramidal tract cells and to elucidate the concept of recurrent inhibition in the neocortex. In the presented scheme for motor cortex these collaterals unfortunately have been neglected.

SZENTÁGOTHAI: Recurrent collaterals of pyramidal cells appear to contact mainly dendritic spines of other pyramidal cells. Although this would not rule out the notion completely that they produce recurrent inhibition, it makes it rather unlikely. Extensive quantitative studies would be needed on chronically isolated cortical slabs in order to show, whether pyramidal cell collaterals contact inhibitory interneurons in sufficient numbers to secure substantial recurrent connexions, for example, over basket neurons. But this would be a formidable task with the techniques available today.

SCHERRER: There are nervous structures in which it is easy to evoke a seizure and others in which it is almost impossible. Are there any common features for these two types of structure?

GLOOR: At one time we had the working hypothesis that a structural prerequisite for the development of epileptic discharge was that neuronal generators had to be oriented in parallel and polarized, *i.e.*, the dendritic and the axonal ends of the neuron must be oriented in opposite directions. We carried out some experiments with penicillin to test this hypothesis and these exploded this idea. Among structures satisfying the postulated structural criteria were the neocortex and hippocampus which are indeed very epileptogenic, but so also is the amygdala which does not satisfy these structural criteria. Conversely the lateral geniculate body and the cerebellar cortex do conform to the postulated structural criteria but do not produce epileptic discharge, at least not with local application of penicillin.

BRAITENBERG: Can you give a list of epileptogenic places and non-epileptogenic places?

GLOOR: Not really. The thalamus and the cerebellum seem to be pretty refractory with regard to their propensity to respond with epileptogenic discharge to a variety of convulsant agents.

NAQUET: FRENCH *et al.* (1956) demonstrated that, by electrical stimulation, the occipital cortex has much higher threshold than the motor cortex and that the hippocampus has a lower threshold than the motor cortex and amygdala and the lowest threshold of all the structures of the brain. Amygdala has a threshold higher than the motor cortex but lower than some other cortical areas (GREEN and NAQUET 1957). With MOLLICA (1953) we demonstrated that it is possible to obtain very short after-discharges by stimulation of the cerebellum but only with fairly high voltages.

SZENTÁGOTHAI: From the viewpoint of positive feedback the hippocampus would be the most unlikely place to have epileptogenic properties. That is, of course, if we consider excitatory recurrent connexions in relatively short (local) loops. With longer loops over distant brain regions everything would be possible.

GLOOR: According to LORENTE DE NÓ, there are collaterals to the basal dendrites. We presumed these were excitatory and could mediate a Jacksonian type of march of epileptic discharge in the hippocampus.

PETSCHE: From our neurophysiological findings we have been guided to the conception of a columnar shape of the electrical generators which are probably arranged in a mosaic-like way in the deeper layers of the cortex, as shown in Fig. 1. Their cross diameter may be in the order of 50 $\mu$. My question is whether this radial organization, which comes out very clearly in myelin stainings (as for instance in the old atlas by KAES 1907), may also have any functional significance.

SZENTÁGOTHAI: I think this is simply a "packing problem". If you have to pack large number of cells, with processes having more or less determined arborization patterns, into the smallest possible space you

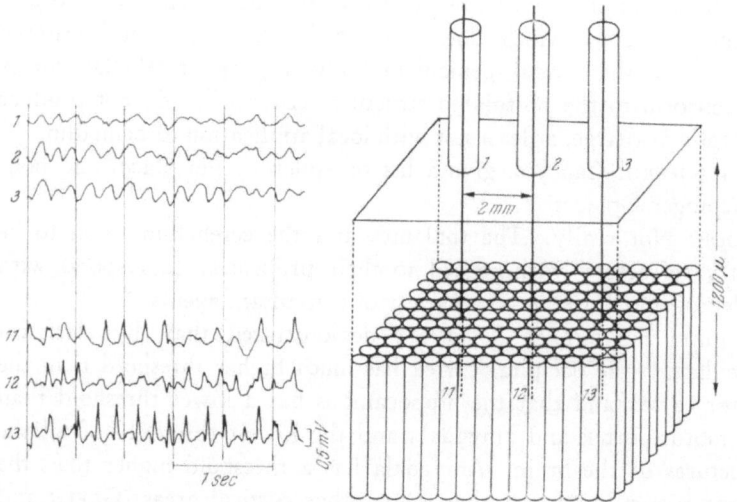

Fig. 1. Seizure activity (Metrazol) recorded from surface (1–3) and intracortical electrodes (11–13), 1200 μ below surface. Common reference recordings.
Epicortical compared with intracortical records look as if passed through a low-pass filter.
Further details see p. 265.

may accept some regularity of arrangement. Additionally you have to consider migration of cells in the neuroblastic stages occurring in different steps. The deeper pyramid cells arrive at their final places somewhat earlier, and the superficial pyramids—coming later—have to move through the rows of the already settled deeper cell layers. This in itself might cause columnar arrangements of cells and bundling of the descending efferent fibres. This may have, of course, electrogenic consequences secondarily but can hardly be of primary functional significance.

PETSCHE: Why are these myelinated fibres then bundled and why aren't they distributed at random?

FLEISCHHAUER: I think it may be worthwhile also considering this problem in an embryological context. As recently shown by MARIN-PADILLA (1971), layers I and VI of the cortex are relatively well developed and already interconnected by the axons or Martinotti cells at a time when the cortical plate giving rise to layers II and V is still very immature. It is conceivable that the early connexions between layers VI and I reveal a frame work, the structure of which influences the further arrangements of axons as well as of dendrites arising from the other layers. And this could perhaps be one mechanism of forming certain vertical packages or columns of functionally related neurones [1].

PETSCHE: It is surprising for me to learn from Dr. TÖMBÖL's interesting findings that there are important fibre systems connecting the cortical surface with the deep layers of the cortex. I wonder if these fibre system may be of any significance as to the phenomenon we have been dealing here. I didn't know either that, in rabbits, there are so many horizontally running fibres in layers V and VI. This may be one reason for the observation that a region with a coherent pattern of activity can be split into two areas with different activities by cutting the cortex down to the deep layers.

TÖMBÖL: As far as the first question is concerned it is extremely difficult to draw, based on the great variety of findings in Golgi sections, one figure that would illustrate all these observations. The axons of the Martinotti cells branch in the first layer horizontally. The other connection is a suggestion. I found multiangular cells in the first layer and was able to follow the axons as far as to the fourth layer, but they do not terminate there. They could not be traced because they leave the plane of the section. The diameters of these axons are of the same size here as in the first layer. These axons, therefore, probably go further down and may contact even the cells of the sixth layer. I hope one day to be able to find the termination of these axons.

As far as your other question on the horizontally running fibres is concerned, most of them are very fine collaterals. They are found in practically all layers, since all neurons having corticofugal axons send

---

1 In the meantime it has been found that the apical dendrites of pyramidal cells are not distributed at random but are organized in vertical bundles consisting of ten or more apical shafts [FLEISCHHAUER, K., H. PETSCHE, and W. WITTKOWSKI: Vertical bundles of dendrites in the neocortex. Z. Anat. Entw.-Gesch. 136, 213—223 (1972)].

collaterals. We were, however, not able to follow these collaterals throughout their entire length, only over about a distance of 500 $\mu$; this indeed is a relatively large distance but they did not yet terminate there. It seems necessary to remark, though, that the horizontal fibre system already mentioned is most prominent in the first and the sixth layers, and this horizontal fibre system may be the cause of the experimental observation by Dr. PETSCHE after cuts into the cortex at different depths.

GLOOR: What is the proportion of these Martinotti cells among the other cells?

TÖMBÖL: The Golgi method does not give any data on the proportion of cells, since the impregnation is too selective. In a Nissl preparation, on the other hand, it is difficult to determine the different cell types. The Martinotti cells are only rarely impregnated in Golgi specimens.

SCHERRER: Indeed it is important to have a precise idea about the relative number of different types of cells. Wouldn't it be possible to combine the Golgi method which does not stain homogeneously with other techniques?

SZENTÁGOTHAI: There are such methods available especially for neuron population analysis. Golgi preparations cannot be used for quantitative purposes, apart from determining the numbers, geometrical, and topological properties of the dendrite branchings. Numerical relations of different cell types can be established from statistical population analytic studies in the Nissl picture on the basis of cell size, shape, nuclear size, nuclear shape, etc. The measurements made on the dendritic arborizations of representative Golgi cells can then be combined with the cell numbers (and densities) established from the Nissl picture.

BRAITENBERG: It may of course be that once we have grasped the variety of neurons in such splendid Golgi material as Dr. TÖMBÖL has just shown us, we may revert to much simpler questions for which the quantitative study with combined Nissl, myelin and Golgi methods is very well possible: how many of the axons leave the cell in upward direction and how many downward, how long the axonal collaterals are on the average and how numerous etc. It is only fair to admit that we do not know where the species specific plan ends and where the individual variation begins: it may be that the bridge between anatomy and physiology, which for some time to come will have to neglect individual variation and ontogenetic effects, can be built between bridgeheads which are much simpler, but perhaps more solid, than the maximum variety which the two sciences are able to demonstrate at present.

BRAZIER: Surely we have to recognize that there are many species differences.

SZENTÁGOTHAI: Although the richness and also the size (spread) of the dendritic arborization—for example in the pyramidal cells—show considerable differences in mammalian phylogenesis, it is remarkable how little the basic pattern of the arborization changes. Some very characteristic types of Golgi Type 2 cells can be found in certain cortical regions (visual cortex) virtually unchanged and easily recognizable as specific cell types from mouse and man.

BRAITENBERG: In answer to Dr. BRAZIER's question I would like to project a few slides of neurons from the cerebral cortex, the seat of the higher mental functions, the sensorium commune, of the cow. Personally, I find it very difficult so distinguish these slides from the others in my collection which are taken from the human cortex.

## References

FRENCH, J. D., B. E. GERNANDT, and R. B. LIVINGSTON: Regional differences in seizure susceptibility in monkey cortex. Arch. Neurol. Psychiat. (Chic.) 75, 260—274 (1956).

GREEN, J. D., and R. NAQUET: Etude de la propagation locale et à distance des démarches épileptiques. VIIIe Réunion de la Ligue Internationale contre l'Epilepsie. Réunions plénières, Acta medica belgica 1957, 225—249.

KAES, T.: Die Großhirnrinde des Menschen in ihren Maßen und in ihrem Fasergehalt. Jena: Fischer. 1907.

MARIN-PADILLA, M.: Early prenatal ontogenesis of the cerebral cortex (neocortex) of the cat (felix domestica). A Golgi study. Z. Anat. Entw.-Gesch. 134, 117—134 (1971).

MOLLICA, A., and R. NAQUET: Activité convulsive et silence électrique dans l'écorce cérébelleuse. Electroenceph. clin. Neurophysiol. 5, 585—587 (1953).

# Studies on Extracellular Potentials Generated by Synaptic Activity on Single Cat Motor Cortex Neurons[1]

W. RAABE and H. D. LUX

Max-Planck-Institut für Psychiatrie, München, Germany

## Introduction

When intracellular recordings from cortical neurons became available it was obvious to most workers in this field that a relationship exists especially between the postsynaptic potentials and the potential recorded from the cortical surface (KANDEL and SPENCER 1961, LI and CHOU 1962, LUX and KLEE 1962, CREUTZFELDT et al. 1964, POLLEN 1964, PURPURA et al. 1964, JASPER and STEFANIS 1965) in accordance with an earlier hypothesis of ECCLES (1951). In the neocortex, the postsynaptic activity of pyramidal tract (PT) cells was assumed to govern the generation of the electrocorticogram and special consideration in this respect was given to the localization of excitatory and inhibitory synapses on the nerve cell soma and apical dendrite (ECCLES 1951, CREUTZFELDT et al. 1966). In a more recent study by HUMPHREY (1968), evidence was presented for a causal relationship between the postsynaptic effects of synchronous recurrent activity of PT cells with the antidromically evoked surface potential. A PT cell model was proposed which rather effectively combined electrophysiological and morphological data. It allowed a quantitative interpretation of this evoked potential based on synaptic activity on single PT neurons. For a check of the model it was thought worthwhile to test experimentally to what amount synaptic currents contribute to the potential field in the extracellular adjacent surround of this neuron.

Since the basal dendritic complex of a PT cell provides the largest amount of membrane area per cortex volume, the density of current generated by synaptic activity was expected to be relatively high or

---

1 Dedicated to the 60th birthday of Professor RICHARD JUNG, Freiburg i. Br.

predominant in the plane of the basal dendritic complex, *i.e.*, parallel to the cortex surface. Accordingly, measurements were begun in the horizontal surround of single cortical nerve cells.

## Methods

For simultaneous recording of intra- and extracellular potentials a microelectrode array was used consisting of three glass-micropipettes

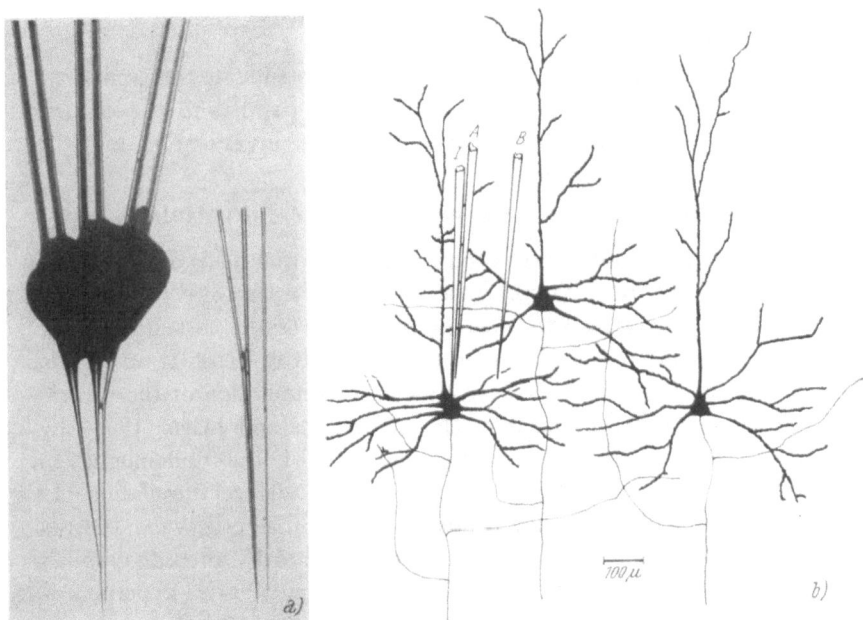

Fig. 1. *a*) Microphotograph of a triple microelectrode, *b*) schematic drawing of the array of the electrode tips, I = electrode for intracellular recording, A = electrode for extracellular recording close to the cell penetrated by electrode I, B = extracellular reference electrode for electrode A

mounted in parallel (Fig. 1). One electrode was for intracellular recording the other two for extracellular recording.

One of the pipettes for extracellular recording was glued over a length of 3 to 4 mm to the shank of the intracellular recording electrode with the tip located 50 to 70 $\mu$ behind the orifice of the latter. This extracellular pipette (A in the scheme Fig. 1 *b*) had been bent beforehand to allow a convenient apposition and mounting under the microscope. The fixation was done with a polymerizing laquer. To have a reference electrode in the farther environment of a nerve cell, but within a defined intracortical and intralaminar surround, a third microelectrode was connected to the electrode pair. This reference electrode (B in the scheme) was similar in construction to electrode A with the last 7 mm of the shank arranged parallel to the other electrodes at a distance of 120 to 150 $\mu$. The tip of this electrode was at the proper level of that of electrode A. Each electrode was connected to a FET amplifier of similar design, to allow a simultaneous potential recording against a common ground (stereotaxic frame).

With this electrode arrangement it was possible to determine an extracellular potential gradient in a plane parallel to the cortical surface and in close proximity of the cell under investigation.

## Simultaneous Intra- and Extracellular Potentials

It was of special interest to compare the potential fields during IPSPs evoked by different stimuli, since this field was assumed to be represented in some cases by a surface positivity of the cortex and by a surface negativity in other instances (Creutzfeldt et al. 1966, Humphrey 1968). IPSPs were elicited by stimulation of the ventral surface of the pyramidal tract (PT) (Stefanis and Jasper 1964), by stimulation of the N. ventralis lateralis of the thalamus (VL) (Nacimiento et al. 1964), and by cortical surface stimulation (Li and Chou 1962). For this study, PT and other cells were selected which showed an IPSP without contamination by preceding spikes or with action potentials without any prominent after-hyperpolarization. During the intracellular IPSPs the extracellular potential differences between electrode A and B (i.e., gradient A against B), showed considerable variances within the total cell population with the exception of a main type of gradient (A–B) which occurred together with IPSPs in PT cells [Fig. 2, $V_e(A–B)$]. In general, a single cell showed somewhat similar extracellular potential slopes after stimulation of the other inputs (N. ventr. lat. thal.; cortex). This observation points to some similarity of the synaptic arrangement of various inhibitory afferents.

## Studies on the IPSP

It was not possible to deduct a simple relationship between the gradients (A–B) and the simultaneously occurring IPSPs. It was therefore thought worthwhile to change the conditions of the synaptic current source of only the cell under observation in order to observe eventual simultaneous changes in the external field. This was done by application of hyperpolarizing current steps across the neuronal membrane via the intracellular electrode or by electrophoretic Cl⁻ injection through this electrode. By such a procedure the IPSP can easily be reduced to zero or be inverted to a depolarizing response. By applying this method, changes in the extracellular gradients (A–B) can be attributed exclusively to changes in the amplitudes of the postsynaptic potentials. By comparing both gradients (A–B), *i.e.*, by subtracting the gradient during the nullified IPSP from that of the original IPSP an estimate of that potential is received which corresponds to the changes in the IPSP. If the apparent reversal potential of the IPSP equals the true IPSP equilibrium potential the resulting difference can be considered to be the field potential generated by the IPSP.

## The IPSP Field Potential

The subtraction of the gradients (A–B) always gave positive potentials at electrode A in respect to the remote electrode B (Fig. 2 *a, b*). The slope of the extracellular potential was similar to that of the intracellular potential but slightly shorter in most instances. The maximum amplitude of the external positivity (A against B) of the difference of gradients was reached almost at the time of the IPSP maximum. The results were similar for PT and unidentified neurons.

For antidromic IPSPs of 4 to 8 mV in amplitude the corresponding external field ranged between 30 and 210 $\mu$V with a mean of 92 $\mu$V. It can be safely assumed that the field potentials of these IPSPs of typical PT neurons should result from currents of 0.5 to 2 nA across the activated inhibitory membrane. This estimate is based upon the assumption of a twofold increase of the neuronal input conductance during recurring IPSPs and other data on electric characteristics of PT cells (POLLEN and LUX 1966, LUX and POLLEN 1966). On the other hand, a comparable current strength across the neuronal membrane is necessary to produce a steady state hyperpolarization of 10 to 20 mV in a neuron with 12 M$\Omega$ total input impedance such as in a slow PT cell neuron (TAKAHASHI 1965). This applies to the neuron shown in Fig. 2, a pyramidal cell with an antidromic latency of 3.6 msec.

4  Synchronization

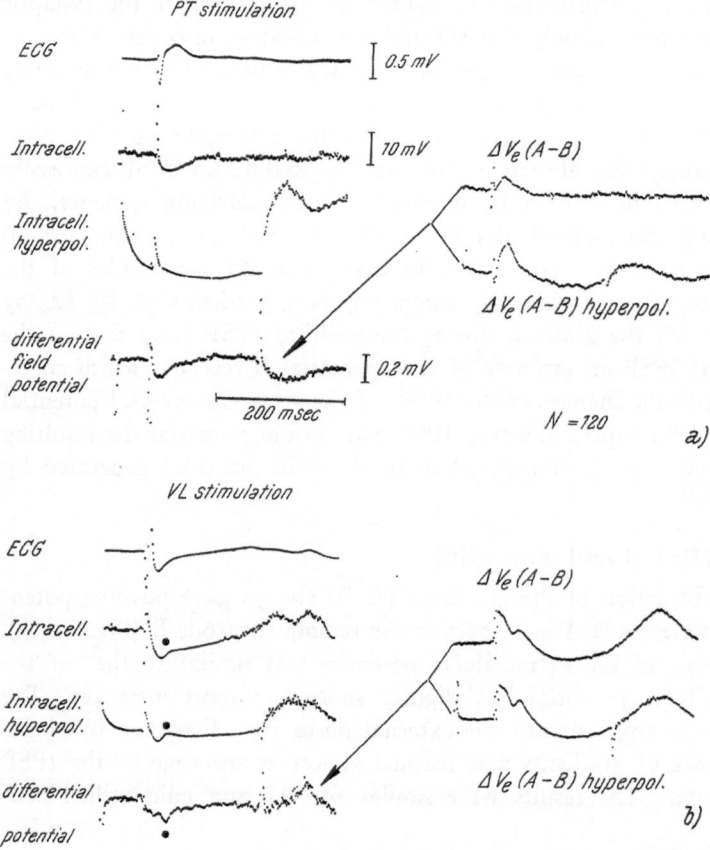

Fig. 2. The IPSP field potential from a pyramidal tract cell after pyramidal tract stimulation (*a*) and thalamic (VL) stimulation (*b*). Averaged records, N = 120. *ECG* = electrocorticogram, intracell. = intracellular recordings, intracell. hyperpol. = intracellular recordings during the passage of hyperpolarizing current steps applied through the intracellular electrode Δ V$_e$(A–B) = potential difference between electrode A and electrode B of Fig. 1 recorded simultaneously with the intracellular recording. Δ V$_e$(A–B) hyperpol. = potential difference (A–B) recorded externally when hyperpolarizing current was applied to the neuron. Differential field potential = computer subtracted data of Δ V$_e$(A–B) records during hyperpolarization of neuronal membrane from data at resting membrane potentials. Note the small size of external voltage change during intracellular current step in comparison with extracellular gradients during IPSPs and EPSPs. The points mark the EPSP mentioned in the text

It can easily be seen in the differential field potential of Fig. 2 *a* that there is only a small if recognizable potential change during such a hyperpolarizing pulse in contrast to a distinct extracellular potential change produced by the IPSP. This difference is to be expected if one considers that the current from the hyperpolarizing pulse is dispersed into many dendritic pathways.

According to estimates of the conductance ratios of dendrite to soma in PT cells (LUX and POLLEN 1966), the current which passes the dendritic inputs is the major part of the applied current and amounts up to 10 times the current passing the soma membrane. The somatic contribution of this current is thus too low to generate a considerable extracellular potential in the environment of the cell soma. The effect of the electrotonic potential must be much lower still than an estimated potential of 2.5 to 20 $\mu$V which should be present during a somatic IPSP if radial potential distribution is assumed. This calculation is based on the assumption that the IPSP current passes a spherical soma, 20 to 40 $\mu$ in diameter (RAMON-MOLINER 1961), assuming furthermore an external impedance of 220 $\Omega$cm (FREYGANG and LANDAU 1955).

The high value of the external field produced by synaptic activity must be ascribed to a considerable deviation from radial potential distribution in this case, *i.e.*, by inhomogeneous local conductances in the external surrounding and by quite localized active areas during inhibitory synaptic activity. Only this can explain the magnitude of the estimated current densities.

## The Antidromic Cortical Response

Quantitatively, the results of these experimental measurements of single cells support the results of HUMPHREY (1968) of a compartmental analysis of the external field generated by typical IPSPs in a dendritic PT neuron model with similar electrotonic characteristics. In the case of the antidromic IPSP, the extracellular potential is in phase and slope quite closely related to the gross intracortical field potential (Fig. 4 PYR). This suggests strongly that this field is generated predominantly by the IPSP. The local current strengths are obviously high enough to account for the interlaminarly extending potential fields and for that recorded at the cortical surface, if only passive electrotonic spread along the apical dendrites is assumed.

The active inhibitory zone is localized in the depth at the level of the PT cell bodies (POLLEN and LUX 1966). Therefore, the greatest amplitude of the antidromic field response is found at the depth of

4*

1.4 mm (layer V) (Fig. 4 PYR). A phase shift of the field response is continuous through the cortical layers. Since the superficial layers with the passively conducting cables of the apical dendritic structures provide the necessary sink, a surface negativity will appear in respect to a remote reference electrode in most experimental situations. It should be stressed that the event of a surface negativity in the usual recording situation is, among other conditions, the result of a high core conductivity of the apical dendrite which spans an external secondary current pathway in which the remote reference is located. The phase reversal can be explained by a stronger influence of the active source on this electrode depending on its location. (This situation is exemplified in the potential divider theory of RALL and SHEPHERD 1968).

This estimate is in full accordance with calculations of laminar and surface gross potentials based upon a model population of pyramidal cells (HUMPHREY 1968). Furthermore, cortical surface potentials as well as potentials in cortical depth respond in parallel to changes of stimulus parameters, for example if a test response is conditioned by other stimuli (HUMPHREY 1968, SASAKI and PRELEVIC 1969). If one compares electrophysiological with anatomical data, there is thus considerable evidence that other current sources than IPSPs are not necessary to explain these intracortical potentials of the antidromic response. In this special case the cortical surface negativity is predominantly the result of postsynaptic inhibitory potentials at the PT cell soma and perhaps with the inclusion of proximal dendrites.

The antidromic cortical evoked response can in purity be elicited with comparably simple techniques. This fact can be especially useful if effects on postsynaptic inhibition, for example by drugs, are tested. In Fig. 3 the action of intravenously applied ammonium acetate solution in a dosage of 2.5 mM/kg bodyweight is shown. This substance reduces inhibitory hyperpolarization as is known from other evidence (LUX 1971) by reducing or eliminating a hyperpolarizing $Cl^-$ gradient across the neuronal membrane. It is interesting here to note that $NH_4$ acts on both the IPSP (after VL-stimulation) and the antidromically evoked cortical surface potential. However, the cortical surface potential evoked by stimulation of specific thalamic afferents (VL) remains at the same time virtually unaffected. This finding suggests that the major part of the surface potential after VL-stimulation comes from other sources than just from the recorded IPSPs.

## Potentials Generated by Thalamic Afferents

After thalamic (VL) stimulation a prominent surface positivity against a remote electrode is present at the time when most records of PT cells and also unidentified neurons show predominantly IPSPs

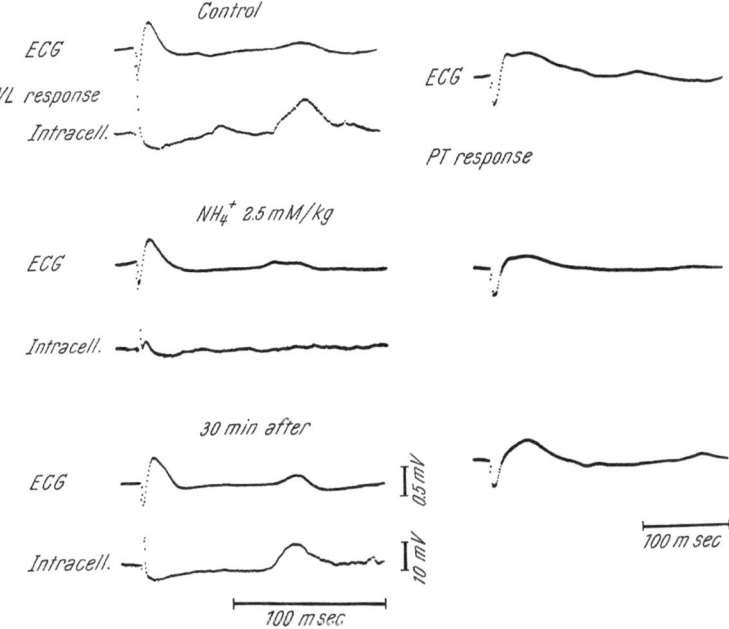

Fig. 3. The action of ammonium ions on the IPSPs and on the cortical surface potentials after pyramidal tract stimulation. Averaged records, N = 10. Left side of figure: *ECG* = electrocorticogram recorded during stimulation (1/sec) of the thalamic N. ventr. lat. (VL), Intracell. = simultaneous intracellular IPSP records, Control = the thalamically evoked IPSP and the electrocorticogram before $NH_4{}^+$ acetate application. Further records of these potentials are shown during slow intravenous infusion of $NH_4{}^+$ acetate solution (2.5 mM/kg bodyweight) and 30 minutes after end of $NH_4{}^+$ application. No effect of $NH_4{}^+$ is obvious on the cortical evoked potential despite a marked decrease in the IPSP amplitude. Right side of the figure: The averaged evoked cortical PT responses recorded immediately after each VL stimulation series (left side of the figure) show (middle trace) a reduction during $NH_4{}^+$ application which is comparable to the effect on the IPSP on the left side

(Figs. 2 *b* and 3). The laminar analysis of this evoked potential is distinctly different from the antidromic response (Fig. 4). There are

prominent differences in both the laminar potentials and intralaminar potential gradients at the time of the intracellular maximal
hyperpolarizing inhibition. It is probable that specific thalamic
afferents provide extra sources in more superficial layers by synaptic
effects on other neurons such as on stellate and spindle cells and,

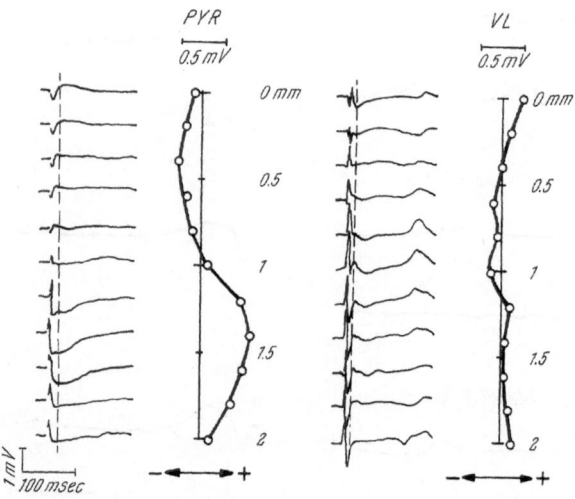

Fig. 4. Intracortical laminar analysis of the antidromic
pyramidal tract response (PYR) and of responses evoked by
thalamic (VL) stimulation. The intracortical evoked potentials are shown on the left of each plot. Evaluations were
made 16 msec after stimuli at about a time when evoked
IPSPs showed an amplitude maximum for both kinds of
stimuli. The plots show a considerable difference in intralaminar potential gradients at this time

especially, by eliciting considerable EPSPs also in apical remote
dendrites of PT cells.
However, one major target for specific thalamic afferents is most
likely the basal dendritic complex of PT cells. A large extracellular
potential gradient along the basal complex is shown in Fig. 2 b. It is
less the poor correspondence of the extracellular record with the
intracellular potentials that suggests the presence of additional
excitatory synaptic drives on this structure than a clearcut prominent
external positivity at the external electrode A which closely corre-

sponds to an apparent EPSP in the intracellular records (see points in Fig. 2 *b*). The polarity of this potential suggests a strong EPSP in the basal dendritic periphery.

## Conclusions

The methods described here make possible the collection of data on extracellular potentials generated by synaptic currents. It can be assumed that high impedance barriers for synaptically produced currents are less significant in the extracellular region, in contrast to the widespread intracellular potential distribution due to the properties of the soma-dendritic membrane. Extracellular synaptic potential should therefore be more restricted which facilitates the localization of synaptic current sources.

Concerning the synaptic genesis of the ECG the suggestion is made to explain cortical surface potentials as resulting from synaptic actions on different parts of the apical dendrites which extend upward to the cortical surface. The obvious phase relations between ECG waves and distinct kinds of postsynaptic potentials elicited from certain inputs were assumed to be the consequence of typical synaptic impingements. As for some examples, surface negative waves were ascribed to EPSPs on the upper parts of apical dendrites. Positive waves or diphasic positive negative wave-forms were explained by EPSPs evoked at the soma and electrotonically "conducted" along the apical dendrites. An inverse relation should apply for IPSPs. It can easily be deduced from available data of neuronal electrotonic characteristics and geometry that, except for the fast transients of evoked potentials, electrotonic conduction of PSPs is not apt to account for the phasic character of the ECG. Within the time domaine of most ECG waves, *i.e.*, the range of a few milliseconds to seconds, it is logical to apply a steady state analysis. Even with this simplification, the results of the present study suggest that general statements of the ECG electrogenesis are unwarranted until detailed knowledge is gained of the neuronal populatoin involved and of the geometry of typical activated cells, together with type of PSP, location and spatial extent of the evoked conductance change.

In one of the reported cases, the pyramidal tract response, it appears possible to reconstruct the slow surface response from locally determined synaptic activity. In this special case of a surface negativity accompanied by deep somatic IPSPs, considerable quantitative agreement is reached towards the hypothesis that recurrent somatic IPSPs mainly produce this surface negativity in respect to a remote electrode. This conclusion cannot be reversed for other evoked surface potentials of the same polarity or of negative-positive waveforms such as those resulting from stimulation of specific thalamic afferents. This is true even if in this situation IPSPs are predominantly recorded in most PT neurons and unidentified cells. It is very likely that intracellular records are selective and that as yet undetermined neuronal aggregates are involved in producing this surface response. Together with other probable sources, EPSPs on remote apical dendrites, secondary to the activity on the basal dendritic complex may also be considered in explaining the cortical surface response. In general, this evoked surface potentials cannot be taken to be representative for a significant part of synaptic activity evoked by thalamic afferents.

It appears necessary to extend and refine methods, such as have been applied in this study, for an intracortical determination of synaptic activity and synaptic targets. Only after a considerable collection of the electrophysiological and anatomical data will it be possible to analyze the variety of wave-forms recorded with a cortical surface electrode.

## Summary

A method has been developed to measure simultaneously intra- and extracellular potentials of single cortical nerve cells. With two extracellular electrodes at defined distances from an intracellular electrode (about 50 and 150 $\mu$) local potential gradients were measured which should be attributed to synaptic activity of a given cell.

Considerable local currents in the horizontal plane of the cell soma and basal dendrites were found during IPSPs evoked by antidromic pyramidal tract stimulation and by EPSPs and IPSPs sequences elicited by thalamic afferents. For an estimate of the contribution of IPSPs of single cells to the extracellular potential, the extracellular field during IPSPs were compared with that when the IPSPs had been brought to their reversal potentials by injection of $Cl^-$ or by hyperpolarizing currents. Evidence is given that the surface negativity after antidromic pyramidal tract stimulation is due to postsynaptic inhibition in PT cells. Other sources than those from IPSPs in deep cortical layers should be considered for the interpretation of the surface positivity after stimulation of specific thalamic afferents.

# References

CREUTZFELDT, O. D., J. M. FUSTER, H. D. LUX, und A. NACIMIENTO: Experimenteller Nachweis zwischen EEG-Wellen und Aktivität corticaler Nervenzellen. Naturwissenschaften *51*, 166—167 (1964).

— S. WATANABE, and H. D. LUX: Relations between EEG phenomena and potentials of single cortical cells. I. Evoked responses after thalamic and epicortical stimulation. Electroenceph. clin. Neurophysiol. *20*, 1—18 (1966).

ECCLES, J. C.: Interpretation of action potentials evoked in the cerebral cortex. Electroenceph. clin. Neurophysiol. *3*, 449—464 (1951).

FREYGANG, W. H., and W. M. LANDAU: Some relations between resistivity and electrical activity in the cerebral cortex of the cat. J. Cell. Comp. Physiol. *45*, 377—392 (1955).

HUMPHREY, D. R.: Re-analysis of the antidromic cortical response. II. On the contribution of cell discharge and PSPs to the evoked potentials. Electroenceph. clin. Neurophysiol. *25*, 421—442 (1968).

JASPER, H. H., and C. STEFANIS: Intracellular oscillatory rhythms in pyramidal tract neurons in the cat. Electroenceph. clin. Neurophysiol. *18*, 541—553 (1965).

KANDEL, E. R., and W. A. SPENCER: Excitation and inhibition of single pyramidal cells during hippocampal seizure. Exptl. Neurol. *4*, 162—170 (1961).

LI, C.-L., and S. N. CHOU: Cortical intracellular synaptic potentials and direct cortical stimulation. J. Cell. Comp. Physiol. *60*, 1—16 (1962).

LUX, H. D.: Ammonium and chloride extrusion: Hyperpolarizing synaptic inhibition in spinal motoneurons. Science *173*, 555—557 (1971).

— und M. R. KLEE: Intrazelluläre Untersuchungen über den Einfluß hemmender Potentiale im motorischen Cortex. I. Die Wirkung elektrischer Reizung unspezifischer Thalamuskerne. Arch.-Psychiatr. Nervenkr. *203*, 648—666 (1962).

— and D. A. POLLEN: Electrical constants of neurons in the motor cortex of the cat. J. Neurophysiol. *29*, 207—220 (1966).

NACIMIENTO, A. C., H. D. LUX und O. D. CREUTZFELDT: Postsynaptische Potentiale von Nervenzellen des motorischen Cortex nach elektrischer Reizung spezifischer und unspezifischer Thalamuskerne. Pflüg. Arch. Ges. Physiol. *281*, 152—169 (1964).

POLLEN, D. A.: Intracellular studies of cortical neurons during thalamic induced wave and spike. Electroenceph. clin. Neurophysiol. *17*, 398—404 (1964).

— and H. D. LUX: Conductance changes during inhibitory postsynaptic potentials in normal and strychninized neurons. J. Neurophysiol. *29*, 369—381 (1966).

PURPURA, D. P., R. J. SHOFER, and F. S. MUSGRAVE: Cortical intracellular potentials during augmenting and recruiting responses. II. Patterns of synaptic activities in pyramidal and non-pyramidal tract neurons. J. Neurophysiol. *27*, 133—151 (1964).

RALL, W., and G. M. SHEPHERD: Theoretical reconstruction of field potentials and dendrodendritic synaptic interactions in olfactory bulb. J. Neurophysiol. *31*, 884—915 (1968).

RAMON-MOLINER, H.: The histology of the postcruciate gyrus in the cat. I. Quantitative studies. J. comp. Neurol. *117*, 43—62 (1961).

SASAKI, K., and S. PRELEVIC: Antidromic invasion of impulses and recurrent collateral inhibition in pyramidal tract neurons. Brain Res. *17*, 355—359 (1970).

STEFANIS, C., and H. H. JASPER: Intracellular microelectrode studies of antidromic responses in cortical pyramidal tract neurons. J. Neurophysiol. *27*, 828—854 (1964).

TAKAHASHI, K.: Slow and fast groups of pyramidal tract cells and their respective membrane properties. J. Neurophysiol. *28*, 908—924 (1965).

# Randomness and Synchrony in the Generation of the Electroencephalogram [1]

R. ELUL

Department of Anatomy and Brain Research Institute, School of Medicine,
University of California, Los Angeles, California

## Subcortical Inputs in Generation of the EEG

The question whether the EEG is dependent upon subcortical inputs or control has never been completely clarified. BURNS (1950, 1951) demonstrated that cortical slabs which have been completely undercut are devoid of spontaneous activity and respond to direct shocks only by brief bursts of activity. However, there have also been conflicting experiments in which some residual activity appeared to have remained in undercut cortex (KRISTIANSEN and COURTOIS 1949, ECHLIN et al. 1952, HENRY and SCOVILLE 1952, INGVAR 1955), although the possibility that this residual activity may be in the nature of injury discharge cannot be ignored. Essentially the same considerations apply to recordings from the frog brain. GERARD was able to record slow potential activity from small fragments of the olfactory bulb of the frog (GERARD 1936, GERARD and LIBET 1940). Such potentials can indeed be recorded but, whereas the amplitude of EEG recorded from the intact frog brain even under the best of conditions does not appear to exceed 20–30 $\mu$V, the potentials recorded from small fragments of the brain are of the order of 200 $\mu$V and very significantly differ in their appearance from the normal EEG of the frog. (Indeed this seizure-like appearance is evident in the original records published by GERARD.) Thus the potentials recorded from the isolated olfactory lobe in the frog unquestionably represent bioelectric activity, but it seems much more dubious that these potentials may legitimately be identified as "EEG".

1 This research has been supported by NIH grants NS-8012 and -8498, and by NSF grant GB-30498.

There exist very extensive anatomical pathways linking the cortex directly with the thalamus and, via thalamic relays, also with the brain stem reticular formation. Functionally, the stimulation of these subcortical centers induces specific changes in the EEG recorded from the cortical hemispheres. As is well known, high-rate stimulation of the reticular formation results in an increase in frequency of cortical EEG associated with reduction in its amplitude. Stimulation of non-specific thalamic nuclei, on the other hand, produces recruiting as well as augmenting responses at the cortical surface. More recently, evidence has been produced also on thalamocortical connections from the specific thalamic nuclei (ANDERSEN and SEARS 1964) and the suggestion has been made that the specific thalamic nuclei act as pacemakers for the EEG (ANDERSEN and SEARS 1964, ANDERSEN and ANDERSSON 1968).

Parenthetically, reviewing the arguments cited by ANDERSEN and ANDERSSON (1968), one may retain certain skepticism in face of the evidence adduced by these authors. For one, it appears rather unlikely that spindle sleep in the cat may be considered an analogue of alpha activity in man—a claim which is one of the cornerstones of ANDERSEN and ANDERSSON's monograph. If the cortex plays any significant role in controlling behavior, it is unlikely that two diverse activities in two mammals, such as sleep (in case of spindle waves in the cat) and restful wakefulness (alpha activity in man), would be characterized by similar EEG patterns. Undeniably, there are functional correlations between unitary neuronal activity in the specific thalamic nuclei and in gross cortical activity (ANDERSEN and SEARS 1964); but these may be in response to some input common to the cortex and the thalamus rather than the origin of cortical wave sequences. Another argument presented by ANDERSEN and ANDERSSON relies on computer simulation of the EEG: in an artificial nerve net of interconnected neurons, these authors have obtained rhythmic activity subsequent to application of repetitive extraneous input (even in these conditions rhythmicity of activity in this model was not very evident, so that autocorrelation had to be used to bring out the weak rhythmic component). However, ANNINOS, who has made a thorough study of cyclic activity in nervous nets, has shown in theoretical analysis (ANNINOS 1969) as well as in simulation studies (ANNINOS 1971, ANNINOS and ELUL 1972) that very clear rhythmicity (indeed, much clearer than in ANDERSEN's model) may quite commonly arise in specific classes of nerve nets even when completely devoid of any extraneous input whatsoever.

Summarizing the information reviewed in this section, the available evidence speaks for, as well as against, the requirement for subcortical pacemakers in generation of the EEG. Experiments involving iso-lation of cortical slabs, or analysis of corticothalamic interactions, are suggestive of the necessity of subcortical inputs, but it is not possible to reach a firm decision based on these data.

## Synchronizing and Randomizing Trends in the EEG

At this point it may be useful to pause and consider the problem of a subcortical pacemaker for the EEG from a somewhat different perspective.

Earlier work by the present author and by others (ELUL 1964, 1966, 1968, CREUTZFELDT et al. 1964, 1966, 1969, JASPER and STEFANIS 1965, POLLEN and MARSAN 1965) has shown that cortical neurons undergo sustained oscillations of their membrane potential. An

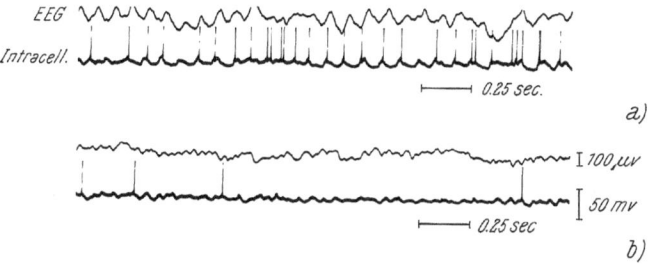

Fig. 1. Wave activity in cortical nerve cells. Intracellular recording from a cortical neuron, 600 $\mu$ depth (lower tracing in a and b). EEG activity recorded simultaneously from the cortical surface by a large electrode (2 mm$^2$) is given in the top tracing. a and b are from the same cell, but animal is sleeping in a and awake in b. Note the similarity between intracellular wave activity and EEG; the two activities change in a similar way between a and b

example of this characteristic activity of cortical neurons is shown in Fig. 1. This figure is taken from an experiment involving intracellular penetration of a cortical neuron. The animal is unanesthetized and no artificial stimuli have been used. This is therefore a recording of spontaneous activity. The top tracing is the EEG recorded from the cortical surface with a large electrode (approximately 2 mm$^2$) located close to the site of intracellular impalement. The lower tracing, which was obtained from a glass micropipette (20 megohm resistance), represents activity inside a single cortical nerve cell, about 600 $\mu$ below the surface. Note that while the calibration for the EEG is in the microvolt range, the intracellular wave potentials reach several tens of millivolts. Intracellular activity also is spontaneous and consists of continuous subthreshold waves which occasionally lead to the discharge of a spike. Discussion of the nature of these waves, which are the hallmark of cortical neurons, is beyond the scope of this paper, but most likely they represent synaptic potentials, possibly with an added endogenous rhythmic component originating in remote dendrites (*cf.*, ELUL and ADEY 1972). It is quite evident from Fig. 1 that the unitary wave potentials closely resemble the EEG in their general appearance (see also ELUL 1968). Glial cells in the cerebral cortex do not exhibit similar oscillations (ELUL 1968). Moreover, since the potentials recorded by intracellular electrodes in cortical neurons on the average are 100 times larger than the EEG recorded from the cortical surface with a gross electrode (or for that matter, from the depth with extracellular microelectrodes; see LI and JASPER 1953, VERZEANO and NEGISHI 1960, ELUL 1962, 1968), clearly the extracellular EEG cannot be the source of the intracellular potentials.

This is because the intracellular potential, as well as the EEG (either from an extracellular microelectrode or from a "gross" electrode) are generally recorded against a common reference point. In this situation, a large intracellular potential might arise from a primary extracellular source, only if the extracellular potential is even larger. Since extracellular exploration never reveals sources of the required magnitude (see references cited above), the possibility that intraneuronal waves reflect extracellular EEG activity cannot be accepted.

In volume conductors such as the brain, the potentials recorded between an intracellular electrode and a remote reference point can only arise as voltage drops caused by the flow of current between these two electrodes. Inasmuch as the reference electrode against which the potentials are recorded is extracellular, it is evident that recordings of intracellular waves, such as in Fig. 1, must be associated with transmembrane currents flowing from the interior of the cell to the extracellular space. In reality, the destination of such currents is not the indifferent electrode. Neither do they arise at the focal microelectrode. Rather, these two electrodes merely act as probes of the current flows associated with synaptic potentials and possibly also with other mechanisms which affect localized patches of the neuronal membrane. Temporary changes of membrane conductance at these locations result in ion movement into, or out of, the neuron. Since such ion movements lead to deviations from the resting potential, the normal ionic selectivity and pumping mechanisms acting on the rest of the cell surface at once restore the resting potential to its level prior to the arrival of the synaptic potential. If, for example, a certain number of anions are lost at a synapse, an essentially equal amount of anions is introduced across the nonsynaptic membrane.

In practice, such movements of charge constitute an electric current, in the intra- as well as extracellular compartments. In turn, these currents induce voltage drops across the cytoplasmic and extracellular space resistances, respectively. It is these voltage drops which are registered by the micropipette and by the indifferent electrode.

Thus, the extracellular space is permeated with current flow originating in and returning to given individual neurons. The resistances of specific extracellular clefts surrounding these neurons combine to perform as a summing resistive network. It is well to recall at this point that, as seen in Fig. 1, the activity recorded from each cell and therefore by necessity also the extracellular current generated by it, is EEG-like in appearance. It is therefore reasonable to assume that the summation of a multitude of neuronal currents could result in EEG activity in the extracellular space.

The significant question at this point relates to the manner of summation of these currents. On an *a priori* basis one may recognize two distinct modes: summation of a random populations of generators, and summation of a synchronized generator population. The latter case is quite obvious: the summed activity by necessity would represent a close replica of the wave activity of any given cell. Insofar as the wave activity of individual neurons has EEG-like characteristics (Fig. 1), this model at first sight does not appear to present any difficulty.

That a population of randomly-related generators also would yield activity essentially indistinguishable from that of a synchronized population is somewhat more surprising. Yet this is the case. To understand the underlying mechanism, consider the sum of two sinusoid potentials of the same frequency (say, 6 c/sec). Unless these waves are exactly opposed in phase, they will sum to form a sinusoid at the same frequency of 6 c/sec. If a third sinusoid at 6 c/sec is now added to the summed waveform, again the output will be at 6 c/sec, and so on for any number of independent sinusoids. Although cortical neurons do not produce pure sinusoidal waves the same principle applies, because, by Fourier's theorem, each complex waveform can be broken down to a number of pure sinusoids. These sinusoids, taken one frequency at a time, can be combined from the respective cells in like manner, yielding a summed output with the same frequency content as the component complex waveforms.

The conclusion, then, is that a synchronized group of neurons as well as a randomly-related generator population are potentially capable of producing EEG activity, and summed activities from these two systems may well appear closely similar. The principal difference between the two systems is in relative amplitude and, as will be seen below, this difference provides a most important clue to the actual mechanism of generation of the EEG. However, inasmuch as the *a priori* considerations equally favor both models of the EEG (namely, origin in a synchronized or in a random population), it is necessary to inquire whether either of these two models is supported by experimental findings.

### Experimental Evidence for Synchrony of EEG and Neuronal Wave Activity

In comparing intracellular recordings with the surface EEG recorded nearby at the same time from a gross electrode, quite frequently one may observe brief periods of apparent correlation (Fig. 2). Here, again, the recording convention is the same as in Fig. 1, with the gross EEG appearing as the uppermost tracing and the intracellular activity below it. As can be seen, the area between the dashed lines characterized by a close correspondence of peaks in the EEG with peaks in the intracellular record. At first sight, such correlations appear to be highly significant; yet on continuing inspection of longer stretches

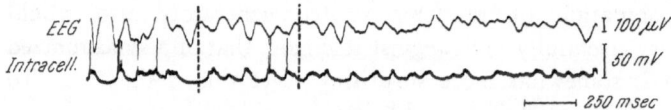

Fig. 2. Correlation of neuronal waves and EEG. Intracellular recording from another cortical neuron, 500 $\mu$ depth. Other details as in Fig. 1. Activity between dashed lines shows clear correlation of EEG with unitary neuronal waves. However, immediately following this segment, an opposite relationship is evident, and later it is difficult to see any constant relationship whatsoever

of activity from the same cell, side-by-side with the surface EEG, invariably it is found that the particular relationship observed is not replicated. (For instance, a negative shift in the EEG may correlate with cellular depolarization at one point and with an hyperpolarization several tens of seconds later.) In fact, analyses of recordings of long duration reveal the total count of correlations over several minutes to be very close to that expected by chance alone.

A quantitative test of the relationship between the EEG and unitary wave activity may be made using the coherence function. This function (GOODMAN 1957, WALTER 1963, WALTER and ADEY 1964) is essentially a measure of the conditional probability of process B attaining a certain value $X_B$ at the same time that process A has the value of $X_A$. An additional feature of coherence calculations is that they are performed on the pure sinusoids obtained from Fourier decomposition of the complex intracellular and surface waveforms. Fig. 3 summarizes coherence measurements made over a period of 200 sec. Each of these measurements yields a probability figure, giving a measure of the likelihood of positive correlation between the intracellular wave activity and the

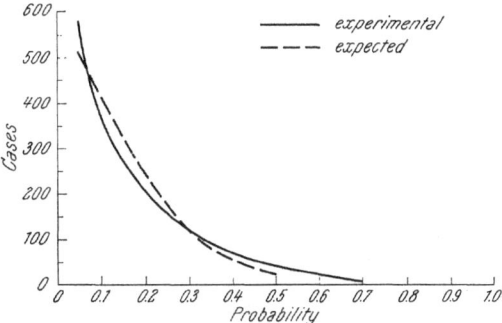

Fig. 3. Correlation between EEG and unitary neuronal waves (solid line) compared to correlation between two independent processes (interrupted line). Further explanation is in text

EEG. The spectral analysis procedure used here involves 30 filters (that is, the complex waveform is decomposed into 30 discrete frequencies), so that there are 30 coherence figures for each 10 sec epoch. There are thus a total of $20 \times 30 = 600$ cases, and their distribution is given as a plot of the number of cases versus the probability figure for each case. This graph is presented in Fig. 3 as a solid line. The distribution of results which would have been obtained if the same coherence test, using the same filters, etc., were applied to examine the coherence between two entirely unrelated processes is given in this figure as a broken line. As may be seen in Fig. 3, the cumulative distribution of correlations in the experimental data is close to that found for two wave processes which are entirely unrelated. However, to gain statistical reliability, averages of over 10 sec of the coherence calculations for any particular frequency must be used in this test. Since the correlation shown in Fig. 2 lasted only 1–2 sec, it is evident that this approach, while yielding reliable results, is not particularly sensitive to brief events.

The issue is complicated even further when it is realized that even striking correlations such as in Fig. 2 may arise by chance alone. To illustrate this point, the intracellular activity from one cell recorded on magnetic tape has been played back onto a chart recorder. Alongside, as the top tracing, EEG from another cat has been recorded (Fig. 4). It is evident that activities from these two animals, which were recorded at different times, show quite striking correlations.

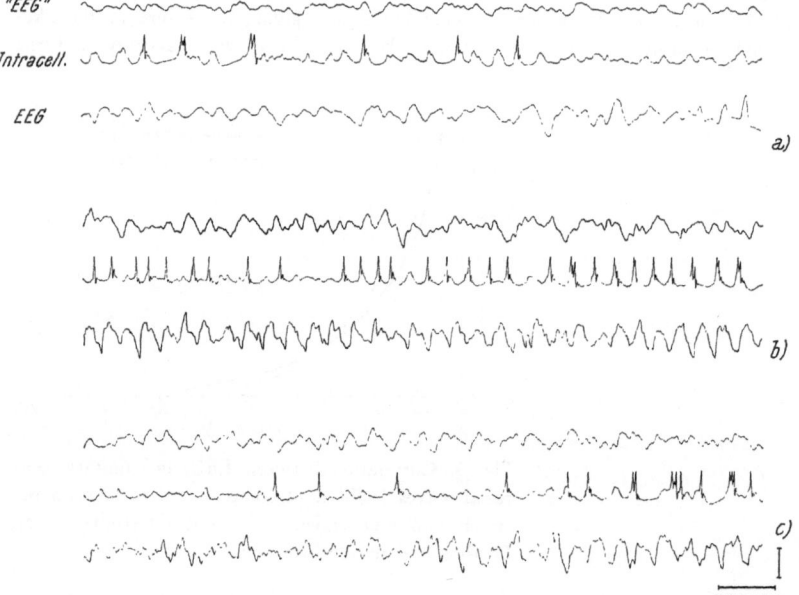

*"EEG"*

*Intracell.*

*EEG*

*a)*

*b)*

*c)*

*250 msec*

Fig. 4. Apparent correlation between EEG from one cat (top tracing in *a–c*) and intracellular activity from a second cat (middle tracing). EEG of the second animal is included for comparison (lower tracing). *a–c* are three different examples from same nerve cell. Activity of nerve cell and EEG (middle and lower tracings) were registered with a chart recorder. EEG activity from the other cat (top tracing) was played back off magnetic tape and recorded subsequently. Note that the top tracing often shows better correlation with the intracellular activity than the EEG activity from the same animal. This is a consequence of the narrow frequency band of the EEG. See text for further discussion. (Modified from ELUL 1972)

EEG activity from the same animal which provided the intracellular recording is given for comparison in the bottom tracing of Fig. 4 *a–c*. The explanation for this bizarre coincidence is not in any form of contact between the two animals, but simply in the fact that the EEG, phenomenologically, constitutes what is known in communication-theory terms as "narrow-band process". The maximal amount of information contained in a complex wave-form in general is inversely proportional to its bandwidth (SHANNON 1948 a, b). With a process such as the EEG, which has a relatively narrow frequency band, this implies in practical terms that an observer may often find it difficult to distinguish between EEG tracings from two sources. Likewise, when two unrelated tracings are placed side-by-side, not uncommonly they may appear correlated, but this is merely because the waveforms, due to their narrow frequency band, do not exhibit sufficient variability of deflection patterns.

Thus, one is able only to summarize the data presented here in saying that the evidence for correlation between the EEG and intracellular waves is inconclusive; if there is any correlation, it must be only of short duration (*i.e.*, less than 10 seconds) and may occur only rather infrequently, for otherwise it would have been apparent in the sensitive "coherence" test.

## Do the Experimental Results Support the Possibility of Random Relationship between the EEG and Unitary Neuronal Wave Activity?

The present author has carried out a detailed investigation of this question (ELUL 1966, 1968, 1969, 1972). These studies lead to the conclusion that, when viewed over relatively long periods of time (several minutes), the EEG exhibits the statistical properties of a normal (Gaussian) random process. Intracellular wave activity, in contrast, does not follow a normal distribution. Recalling that the EEG represents a summation of these non-Gaussian cellular waves, it becomes apparent that in order to produce the randomness observed in the gross EEG, their individual generators should be randomly related to one another (ELUL 1966, 1968).

This statement merely constitutes a rephrasing of the Central Limit Theorem of statistics (*e.g.*, CRAMÉR 1955). Evidently, if the cortical generators all were synchronized among themselves, the resultant sum would constitute a replica of unitary activity and hence must be non-Gaussian. It is only owing to the random variation of the phase relationship between non-Gaussian generators which are not synchronized that the summed activity becomes Gaussian.

In conclusion, two conflicting lines of evidence have been presented. Brief periods of synchrony between the EEG and individual neuronal wave activity can be detected, but these as a rule last only a few seconds. When one is able to follow over longer time the activity of any given neuron, some periods of synchrony may be seen, but most of the time the relationship between the unitary events and the EEG appears to be random. Moreover, when this problem is approached from a different angle, investigating the type of interneuronal relationships which must exist in order to satisfy the statistical properties found for the gross EEG, one is led to conclude that the EEG cannot remain synchronized with any given cell over any extended period of time.

These conflicting results seem to exclude any model representing the EEG as the product of a synchronized neuronal population. On the other hand, the EEG might result from random interactions between activities of essentially independent neurons; this, however, would make rather high the odds for encountering synchronous events such as seen in Fig. 2. It appears that the ultimate explanation should allow for synchrony as well as for randomness in interactions between neurons. Whereas randomness may well thrive on the absence of neuronal connections, synchrony by necessity implies some form of interneuronal communication. Whether this communication is at the corticocortical level or depends on corticothalamic communication is not clear at this stage, but the experiments described in the following section shed some light on the role played by corticothalamic connections in synchronization of cortical neuronal wave activity.

## Subcortical Synchronizing of Neuronal Wave Activity in the Cerebral Cortex

An important new approach to the mechanisms of generation of the EEG is through use of tetrodotoxin. The chief effect of this drug is to block spike activity. Secondarily, it therefore blocks synaptic potentials, since the latter arise only following depolarization of the presynaptic terminal by a spike. It has been found that tetrodotoxin injected into the lateral brain ventricle in small dose (0.1–0.3 $\mu$g) induces within 15–20 minutes intermittent "pauses" in the EEG (ELUL 1971). These pauses are periods of complete (or sometimes only relative) isoelectricity (Fig. 5). This drug thus offers an opportunity for direct comparison of "on" and "off" states in the EEG. One may make *a priori* predictions as to the mechanism of the EEG pauses of tetrodotoxin, depending on whether the EEG is produced from a synchronized or random population. With a mechanism based on random interneuronal relations, the EEG is the summed activity of all

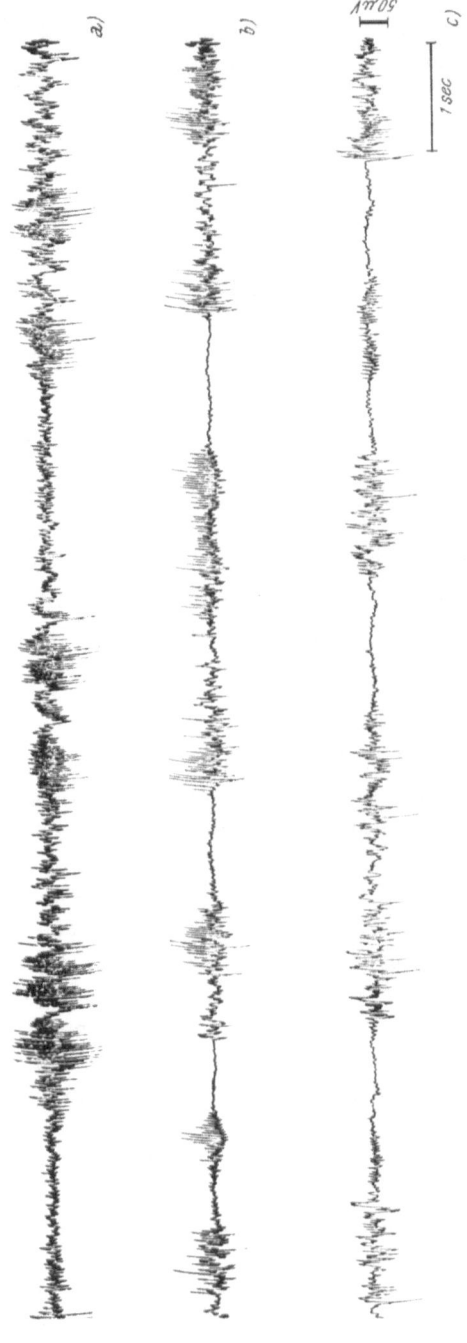

Fig. 5. Effect of intraventricular tetrodotoxin on EEG. *a*) taken just before injection of the drug, *b*) 10 minutes later, and *c*) 20 minutes following the injection. Bipolar recording from surface electrodes making contact with the dura (parietal-parietal). Calibration: 1 second, 100 μV. (Modified from ELUL 1972)

neurons present in the tissue. This summed activity will be reduced in amplitude if, and only if, the wave output of individual neurons also is reduced to the same degree; for example, if the amplitude of the gross EEG is measured in two experimental states and found to have been changed by a factor of 10, a similar tenfold change should be observable, on the average, also in the wave activity of individual neurons.

It is relevant to stress the importance in the preceding sentence of the phrase "on the average." A tenfold change in amplitude of the gross EEG could be achieved also by "dropping out" of 90% of the nerve cell population. However, in this situation one would be quite likely to observe cells which have previously been active to cease wave activity in synchrony with the change observed in the gross EEG. Indeed, such observations should be the rule since the probability of impaling one of these cells would be 0.9 as against a probability of 0.1 of impaling a cell which persists in being active during the "pause."

With a synchronized population, a tenfold change in EEG amplitude also may ensue from reduction in activity in all cortical neurons involved. However, an additional alternate mechanism for reduction in amplitude exists in this situation. This mechanism involves desynchronization. It is quite obvious that the maximal output from a group of generators is attained when they are all in synchrony; even if the output from each generator is not decreased, the summed output will diminish when the generators lose synchrony.

The magnitude of the change in output from a group of synchronized generators due to desynchronization is often not appreciated. In general, if the output from a single generator is V, then given a population of N generators which are synchronized, the summed output will be NV. When the generators lose synchrony among themselves, the output will decrease to $\sqrt{N}V$ (cf., Elul 1966, 1968). For any number of generators N which is realistic in terms of cortical nerve cell populations, the changes in amplitude involved in desynchronization can be quite staggering. As an example, consider a population of one million generators ($10^6$); the summed output from this population would decrease in amplitude 1000 times in the transition from synchrony to randomness. Smaller changes (e.g., tenfold) would require only very minimal divergence from total synchrony.

It is important to stress that with this mechanism, an observer need not note a change in activity of any given individual generator. All that would change is the interneuronal relationship. Thus we have a critical test between the two mechanisms: if the EEG is produced

by a randomly-organized population, the pauses induced by tetrodotoxin must be accompanied by corresponding reductions in wave
activity on the single-neuron level. Furthermore, these reductions
should be synchronized with the pauses appearing in the EEG. In
contrast, if the EEG represents the product of a synchronized population of unitary generators, there need not be any change in the
output from individual cells examined in intracellular recording, as

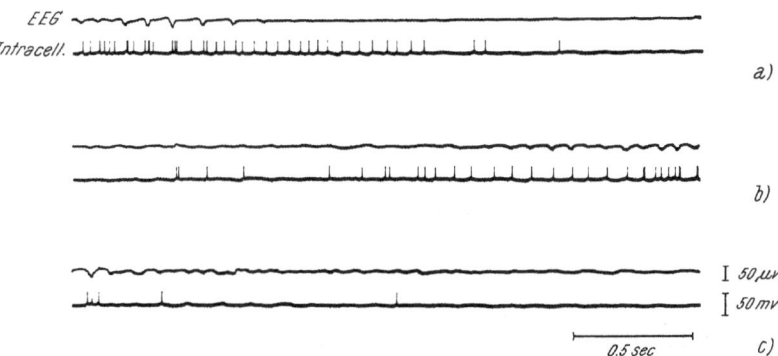

Fig. 6. Effect of tetrodotoxin on unitary neuronal activity. *a–c* is a continuous
record. Top tracing—surface EEG from a gross electrode. Lower tracing—intracellular recording from cortical neuron $1100\,\mu$ depth. *a* "pause" in the gross
EEG starts in the 2nd third of *a* and continues to *b*. Note that neuronal firing
and subthreshold wave activity continue well into the "pause", although the
firing decreases and eventually stops altogether. (Modified from ELUL 1972)

their desynchronization by itself would readily bring about flattening
of the EEG.

Fig. 6 presents a continuous recording from a cortical neuron containing segments during an EEG "pause" and preceding and following
it. The top tracing represents the surface EEG recorded by a gross
electrode adjacent to the micropipette. It is quite evident that the
neuron in this figure (which is typical of all cells explored so far) does
not block its firing, nor is its wave activity diminished during the EEG
pause. In the latter half of the pause (Figs. 6 *a–b*), firing is reduced,
but the subthreshold activity which, in the context of the EEG plays
a far more significant role than spike activity, persists throughout.
Thus, flattening of the gross EEG cannot be attributed to a reduction
in wave activity in the entire neuronal population. This immediately
precludes generation of the EEG by a randomly-organized popu-

lation. Rather, the results shown here support the concept of the EEG as the output from a synchronized population. That this in fact may be the case is suggested from the reduction in spike discharge associated with EEG pauses, for these spikes provide the vehicle for communication among cells and any reduction in spike output is therefore likely to have a desynchronizing effect.

How can these results be reconciled with the evidence presented earlier that no cell maintains a constant synchronized relationship with the EEG? (As a matter of fact, the absence of sustained synchrony is evident even in Fig. 6.) This difficulty may be circumvented by two elementary and plausible assumptions: first, that only a small fraction of the total cell population need be synchronized to produce the gross EEG.

To make this assumption more concrete, let us consider again a hypothetical population of one million generators. If only one percent of the cells are synchronized, and the contribution from each cell to the gross electrode is, say, 0.01 $\mu$V, the total voltage produced by this small fraction of the population will be $10^4 \times 0.01 = 100 \ \mu$V. The rest of the population (*i.e.*, 99% of it), being desynchronized, will produce only $0.99 \times 10^6 \times 0.01 \approx 10^3 \times 0.01 = 10 \ \mu$V. Thus, even one percent of the total population, when synchronized, overshadows the combined activity of the rest of the neuronal population.

The second assumption simply is that different groups of cells may be synchronized to produce the EEG in successive instants in time. Such "sweeping" through the population would cause each individual cell to be synchronized with the EEG for only brief, fleeting intervals, thus accounting for example, for the short but striking synchronization of intracellular wave activity with the EEG seen in Fig. 2.

The physiological and functional implications of this model are examined in greater detail elsewhere (ELUL 1971, 1972). The present discussion has opened with the inquiry whether production of the EEG may require subcortical "pacemaking". We have already mentioned that synchrony most likely would involve a subcortical pacemaker. Fig. 6 provides more direct information on this question. Spikes are present in the intracellular recording both before and during EEG pauses. Now tetrodotoxin blocks the sodium current responsible for the depolarizing part of the spike (NARAHASHI *et al.* 1964, HUBBARD *et al.* 1967, BLANKENSHIP 1968, COLOMO and ERULKAR 1968). Cortical neurons are susceptible to this drug to the same degree that spinal and peripheral neurons are, as can be demonstrated through topical application of tetrodotoxin on the cortical surface; within a few minutes the drug applied in this manner abolishes all spike discharges (ELUL 1971, 1972). Thus the presence of spikes in Fig. 6 indicates that tetrodotoxin applied into the ventricle has not in these experiments reached the cortex in any effective concentration. (Indeed, the diffusion of tetrodotoxin in the brain is exceedingly slow. See HAFEMAN et al. 1969, HAFEMAN and HOUSTON 1971.) Hence, we must conclude that the presence of "pauses" in the EEG is due to a subcortical effect of tetrodotoxin. There must therefore be some subcortical center which, when depressed by tetrodotoxin, may cause cortical neurons to desynchronize, resulting in flattening of the gross EEG. According to this view, the EEG is entirely dependent upon subcortical input, and in the absence of such input, the desynchronized activity falls below the resolution level of available recording instruments.

## Neuronal Behavior in Epileptic Discharge

*Is there any increase in interneuronal synchronization during epileptic discharges?* Probably the most important feature of neurons in an experimental epileptic focus, such as caused by application of penicillin, is the presence of "Paroxysmal Depolarization Shifts". This potential, described by MATSUMOTO and MARSAN (1964 a, b, see also MATSUMOTO et al. 1969, PRINCE 1966, 1967, 1968, DICHTER and SPENCER 1969), is similar in many respects to an EPSP, having a variable duration of several tens of msec and variable amplitude. However, it is distinguished from EPSP's in normal neurons chiefly by its amplitude which often reaches 20–30 mV. Assuming no change during seizure in the basic mechanisms of EEG generation discussed in the preceding sections, it is clear that the increase in the output of each neuron due to this large depolarizing potential would act to increase the amplitude of the gross EEG to the same degree, *i.e.*, 2–20 fold. This by itself would be sufficient to account for the larger amplitude of many, but not all, epileptic potentials in the gross EEG. The characteristic increase in amplitude of the gross EEG to around 1 mV during epileptic seizure may adequately be accounted for by Paroxysmal Depolarization Shifts. However, in many generalized seizures one encounters gross EEG potentials of the order of 5–10 mV. Since at the most each neuron may depolarize down to the EPSP equilibrium potential, *i.e.*, a maximal shift of about 50 mV, clearly the huge potentials in the gross EEG cannot be explained simply in terms of increased output of each neuron without a change in size of the synchronized neuronal pool participating in the generation of gross brain potentials. Rather, EEG potentials in excess of 1–2 mV probably require hypersynchrony of the cortical neuronal population and, specifically, such large EEG potentials could only be possible if the percentage of synchronized generators in the cortex were increased.

The cellular mechanism responsible for generation of the Paroxysmal Depolarization Shifts is not clear, but inasmuch as the properties of the membrane do not seem to be altered, it has been suggested that the Paroxysmal Depolarization Shift represents the synchronized activation of a large number of excitatory synapses on the same neuron (MATSUMOTO et al. 1969). This mechanism may provide also the means for synchronization of a larger number of cortical neurons. The early phase of the Paroxysmal Depolarization Shift is characterized by a rapid sequence of spikes. Such spike trains, arriving in close proximity, would tend to depolarize other, efferent, nerve cells and induce in them firing in synchrony with the Paroxysmal Depolarization Shift. The fundamental neuronal mechanism of temporal summation may in this way operate to increase gradually the size of the synchronized generator pool, recruiting to paroxysmal activity an increasingly large portion of the cortical neuronal population. This recruitment process would lead in the gross EEG to a continuing increase in amplitude of the paroxysmal potentials, a phenomenon which characterizes the generalization of many types of seizure activity.

## Summary

This presentation is concerned with the role of subcortical inputs in generation of the EEG. Previous evidence on the role of subcortical centers in control of cortical rhythms is briefly reviewed. Since subcortical control is likely to have a synchronizing effect, evidence in favor of, as well as against synchrony of unitary neuronal wave activity with the EEG is reviewed. New data from experiments involving intraventricular injections of tetrodotoxin are presented. From these and previous results it is concluded that the EEG represents the output of a relatively small group of cortical neurons which are synchronized by subcortical inputs. Activity of the bulk of the neuronal population is not synchronized and is below the resolution level of EEG recording machines. The evidence from the tetrodotoxin experiments indicates that the groups of synchronized neurons which generate the EEG are not fixed; rather, they are temporarily selected from the total neuronal pool in the cerebral cortex, so that any given nerve cell does not remain synchronized with other cells—and with the EEG—over any extended period of time.

## References

ANDERSEN, P., and S. A. ANDERSSON: Physiological Basis of the Alpha Rhythm. New York: Appleton-Century-Crofts. 1968.
— and T. A. SEARS: The role of inhibition in phasing of spontaneous thalamo-cortical discharge. J. Physiol. *173*, 459—480 (1964).
ANNINOS, P. A.: Dynamics and function of neural structures. Doctoral dissertation, Syracuse University, Syracuse, N.Y., 1969.
— Cyclic modes in probabilistic neural nets. Submitted for publication (1971).
— and R. ELUL: A neural net model of the alpha rhythm. In preparation (1972).
BLANKENSHIP, J.: Action of tetrodotoxin on spinal motoneurons of the cat. J. Neurophysiol. *31*, 186—194 (1968).
BURNS, B. D.: Some properties of the cat's isolated cerebral cortex. J. Physiol. *111*, 50—68 (1950).
— Some properties of isolated cerebral cortex in the unanesthetized cat. J. Physiol. *112*, 156—175 (1951).
CRAMER, E. G. H.: The Elements of Probability Theory, pp. 168—171. New York: Wiley. 1955.
CREUTZFELDT, O. D., J. M. FUSTER, H. D. LUX, und A. C. NACIMIENTO: Experimenteller Nachweis von Beziehungen zwischen EEG-Wellen und Aktivität corticaler Nervenzellen. Naturwissenschaften *51*, 166—167 (1964).
— S. WATANABE, and H. D. LUX: Relations between EEG phenomena and potentials of single cortical cells, II. Spontaneous and convulsoid activity. Electroenceph. clin. Neurophysiol. *20*, 19—37 (1966).
— H. ROSINA, M. ITO, and W. PROBST: Visual evoked response of single cells and of the EEG in primary visual area of the cat. J. Neurophysiol. *32*, 127—139 (1969).
COLOMO, F., and S. D. ERULKAR: Miniature synaptic potentials at frog spinal neurones in the presence of tetrodotoxin. J. Physiol. (London) *199*, 205—221 (1968).
DICHTER, M., and W. A. SPENCER: Penicillin-induced interictal discharges from the cat hippocampus, I. Characteristics and topographical features. J. Neurophysiol. *5*, 649—663 (1969).
ECHLIN, F. A., V. ARNETT, and J. ZOLL: Paroxysmal high voltage discharges from isolated and partially isolated human and animal cerebral cortex. Electroenceph. clin. Neurophysiol. *4*, 147—164 (1952).
ELUL, R.: Dipoles of spontaneous activity in the cerebral cortex. Exptl. Neurol. *6*, 285—299 (1962).

ELUL, R.: Specific site of generation of brain waves. The Physiologist 7, 125 (1964).

— Statistical mechanisms in generation of the EEG. Progr. Biomed. Eng. 1, 131—150 (1966).

— Brain waves: Intracellular recordings and the origin of the EEG. In: Data Acquisition and Processing in Biology and Medicine, Vol. V, pp. 93—115. Oxford: Pergamon Press. 1968.

— Gaussian behavior of the electroencephalogram: Changes during performance of mental task. Science 164, 328—331 (1969).

— Effect of tetrodotoxin on spontaneous electrical activity of the cerebral cortex: Intracellular evidence for subcortical control of the electroencephalogram. Submitted for publication (1971).

— Cellular sources of the EEG, II. Summation of unitary potentials in the gross electrocorticogram. Submitted for publication (1972).

— and W. R. ADEY: Cellular sources of the EEG, I. Neuronal mechanisms. Submitted for publication (1972).

GERARD, R. W.: Factors controlling brain potentials. Cold Spring Harbor Symp. Quant. Biol. 4, 292—298 (1936).

— and B. LIBET: The control of normal and "convulsive" brain potentials. Amer. J. Psychiat. 96, 1125—1151 (1940).

GOODMAN, N. R.: On the joint estimation of the spectra, cospectrum and quandrature spectrum of a two-dimensional stationary Gaussian process. Ph. D. dissertation, Princeton University, 1957.

HAFEMANN, D. R., A. COSTIN, and T. J. TARBY: Neurophysiological effects of tetrodotoxin in lateral geniculate body and dorsal hippocampus. Brain Res. 12, 363—373 (1969).

— and A. H. HOUSTON: Specificity of binding of radioactive tetrodotoxin. Fed. Proc. 30 (2), 349 (1971).

HENRY, C. E., and W. B. SCOVILLE: Suppression-burst activity from isolated cerebral cortex in man. Electroenceph. clin. Neurophysiol. 4, 1—22 (1952).

HUBBARD, J. E., D. STENHOUSE, and R. M. ECCLES: Origin of synaptic noise. Science 157, 330—331 (1967).

INGVAR, D. H.: Electrical activity of isolated cortex in unanesthetized cat with intact brain stem. Acta physiol. scand. 33, 151—168 (1955).

JASPER, H., and C. STEFANIS: Intracellular oscillatory rhythms in pyramidal tract neurones in the cat. Electroenceph. clin. Neurophysiol. 18, 541—553 (1965).

KRISTIANSEN, D., and G. COURTOIS: Rhythmic electrical activity from isolated cerebral cortex. Electroenceph. clin. Neurophysiol. 1, 265—272 (1949).

LI, C. L., and H. JASPER: Micro-electrode studies of the electrical activity of the cerebral cortex in the cat. J. Physiol. 121, 117—140 (1953).

MATSUMOTO, H., and C. AJMONE MARSAN: Cortical cellular phenomena in experimental epilepsy: Interictal manifestations. Exp. Neurol. 9, 286—304 (1964).

— Cortical cellular phenomena in experimental epilepsy: Ictal manifestations. Exp. Neurol. 9, 305—326 (1964).

— G. F. AYALA, and R. J. GUMNIT: Neuronal behavior and triggering mechanism in cortical epileptic focus. J. Neurophysiol. 5, 688—704 (1969).

NARAHASHI, T., J. W. MOORE, and W. R. SCOTT: Tetrodotoxin blockage of sodium conductance increase in lobster giant axons. J. gen. Physiol. *47*, 965—974 (1964).

POLLEN, D. A., and C. AJMONE MARSAN: Cortical inhibitory postsynaptic potentials and strychninization. J. Neurophysiol. *28*, 342—358 (1965).

PRINCE, D. A.: Modification of focal cortical epileptiform discharge by afferent influences. Epilepsia *7*, 181—201 (1966).

— Electrophysiology of "epileptic" neurons. Electroenceph. clin. Neurophysiol. *23*, 83—84 (1967).

— The depolarization shift in "epileptic" neurons. Exptl. Neurol. *21*, 467—485. (1968).

SHANNON, C. E.: A mathematical theory of communications, Part I. Bell Syst. Tech. J. *27*, 379—423 (1948 a).

— A mathematical theory of communications, Part II. Bell Syst. Tech. J. *27*, 623—656 (1948 b).

VERZEANO, M., and K. NEGISHI: Neuronal activity in cortical and thalamic networks. A study with multiple microelectrodes. J. gen. Physiol. *43*, suppl., 177—195 (1960).

WALTER, D. O.: Spectral analysis for electroencephalograms: Mathematical determination of neurophysiological relationships from records of limited duration. Exptl. Neurol. *8*, 155—181 (1963).

— and W. R. ADEY: Analysis of brain-wave generators as multiple statistical time series. IEEE Trans. Bio. Med. Engin. *BME-12*, 8—13 (1965).

# Neuronal Correlates of the Visual Evoked Response and Disinhibition in the Visual Cortex during Flicker Stimulation

U. KUHNT

Max-Planck-Institut für Biophysikalische Chemie, Abteilung Neurobiologie, Göttingen, Germany

## Introduction

The hypothesis was put forward more than 20 years ago that postsynaptic potentials (PSPs) might be the generators of the surface EEG (BREMER 1949, CHANG and KAADA 1950, ECCLES 1951). In the visual cortex, temporal relationships between the evoked response to a flash of light recorded from the surface (VEP) and extracellularly were reported (FROMM and BOND 1967, DILL et al. 1968). A close correspondence between evoked spike activity and the positive wave of the VEP was found as well as a marked reduction of the spike activity during the negative wave recorded from the surface. Different results from intracellular recordings to light stimulation were reported by CREUTZFELDT et al. (1969). They found a correspondence between the positive wave of the VEP and the evoked inhibitory postsynaptic potential (IPSP). With electrical stimulation of the specific afferent pathway the relationship between the evoked cellular potential and the surface potential becomes clearer: here the negative wave of the VEP runs parallel with the evoked IPSP (WATANABE et al. 1966).

The different results obtained by light stimulation justify a further approach to this problem. As will be shown, either constant relationships between the negative wave recorded from the surface and the evoked IPSP or between the positive wave of the VEP and the evoked IPSP have been found. Possible reasons for these different findings are going to be discussed.

Furthermore the cellular response to a light flash or an electrical stimulus is to be described. The cellular response contains excitatory postsynaptic potentials (EPSPs) and inhibitory postsynaptic potentials (IPSPs). EPSPs and IPSPs react differently according to stimulus rate and intensity. The disinhibition found at higher stimulus rates is to be related to photosensitive epilepsy in humans.

## Methods

The animals (adult cats) were initially anesthetised (30 mg Nembutal/kg bodyweight) and after the preparation immobilised (Flaxedil) and artificially respirated. All pressure points were carefully anesthetised (Carbostesin). No further Nembutal was given during the experiment. The VEPs were recorded by a chlorided silverball electrode situated near the insertion point of the microelectrode. The neckmuscle served as reference point. The glass-micropipettes were filled with 1.5 M potassium citrate (30 to 60 Mohm). Stimuli were either shortlasting flashes from a stroboscope, longlasting diffuse flashes or electrical stimuli.

## Results

The neurons to be reported on were all in area 17 or at the border between area 17 and 18. To a diffuse light stimulus, they react with a short excitation followed by a strong inhibition. These units form the largest group of cells recorded (about 65%). Most of the neurons of this group were found between 600 and 1100 $\mu$m below surface. According to Otsuka and Hassler (1962) this corresponds to layers 3 and 4. In both layers there are small pyramidal cells, stellate cells and, on the border to layer 5, large pyramidal cells (Garey 1971). So it might be that most neurons of this group are pyramidal cells which, because of their large soma, are not so easily irritated by the penetration of the electrode.

### A. The Cellular Response

In the visual cortex the typical response of this type of cell when stimulated with 1/sec light flashes consists of a quickly rising EPSP starting 20–70 msec after the onset of the stimulus (Fig. 1 *b*, arrow). The latency depends on the intensity of the light flash over a range of 4–5 log units. The evoked primary EPSP has a normal time course at low intensities only. In this case the evoked EPSP often does not reach the firing threshold. At stronger intensities, however, an evoked IPSP starts either directly after the decay of the EPSP or in the late part of the decay and therefore cuts off the usually slow decay of the EPSP. Both the amplitude and latency of the IPSP depend on the stimulus intensity. The decay of the IPSP which lasts up to 200 msec is often superimposed by a late excitation which rises more slowly, but often reaches the firing threshold. Since the shifting of the latencies of the IPSP is steeper than that of the EPSP, the evoked IPSP starts relatively earlier than the EPSP when the intensity is increased. Therefore, with higher intensities, the compound IPSP partly overlaps the EPSP; the primary EPSP is cut off and the firing rate is lowered. But the IPSP never starts before the EPSP. Fig. 2 shows the typical response to different light intensities. The upper traces (Fig. 2 a) were recorded with low intensity; the broad EPSP is indicated by an arrow. The traces below (Figs. 2 b, c) show the responses to higher intensities. Here the EPSP becomes narrower and the IPSP is more marked.

*Different stimulus frequencies*: A slight increase of both the firing rate and the maximum amplitude of the IPSP is seen in some cells with 2/sec stimulation. By changing the repetition rate to 4/sec the amplitude of the IPSP is reduced and the EPSP is widened (Fig. 3). The marked decrease of the amplitude of the IPSP is often accompanied by a higher firing rate of the primary EPSP. In many cells the IPSP disappears with 8/sec stimulation, whereas the primary EPSP is still visible. It is always longer than with lower repetition

---

Fig. 1. Surface and cellular evoked response to a short diffuse flash of light.
a) The averaged surface VEP is shown in the upper trace, the averaged membrane potential in the middle trace and the poststimulus histogram (PSTH) in the lower trace (25 sweeps, binwidth 2 msec for the upper traces, 4 msec for the PSTH). The stimulus is at the beginning of the traces. Negativity of the VEP and depolarization of the cell response in this and all subsequent figures upward. The coincidence between the node of the initially positive deflection of the VEP and the evoked EPSP is indicated by the first dashed line. The close time relationship between the peak of the positive deflection of the VEP and the peak of the IPSP is shown by the second dashed line. A strong delayed excitation with discharges (PSTH) occurs simultaneously with a surface negative deflection.
b) Two original records of the same cell with different amplification. The evoked IPSP starts during the positive deflection of the VEP (arrow in the upper record). In the lower record the short, quickly rising evoked EPSP is indicated by an arrow. The stimulus is marked by artefacts on the traces

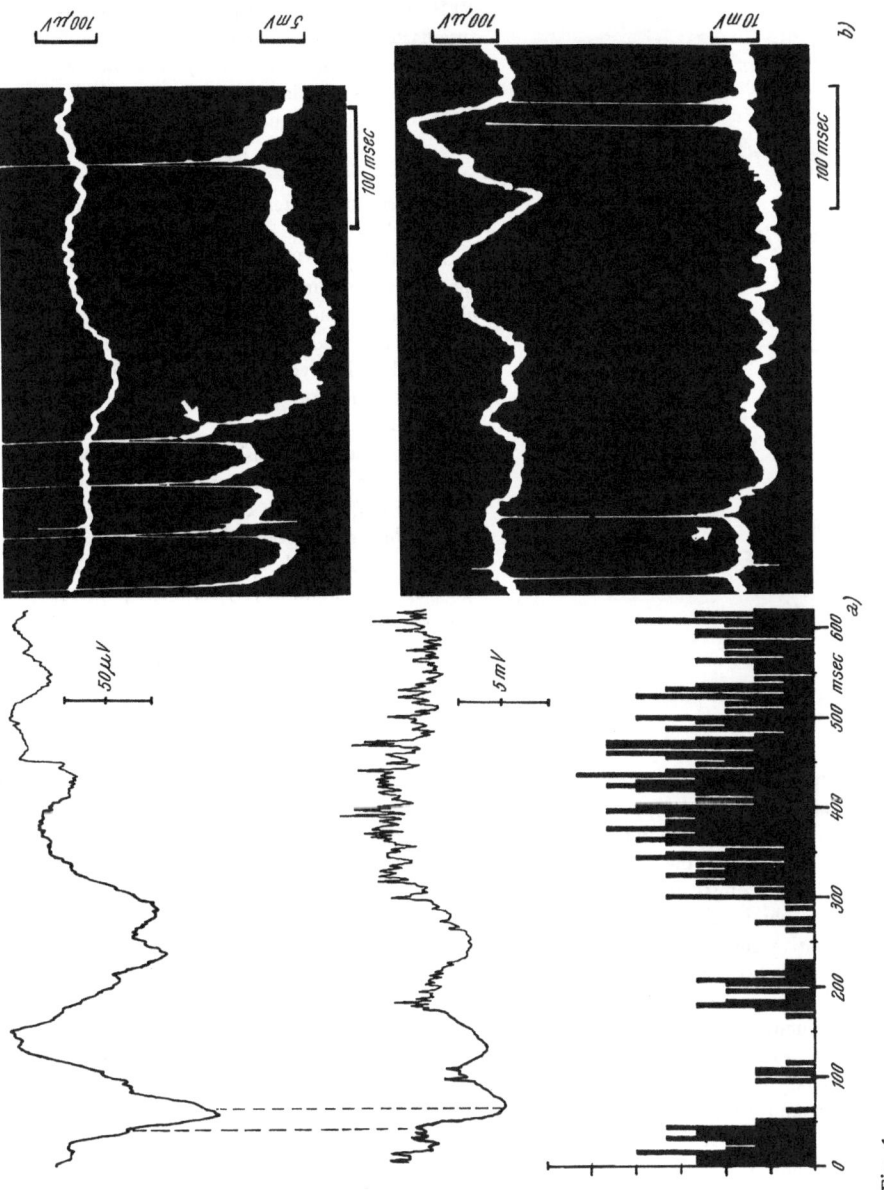

Fig. 1

6  Synchronization

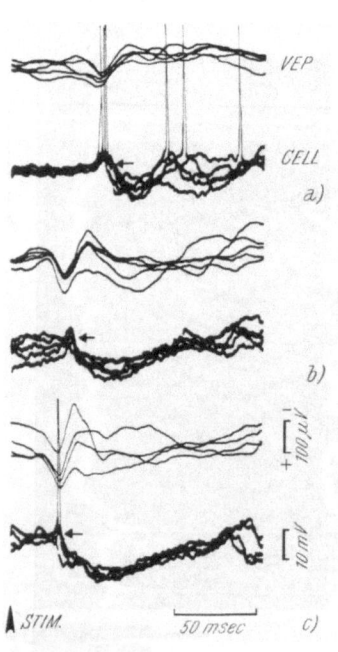

a)

b)

c)

STIM.    50 msec

Fig. 2. Superimposed original records of the VEP and the cellular responses to short diffuse flashes of different intensity.

The upper records of each pair show the surface VEPs, the lower records the cellular responses. The stimulus is marked by an arrow below c. Same calibration for a, b and c.

a) Low stimulus intensity. The cellular responses show broad evoked EPSPs (arrow) with discharges. The evoked IPSPs correspond with the surface negative waves.

b) Medium stimulus intensity. The latencies of the VEPs and the cellular responses are shortened. The evoked primary EPSPs are narrower (arrow). The duration of the IPSPs is increased.

c) High stimulus intensity. A further decrease of the latencies is seen. The primary evoked EPSPs are cut off by the following IPSPs, during 5 sweeps the evoked EPSPs only once reached the firing threshold. The spike occurs simultaneously with the positive peak of the VEP

Fig. 3. Effect on the cellular response of changing flash intensity and repetition rate.

In each group the upper record represents the averaged intracellular potential, the lower one the PSTH. The curves represent averages of 30 sweeps, the bin width is 2 msec for the intracellular record, 4 msec for the PSTH. The top row was obtained with lowest, the bottom row with highest intensities. The log relative intensity is written in the right column. The relative position of the averaged cellular potentials with respect to the ordinate does not reflect the absolute membrane potential level.

Column A: 1/sec stimulation. The early IPSP is well developed at all intensities. The early EPSP is smaller and shorter at higher intensities since the IPSP increases. At the highest intensity the early spike response is suppressed.

Column B: 4/sec stimulation. The IPSP is diminished at all intensities, while the early EPSP becomes broader and the late EPSP becomes more prominent.

Column C: 8/sec stimulation. The IPSPs are absent at all except the highest intensity. The early EPSPs are longer and stand out clearly, the firing threshold is not reached.

Column D: 16/sec stimulation. Only with highest intensity some depolarizing potentials are seen

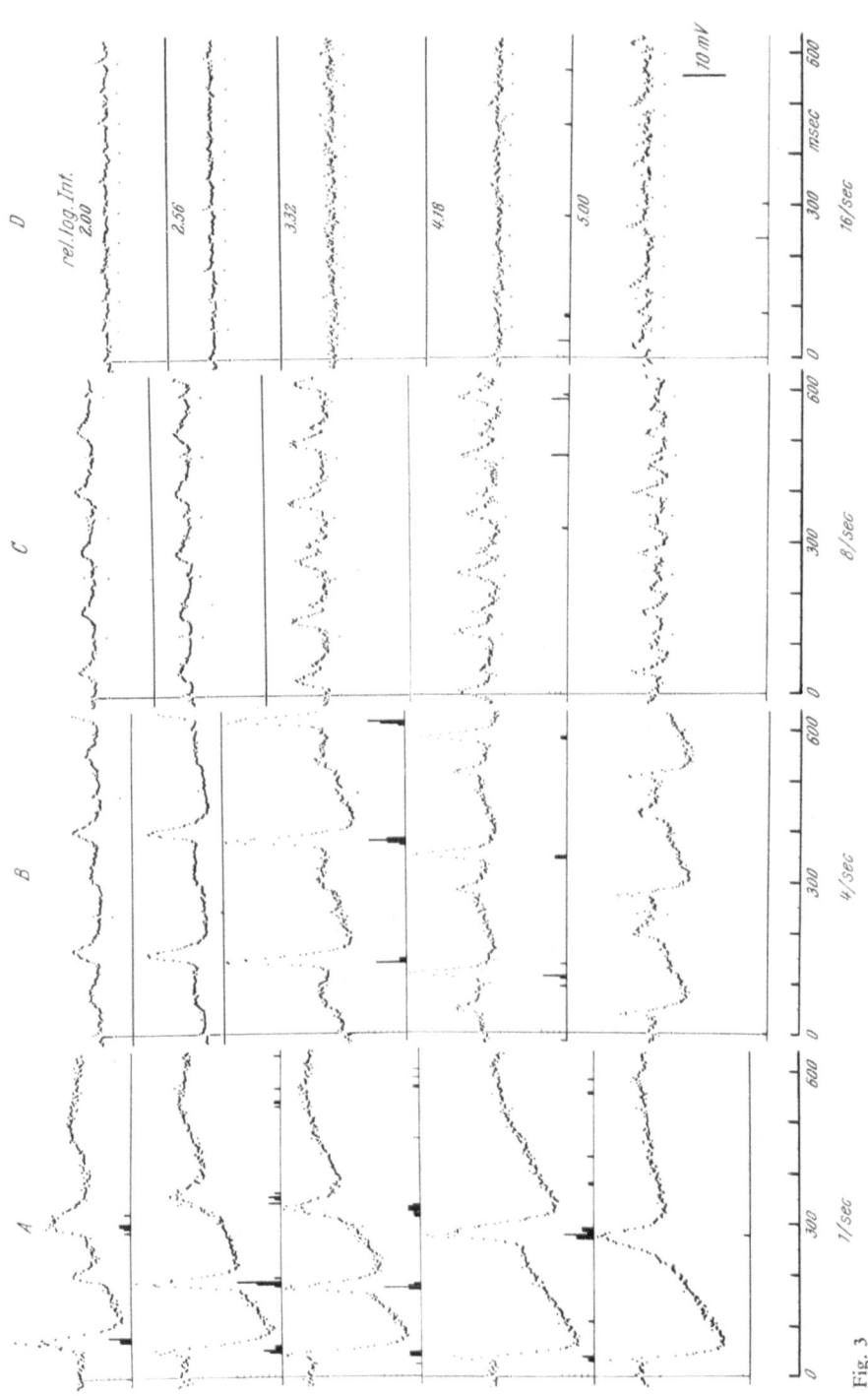

Fig. 3

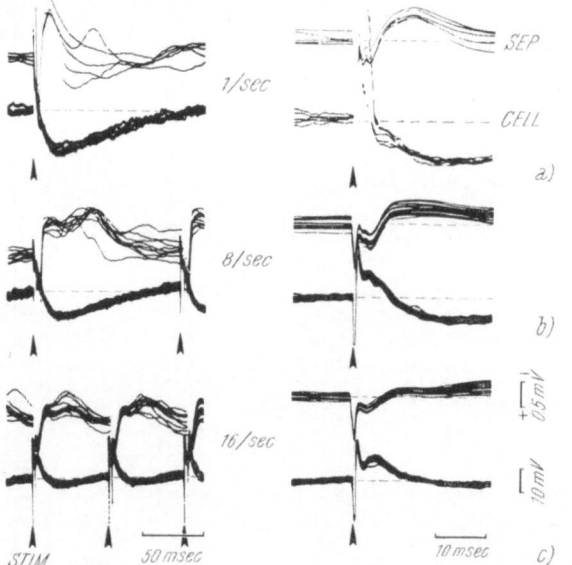

Fig. 4. The effect of the rate of electrical stimulation on intracellular response and surface potential.

The upper trace of each group shows the superimposed surface potentials (SEP), the lower trace the superimposed intracellular responses. The right column shows the same cell with higher sweep speed. The dashed lines indicate the base lines. No filters were used.

*a*) 1/sec stimulation. The classical surface response is obtained. The cell responds with an early EPSP and spikes (this is better seen in the right column). The IPSP corresponds to the surface negative wave.

*b*) 8/sec stimulation. Both the amplitude of the IPSP and of the surface negative wave are diminished. Because of the decreased IPSP the EPSP is seen more clearly. It does not reach the firing threshold.

*c*) 16/sec stimulation. The IPSP is absent, the EPSP is broader. Only a slight hyperpolarizing shift occurs after the EPSP, corresponding to a slight negative overshoot of the surface potentials

rate. Usually the EPSP is not high enough to reach the threshold. Between 8 and 16/sec the IPSP disappears in all cells. The primary EPSP is still present and reaches the firing threshold in a few cells (Fig. 5). The stimulus frequency at which disinhibition occurs depends partly on the stimulus intensity. The lower the intensity the lower the critical frequency at which the IPSP disappears. In Fig. 3 the averaged slow membrane responses of a cell and the post-stimulus histograms (PSTH) are shown. Both the intensity of the light flashes (vertical columns) and the frequency of the stimuli were changed. In the cell of Fig. 3, the IPSP had already disappeared at 4/sec stimulation when lowest light intensity was used. With 8/sec a slight hyperpolarization was seen only when highest light intensity was applied. The EPSP was only broader, but otherwise remained unchanged. With 16/sec only a few depolarizing potentials were seen with most intense flashes (KUHNT and CREUTZFELDT 1971).

When electrical stimuli are applied to the geniculate body or to the optic radiation a comparable disinhibition is found. The distinct IPSP which is present at 1/sec stimulation diminishes when the frequency is changed to 4/sec (Fig. 4). In this cell a slight hyperpolarization was still seen when the frequency was increased to 8/sec. With 16/sec this slight inhibition had completely disappeared. A comparison of the primary EPSPs at a low and high stimulation rate shows the unmasking very well (Figs. 4 *a, c*): the EPSP becomes broader as the frequency rises. The different cells have different critical frequencies for this disinhibition. But for both kinds of stimulation the upper limit for the disinhibition is about 16/sec.

*Discussion to A*

The intermittent light stimulation of visual cortical neurons at frequencies between 4 and 16/sec results in a diminution and total disappearance of the evoked IPSP. As it was shown, this effect depends partly on the intensity used. Simultaneously the evoked EPSP broadens. Electrical stimulation of the specific afferent visual pathway produces a comparable frequency-dependant disinhibition.

Since there is a difference between the shortening of latencies in EPSPs and IPSPs with increasing intensities the broadening of the evoked EPSP at higher stimulation is probably due to the fact that the excitation is not suppressed any longer by the evoked IPSP. From frequency studies in the afferent visual system it is known that the specific afferent input per stimulus diminishes if the frequency of the light stimulation is increased (GRÜSSER and RABELO 1957, 1958). This might be the reason for the observation that, in most cells, the evoked EPSPs do not reach the firing threshold at higher stimulation rates.

Attenuation and total disappearance of the evoked IPSP similar to that reported here were observed in the motor cortex of the cat after electrical stimulation of the specific thalamic relay nuclei (VL) (NACIMIENTO et al. 1964, PURPURA et al. 1964). A slowly working transmitter system and/or a rapidly exhausting transmitter-receptor system were discussed as possible mechanisms by NACIMIENTO et al. Although PHILLIPS (1961) suggested an interneuron which could account for this disinhibition these authors disregarded this possibility because cells with the necessary qualities had not yet been described. An intracortical mechanism based on inhibitory interneurons may be present. The correlation between latency and light intensity of these interneurons should be different from the one observed in pyramidal cells. (KUHNT in prep.) A higher threshold of such postulated inhibitory interneurons would also explain the masking phenomenon at low and higher frequencies. The possible anatomical sites for this intracortical inhibition are the symmetrical synapses which survive undercutting (SZENTÁGOTHAI 1965) and thalamic lesions (GAREY 1970, GAREY and POWELL 1971). It is assumed that these synapses are the terminals of intracortical neurons which may be inhibitory (GAREY and POWELL 1971).

The disinhibition obtained in the visual cortex of the cat deserves consideration with respect to the photosensitive epilepsy in humans. It is known that repetitive light flashes can elicit paroxysmal activity and spike—and—wave patterns in the EEG of humans suffering from epilepsy. Although in some cases even single flashes may elicit such EEG-patterns (BICKFORD and KLASS 1969), the effective frequency range is usually between 4 and 20/sec (GASTAUT et al. 1948, BICKFORD and KLASS 1969); the critical frequency often depends on the intensity of the light flashes (BICKFORD and KLASS 1969). The same is true of the disinhibition in the visual cortex mentioned above.

To explain the epileptic discharges and attacks, a disturbed inhibitory mechanism has been proposed to account for the excitatory state of the cortex (JUNG and TÖNNIES 1950, ECCLES 1969). The results dealt with here only partially support this hypothesis since disinhibition itself is not sufficient to elicit paroxysmal activity in a healthy animal. It may be, however, that subjects suffering from photosensitivity may have—in their visual cortex—a basic disturbance of the excitatory pathways which becomes evident when the inhibition is suppressed.

## B. Correlation between Evoked Potential Recorded from the Surface (VEP) and Cellular Response

Two types of VEP may be found in one animal. The first type consists of a sharp and short-lasting positive wave followed by a longer lasting negative deflection (Fig. 2). The negative wave often passes into a long-lasting positive wave of low amplitude. In the second type of the VEP, the initially positive deflection may be immediately followed by a second positive wave the onset of which appears as a little node (Fig. 1 a). In this case the peak of the negative wave often does not reach the baseline or just crosses it (Fig. 1 a). Both types of VEP can be seen in the same experiment, gradually changing from one type to the other.

In all animals the initially positive deflection of the VEP started some msec earlier than the cellular response. The peak of the evoked EPSP (and, if the threshold was reached, of the spike) occurred at the positive deflection of the initially positive wave, simultaneously with the little node of the type II response (Fig. 1 a, first dashed line) and with the initially positive wave of type I response (Figs. 2 a, c). The time relations between the first positive deflection of the surface response and the unit response are therefore almost alike in the type I and type II response.

In type I response the evoked IPSP starts during the negative deflection, sometimes near the negative peak (Fig. 2 a), sometimes near the positive peak (Fig. 2 c). The time relation between the light intensity, the time course of the IPSP and the rising phase of the surface negative deflection as seen in Fig. 2 is somewhat inconsistent. In a few cases the peak of the negative wave and of the IPSP occurred simultaneously, but in most cases the IPSP reached its full amplitude later than the negative peak of the VEP. The other group of VEPs (type II response) shows different relations as shown in Fig. 1 a where the IPSP starts during the positive deflection of the initially positive wave. The full amplitude of the evoked IPSP is reached shortly after the positive peak. The decay of the evoked IPSP usually corresponds with the longlasting positive wave which follows the negative wave. Late and always broad EPSPs show a close correspondence with negative deflections, they seem to be superimposed on the longlasting positive deflection of the VEP.

If the repetition rate of the light stimuli is high enough for the cell to be disinhibited, a better relation is obtained. Fig. 5 shows the averaged membrane potential and the PSTH of 16/sec stimulation. The peak of the evoked EPSP occurs at the same time as the peak of the surface potential. In a few experiments the baseline of the EEG and of the membrane potential were monitored, when the frequency of the light stimulation was increased. Neither a shifting

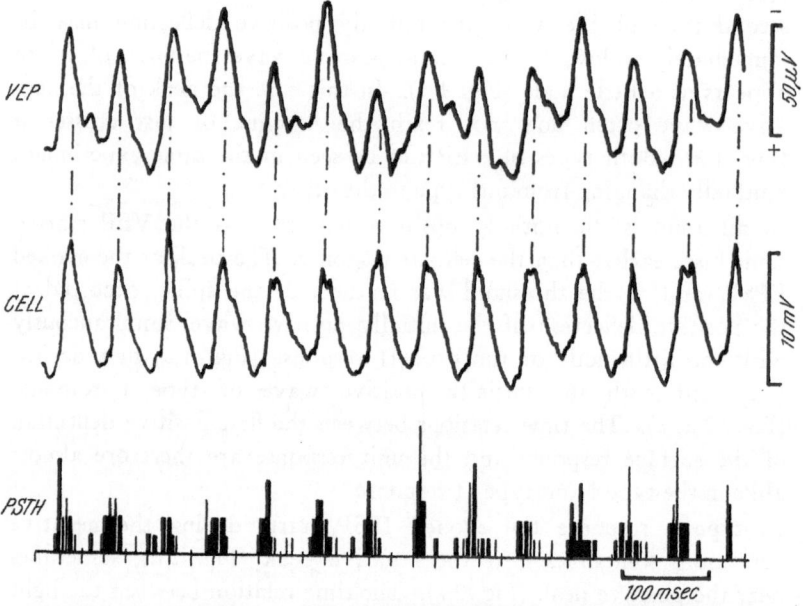

Fig. 5. Averaged VEP and averaged intracellular response at 16/sec light stimulation.

The upper trace shows the averaged surface potential, the lower one the averaged intracellular potential. The PSTH is on the bottom trace. No filters were used. The curves and the PSTH show averages of 30 sweeps of the same cell. The binwidth is 2 msec for the upper two traces, 4 msec for the PSTH. The EPSPs which often reach the firing threshold occur simultaneously with the surface negative deflections. This is indicated by the lines which connect the peaks

of the membrane potential nor a shifting of the baseline of the EEG was seen.

With electrical stimulation only one type of response was seen; this is the classical CLARE-BISHOP response (1952) which is similar to the type I response to light stimulation. The primary EPSP coincides with the positive, and the evoked IPSP with the negative wave of the VEP (Fig. 4 a). At higher frequencies the amplitude of the negative wave diminishes and the amplitude of the evoked IPSP decreases (Fig. 4 *b*). If the stimulus frequency is high enough for disinhibition to occur the unmasked EPSP corresponds with the positive wave recorded from the surface (Fig. 4 *c*), and only a very small negative overshoot of the surface potential is left.

*Discussion to B*

An evident coincidence exists between the initially positive wave of the VEP and the primary evoked EPSP after light stimulation as well as between the evoked IPSP and the negative wave of type I response. The same relation holds good for electrical stimulation. In this respect the results meet with the classical volume conductor concept and earlier observations on the visual cortex after light stimulation (FROMM and BOND 1967, DILL *et al.* 1968) or electrical stimulation (BISHOP and CLARE 1952, WIDÉN and AJMONE MARSAN 1960, WATANABE *et al.* 1966). The type II response, however, shows a more complicated time relationship (CREUTZFELDT *et al.* 1969). In this case the evoked EPSP coincides with the initially positive wave, and the IPSP with the immediately following second positive wave. It is hard to believe that the different correlations between cellular and EEG responses of the type I and II VEP are due to an experimental bias, *i.e.*, that one should have missed a whole population of cells, presumably active during the large surface-positivity. Since both types of VEPs can be found in the same animal it is more reasonable to look for a physiological mechanism. It is possible for example that the form of the VEP depends on the transcortical steady potential which seems to represent different states of cortical excitability (CASPERS 1964, DENNEY and BROOKHART 1962, FROMM and BOND 1964, PURPURA and MCMURTRY 1965, FROMM and BOND 1967).

With 16/sec light stimulation a reversed behaviour was found: the evoked EPSPs correspond with the negative wave of the VEP. Such a relation was also found in the case of the late broad EPSPs and the VEP and has also been observed on barbiturate spindles.

A satisfactory explanation for the various relationships between surface potentials and the sharp primary versus the slower secondary responses cannot be given yet. This problem may be approached by using combined anatomo-physiological methods to detect the location of the recorded cells or the synapses activated during different excitatory states.

## Summary

1. The evoked response to light flashes of a certain group of cells (presumably pyramidal cells) in the visual cortex is composed of an EPSP-IPSP-sequence. With higher light intensities the EPSP is cut off by the following IPSP and the discharge rate of the primary evoked EPSP is therefore lowered.

2. An increase of the stimulus frequency (3–4/sec) results in a diminution of the IPSP amplitude and a broadening of the evoked EPSP. Between 8 and 16/sec the IPSP disappears whereas the EPSP is broadened but mostly does not reach the firing level. A similar disinhibition of the EPSP is seen with 8–16/sec electrical stimulation.

It is assumed that the disinhibition as well as the masking effect is due to an inhibitory interneuron.

The disinhibition of the visual cortex during 16/sec light stimulation is discussed with respect to the photosensitive epilepsy seen in humans.

3. Two types of VEPs were recorded. The time relation between the cellular response and the type I response to light stimulation is similar to the time relation between cellular and surface response to electrical stimulation. For 16/sec light stimulation a reversed relation between VEP and cellular response was found.

## References

BICKFORD, R. G., and D. W. KLASS: Sensory precipitation and reflex mechanisms. In: JASPER, H. H., A. A. WARD, and A. POPE (eds.), Basic mechanisms of the epilepsies, pp. 543—564. Boston: Little, Brown & Co. 1969.

BISHOP, G. H., and M. H. CLARE: Sites of origin of electrical potentials in striate cortex. J. Neurophysiol. *15*, 201—220 (1952).

BREMER, F.: Considérations sur l'origine et la nature des "ondes" cérébrales. Electroenceph. clin. Neurophysiol. *1*, 177—193 (1949).

CASPERS, H.: Über die Beziehung zwischen Dendritenpotential und Gleichspannung an der Hirnrinde. Pflügers Arch. ges. Physiol. *269*, 157—181 (1959).

CHANG, H. T., and B. KAADA: Analysis of primary response of visual cortex to optic nerve stimulation in cats. J. Neurophysiol. *13*, 305—318 (1950).

CREUTZFELDT, O. D., A. ROSINA, M. ITO, and W. PROBST: Visual evoked response of single cells and of the EEG in primary visual area of the cat. J. Neurophysiol. *32*, 127—139 (1969).

DENNEY, D., and J. M. BROOKHART: The effects of applied polarization on evoked electrocortical waves in cat. Electroenceph. clin. Neurophysiol. *14*, 885—897 (1962).

DILL, R. C., E. VALLECALLE, and M. VERZEANO: Evoked potentials, neuronal activity and stimulus intensity in the visual system. Physiology and Behavior *3*, 797—801 (1968).

ECCLES, J. C.: Interpretations of action potentials evoked in the cerebral cortex. Electroenceph. clin. Neurophysiol. *3*, 449—464 (1951).

— Excitatory and inhibitory mechanisms in brain. In: JASPER, H. H., A. A. WARD, and A. POPE (eds.), Basic mechanisms in the epilepsies, pp. 229—252. Boston: Little, Brown & Co. 1969.

FROMM, G. H., and H. W. BOND: Slow changes in the electrocorticogram and the activity of cortical neurons. Electroenceph. clin. Neurophysiol. *17*, 520—523 (1964).

— — The relationship between neuron activity and cortical steady potentials. Electroenceph. clin. Neurophysiol. 22, 159—166 (1967).

GAREY, L. J.: The termination of thalamo-cortical fibres in the visual cortex of the cat and monkey. J. Physiol. *210*, 15P (1970).

— A light and electron microscopic study of the visual cortex of the cat and monkey. Proc. roy. Soc. (Lond.) *B 179*, 21—40 (1971).

— and T. P. S. POWELL: An experimental study of the termination of the lateral geniculo-cortical pathway in the cat and monkey. Proc. roy. Soc. (Lond.) *B 179*, 41—63 (1971).

GASTAUT, H., J. ROGER, and Y. GASTAUT: Les formes éxperimentales de l'épilepsie humaine. Rev. neurol. *80*, 161—183 (1948).

GRÜSSER, O. J., and C. RABELO: Die Wirkung von Flimmerreizen mit Lichtblitzen an einzelnen corticalen Neuronen. I. Int. Congr. Neurol. Sci. Vol. III: IV. Int. EEG congress Brussels 1957, pp. 371—375. London: Pergamon Press. 1959.

— — Reaktionen einzelner retinaler Neurone auf Lichtblitze. I. Einzelblitze und Lichtblitze wechselnder Frequenz. Pflügers Arch. ges. Physiol. *265*, 501—525 (1958).

JUNG, R., and J. F. TÖNNIES: Hirnelektrische Untersuchung über Entstehung und Erhaltung von Krampfentladungen. Arch. Psychiat. Nervenkr. *185*, 701—735 (1950).

KUHNT, U., and O. D. CREUTZFELDT: Decreased postsynaptic inhibition in the visual cortex during flicker stimulation. Electroenceph. clin. Neurophys. *30*, 79—82 (1971).

NACIMIENTO, A. C., H. D. LUX, and O. D. CREUTZFELDT: Postsynaptische Potentiale an Nervenzellen des motorischen Cortex nach elektrischer Reizung spezifischer und unspezifischer Thalamuskerne. Pflügers Arch. *281*, 152—169 (1964).

OTSUKA, R., and R. HASSLER: Über Aufbau und Gliederung der corticalen Sehsphäre bei der Katze. Arch. Psychiat. Nervenkr. *203*, 212—234 (1962).

PHILLIPS, C. G.: Some properties of pyramidal neurones of the motor cortex. In: WOLSTENHOLME, G. E. W., and M. O'CONNOR (eds.), The nature of sleep, pp. 4—24. (Ciba Symposium.) London: Churchill. 1961.

PURPURA, D. P., and J. G. McMURTRY: Intracellular activities and evoked potential changes during polarization of motor cortex. J. Neurophysiol. *28*, 166—185 (1965).

— R. J. SHOFER, and F. S. MUSGRAVE: Cortical intracellular potentials during augmenting and recruiting responses. J. Neurophysiol. *27*, 133—151 (1964).

Szentágothai, L. J.: The use of degeneration methods in the investigation of short neuronal connection. In: Singer, M., and J. P. Schadé (eds.), Progr. in Brain Res., Vol. 14, pp. 1—32. Amsterdam: Elsevier. 1965.

Widén, L., and C. Ajmone Marsan: Unitary analysis of the response elicited in the visual cortex of the cat. Arch. ital. Biol. *98*, 248—274 (1960).

Watanabe, S., M. Konishi, and O. D. Creutzfeldt: Postsynaptic potentials in the cats visual cortex following electrical stimulation of afferent pathways. Exp. Brain Res. *1*, 272—283 (1966).

# Relations between Cortical DC Shifts and Membrane Potential Changes of Cortical Neurons Associated with Seizure Activity

E.-J. Speckmann, H. Caspers, and R. W. Janzen

Institute of Physiology, University of Münster, Germany

It is well established that seizure activity of the cerebral cortex is accompanied by slow potential changes which have been labelled as "steady" or "DC" potential shifts. With generalized seizure discharges or direct recordings from an active focus, these potential deviations are usually surface-negative in polarity and finally turn over to a positive deflection which corresponds to the post-ictal silent period (for references see O'Leary and Goldring 1964, Caspers and Speckmann 1969, Gumnit et al. 1970).

While the occurrence of such slow bioelectric phenomena is beyond discussion, the opinions on their origin are still controversial. There is no doubt that DC shifts in the brain can be produced, in principle, by a variety of generator structures. Potential differences across the blood-brain barrier or the meninges must be taken into account as well as membrane potential changes of neuronal elements and/or of glial cells (Loeschcke 1955, 1971, Tschirgi and Taylor 1958, Grossman and Hampton 1968, Castelucci and Goldring 1970, Somjen 1970). In case of seizure activity many experimental observations point to a *neuronal* origin of the DC shifts. The purpose of this paper is to study this particular problem in greater detail by simultaneous recordings of the membrane potential (MP) of pyramidal tract (PT) cells and of the DC component both from the surface and at various depths of the cortex. All experiments were performed on cats. Generalized seizure activity was most often elicited by intravenous administration of pentylenetetrazol. In some investigations focal convulsive discharges evoked by epicortical application of penicillin served as an additional model.

## Cortical DC Shifts and Activity Changes of PT Neurons during Generalized Seizure Discharges

In the anesthetized animal repeated applications of pentylenetetrazol (PTZ) finally lead to tonic-clonic convulsions which appear periodically at intervals ranging from approximately 1–10 minutes. In any given stage of this process the phase duration of the cyclic excitability

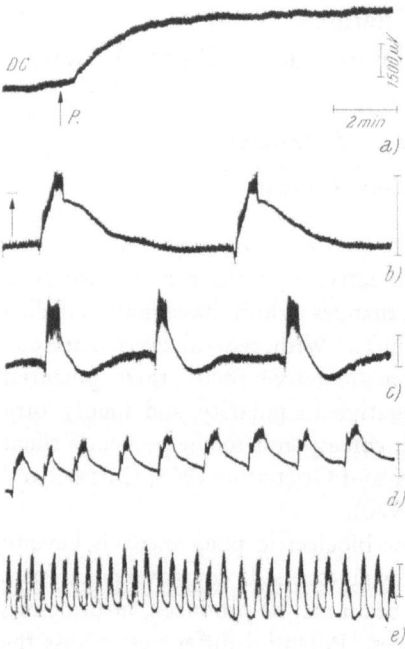

Fig. 1. Cortical DC shifts associated with series of generalized seizure discharges evoked by repeated applications of pentylenetetrazol (P) in the anesthetized cat. The tracings were recorded from the surface of the intact sensorimotor cortex against a reference point in the nasal bone. a) Sustained surface-negative DC deviation following the first administration (arrow) of 50 mg/kg P. b)–e) Surface-negative DC shifts associated with series of seizure discharges recurring periodically with various phase durations after repeated applications of the drug

changes may be influenced by low frequency cortical stimulation (Caspers and Simmich 1966, Caspers and Speckmann 1969). Each single fit occurring within a series of convulsive attacks is accompanied by typical changes of the cortical DC potential and of neuronal activity. These changes will, at first, be described by means of separate recordings, as the most informative display of the two bioelectric phenomena is entirely different.

The tracings in Fig. 1 represent a typical sample of epicortical DC shifts associated with periodical seizure discharges. They were recorded from the surface of the intact motor cortex against a reference point in the nasal bone at a very low paper speed. As shown in Fig. 1 a, the first administration of PTZ releases a sustained surface-negative DC displacement which corresponds to the general arousing capability of the drug. After repeated applications of PTZ, tonic-clonic convulsions appear, the intervals of which may vary, in principle, within a wide range. At any given increase of seizure susceptibility, however, the phase duration of the cyclic excitability changes proves rather constant (Figs. 1 *b–e*). The records furthermore

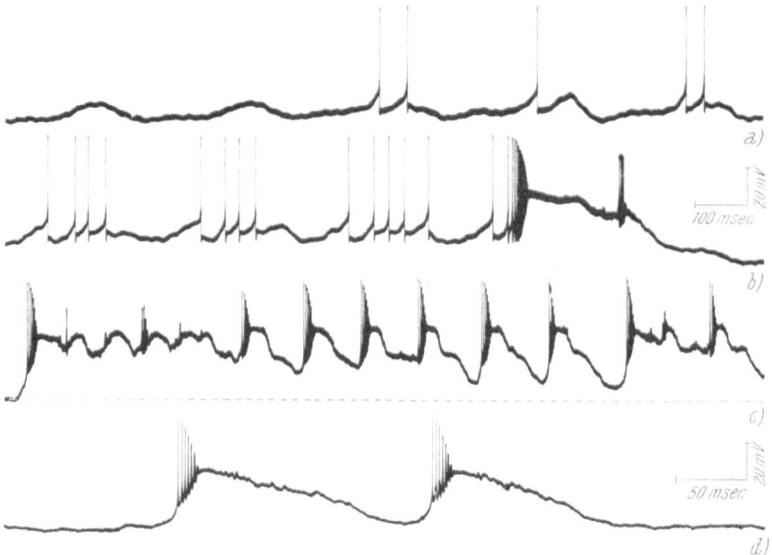

Fig. 2. Intracellular recording of a pyramidal tract (PT) cell in the postcruciate cortex of the cat during the development of seizure discharge released by intravenous application of 50 mg/kg pentylenetetrazol. In *b* the preceding increase in discharge frequency is suddenly interrupted by a paroxysmal depolarization shift (PDS), which tends to reiterate and thus releases a sustained depolarization of the neuron (*c*). In *d* the propagated spikes initiated during the rise of a PDS are displayed at a higher paper speed

indicate that cortical DC shifts associated with generalized seizure activity are essentially surface-negative in polarity. Transient positive deviations usually coincide with a post-ictal silent period. They may occur, moreover, at the immediate beginning of a fit.

The intracellular recording in Fig. 2 shows the activity changes of a PT neuron during the development of a single fit within a series of convulsions characterized by longer lasting surface-negative DC shifts as indicated in Fig. 1 *b*. With rising seizure susceptibility at first polysynaptic EPSPs appear which increase in amplitude and finally evoke groups or series of propagated discharges (Fig. 2 *a* and *b*). According to previous studies using higher DC amplification rates, this

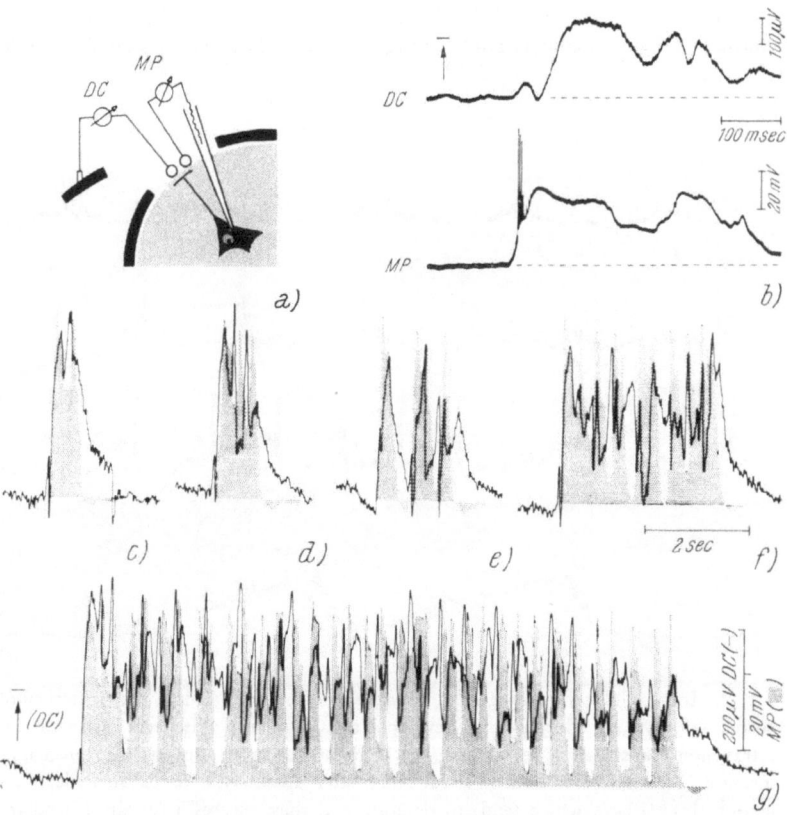

Fig. 3. Simultaneous recordings of the epicortical DC component (DC) and of the membrane potential of a pyramidal neuron (MP) during seizure activity of various duration in the cat. *a*) Schematic presentation of the positions of the DC and MP recording electrodes. *b*) Original tracings of a paroxysmal depolarization shift (PDS) in a PT cell (MP) and of the concomitant DC deviation at the cortical surface during the onset of seizure activity. *c*) Projection of the MP changes of PT cells indicated by screened areas on simultaneous epicortical DC shifts (solid lines) with seizure discharges of increasing duration

pre-convulsive stage of the unit response is usually accompanied by a flat surface-negative DC deviation provided that the shift originates with a sufficient delay after a preceding attack (Caspers and Speck-

MANN 1969). This problem will be discussed in greater detail in a following chapter. At a critical increase of convulsive excitability, the above mentioned activity changes of a PT neuron finally turn to paroxysmal depolarization shifts (PDS).

The transition to such a discharge pattern, the properties of which

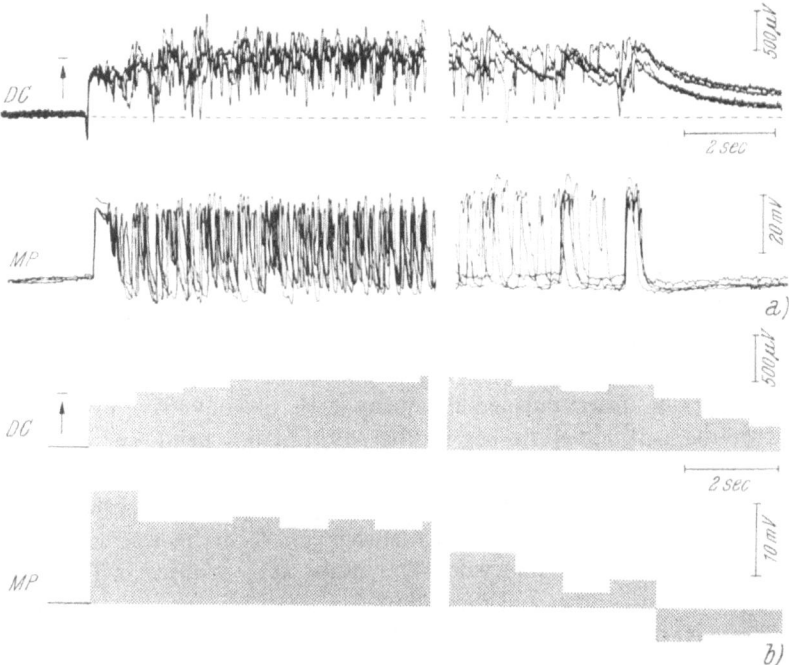

Fig. 4. *a*) Simultaneous recording of the epicortical DC component and of the membrane potential (MP) of a PT cell during series of seizure discharges in the cat. Four consecutive tracings are superimposed. The propagated spikes in the MP curves have been cut off. *b*) Mean epicortical DC shifts and mean MP changes of a PT cell during seizure discharges calculated from consecutive recordings at 1 second intervals

have already been studied in a number of papers (MATSUMOTO and AJMONE-MARSAN 1964 a, b, PURPURA *et al.* 1966, PRINCE 1968, 1969, PRINCE and FUTAMACHI 1970, DICHTER and SPENCER 1969 a, b, MATSUMOTO *et al.* 1969, AYALA *et al.* 1970, SPECKMANN *et al.* 1971), is illustrated in Fig. 2 *b*. As shown in Fig. 2 *c*, the PDS tend to reiterate and thus give rise to a sustained depolarization of PT neurons which corresponds to the surface-negative DC shifts.

7 Synchronization

The actual relations between the two bioelectric phenomena are presented in greater detail in Fig. 3. In this experiment the cortical DC component and the membrane potential of a PT neuron were simultaneously recorded from the same area during seizure discharges of various length. The positions of the active and reference electrodes in either case are illustrated in Fig. 3 a. Part B of the figure shows an original tracing of a PDS in a pyramidal cell and of the concomitant potential fluctuations at the cortical surface during the onset of seizure activity. Voltage, polarity, and time course of the two bioelectric phenomena were evaluated both with single and repetitive PDS of increasing duration. The results are summarized in Figs. 3 c–g. In these diagrams the membrane potential changes of the neuron indicated by screened areas are projected on the concomitant epicortical potential shifts. The findings suggest that the surface-negative DC deviation during seizure activity is closely related to the mean depolarization of cortical neurons.

The assumption is substantiated by Fig. 4. In this experiment the epicortical DC component and the MP of a PT neuron were simultaneously recorded during subsequent series of convulsive discharges with the propagated spikes in the MP tracings being cut off. In Fig. 4 a four consecutive records of the two bioelectric phenomena are superimposed. On the basis of these tracings the mean cortical DC deviations as well as the mean MP changes of the neuron were calculated for 1 second intervals. The results are presented in Fig. 4 b. Apart from the onset and the end of a series of seizure discharges, the diagrams demonstrate a rather close correlation between the two variables. It seems possible, therefore, that the typical DC shifts associated with convulsive activity depend, essentially, on a summation of slow membrane potential changes of neuronal elements (cf., also discussion section).

## Differences between Cortical Field Potentials and MP Changes of Cortical Neurons during Seizure Activity

As was already mentioned above, a more detailed evaluation of simultaneous DC and MP recordings during seizure activity reveals some less impressive, but rather constant differences between the two bioelectric phenomena. This statement refers almost exclusively to the more rapid potential fluctuations superimposed on the slow deviations of the base-line in either case. Distinctions of this kind are found throughout a series of seizure discharges, but appear most

prominent at the onset and at the end of a fit. These experimental
situations have therefore been studied in greater detail.

In intracellular recordings from PT neurons the *initiation* of a series
of seizure discharges is marked by a PDS. As a rule, the steep

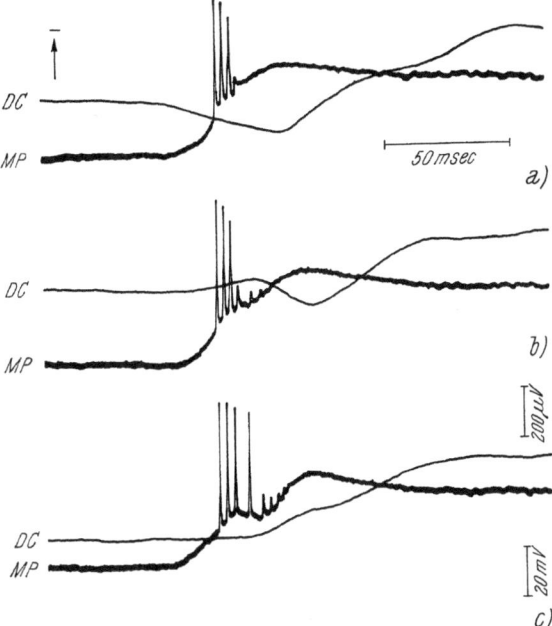

Fig. 5. Simultaneous recording of the epicortical DC
component and of the membrane potential of a PT cell
(MP) at the onset of a series of convulsive discharges. The
first paroxysmal depolarization shift (PDS) in the neuron
is usually accompanied by rapid deflections in the surface
leads (prepotentials) which tend to be positive (*a*) or
negative-positive in polarity (*b*). Only on rare occasions
the admixture of a positive prepotential is found to be
missing (*c*)

decrease of the membrane potential coincides with one or two rapid
deflections in the surface leads preceding the sustained negative DC
shift. Such fast prepotentials have already been shown to occur in
Figs. 3 and 4 *a*. In Fig. 5 they are illustrated at a more informative
display. The tracings indicate that the initial fluctuations in the DC
recordings are positive or negative-positive in polarity (Figs. 5 *a*
and *b*). Occasionally, the admixture of a positive component may

7*

be missing (Fig. 5 c). The sample of records in Fig. 5 furthermore shows that the sustained negative DC deviation associated with a series of seizure discharges commences approximately at the top of the first neuronal PDS, the time lag ranging from 20–50 msec. Provided that the electrodes are kept in the same position, these sequences of MP and DC fluctuations at the onset of a convulsive fit prove very constant.

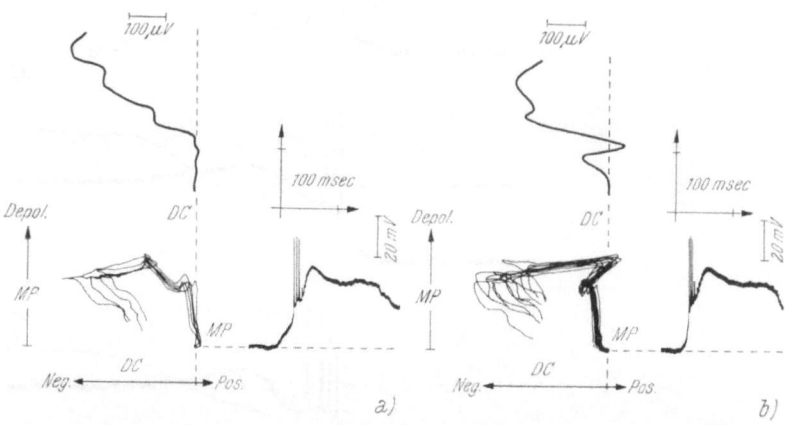

Fig. 6. Display of epicortical DC shifts (abscissa) as a function of membrane potential changes (MP) of a PT neuron (ordinate) during the onset of seizure activity. The tracings in *a* and *b* represent typical examples different by the occurrence of fast prepotentials in the DC records (see also Fig. 5). In each case 8–10 single sweeps taken at consecutive series of seizure discharges are superimposed. The records demonstrate the stability of MP/DC function during subsequent series of convulsion discharges at the same recording site

The above-mentioned relations can be shown to exist still more precisely by plotting the field potentials in the surface leads directly against the MP changes of the PT neurons. For this purpose the outputs of the DC and MP channels of the recording system were connected to the x- and y-axis, respectively, of an oscilloscope and/or of an XY writer. The results obtained with such a technique are presented in Fig. 6 by means of two typical examples corresponding to tracings *b* and *c* in Fig. 5. In each case 8–10 sweeps at the onset of consecutive series of seizure discharges are superimposed. The diagrams at first confirm the time—lag between the initial neuronal PDS and the commencement of the sustained negative DC shift. They furthermore illustrate the high persistence of the various potential sequences at the same recording site during the initiation of successive fits. By simultaneous shifting of the reference coordinates as a function of time, moreover, a 3-dimensional aspect of the curves given in Fig. 5 may be obtained. Examples of such a stereoscopic display are presented in Fig. 7 at hand of two original records. They yield a still more detailed insight into the sequence of potential changes concerned. As a whole, the studies on differences between MP and DC deviations at the *onset* of a series of seizure discharges confirm the

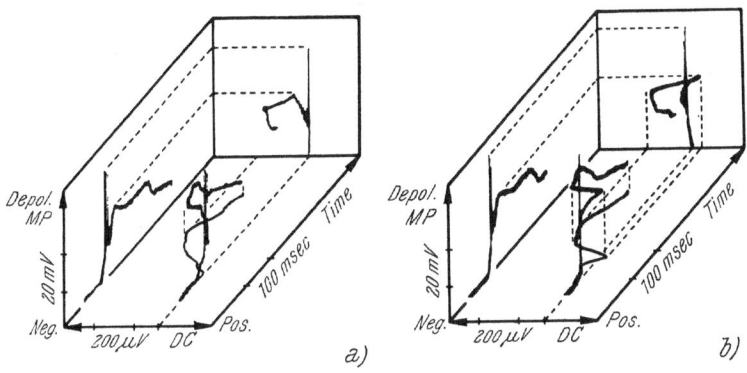

Fig. 7. Presentation of the MP/DC relation as a function of time. *a* and *b* represent two typical single records corresponding to tracings *a* and *b* in Fig. 5. In the 3-dimensional display the original DC and MP components are projected on the corresponding coordinates

assumption that the close correspondence between the mean epicortical DC shift and the mean depolarization of PT neurons cannot be attributed entirely to a direct recording of summated MP fluctuations at the deeper cellular layer by the surface leads. The sustained negative DC displacements are obviously mediated by additional generator structures, the average reactions of which to a synchronized neuronal input must, however, be similar to those of the soma membrane. This problem will be discussed in greater detail in a following chapter.

The relations between DC and MP responses illustrated in Figs. 5, 6, and 7 refer, exclusively, to the *onset* of series of seizure discharges separated by longer phases of non-ictal activity. In the course of sustained convulsive discharges, the above-mentioned correlations between the two bioelectric phenomena are found, in principle, to persist. There is, however, a far greater variability between single neuronal PDS and the concomitant fluctuations of AC field potentials in the surface leads. This statement becomes evident already from the diagrams in Fig. 3. In Fig. 8 it is documented at a more informative display. Such differences between MP and epicortical AC potentials depend, in part, on the actual voltage of the surface-negative DC displacement. In accordance with earlier polarization

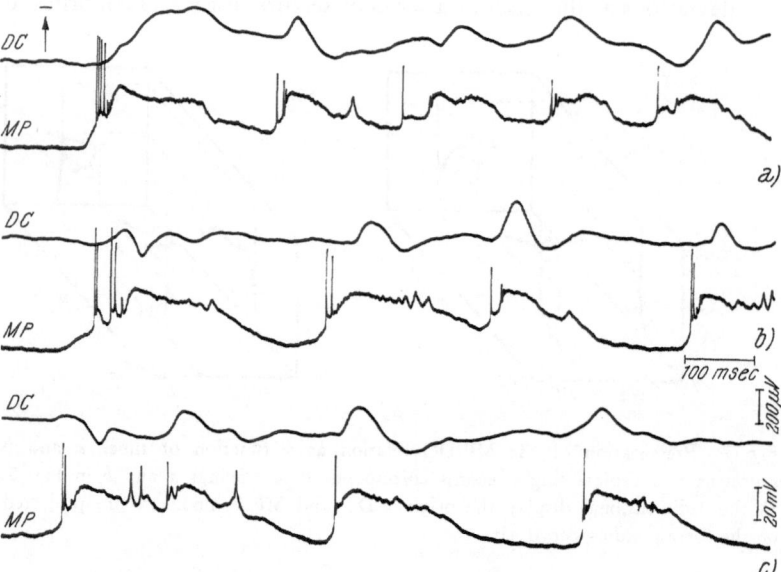

Fig. 8. Simultaneous recordings of the epicortical DC component and of the membrane potential of a PT cell (MP) in the cat sensorimotor cortex during a series of seizure discharges. The tracings *a–c* demonstrate the variability of correlations between neuronal PDS and rapid field potentials superimposed on the sustained negative DC deviation in the surface lead

experiments (Caspers 1959), the negative waves superimposed upon the slow deviation of the base line decrease in amplitude with rising negativity of the cortical surface, whereas positive deflections usually become more prominent. A further discussion of such observations concerning the relationship between MP changes and AC potentials of the brain would exceed the scope of the present paper and will be postponed to a subsequent publication. The general problem has been dealt with, moreover, already in a number of studies by other authors (Matsumoto and Ajmone Marsan 1964 a, b, Sugaya et al. 1964,

KLEE *et al.* 1965, KLEE 1966, CREUTZFELDT *et al.* 1966 a, b, GLÖTZNER and GRÜSSER 1968).
Another difference between MP changes of PT cells and epicortical DC reactions emerges at the *cessation* of a convulsive attack. Such

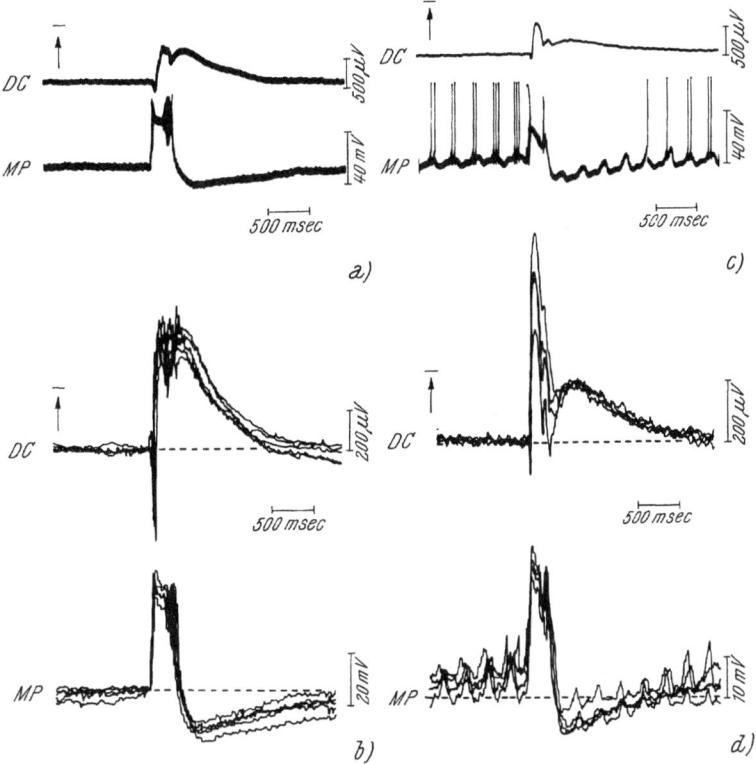

Fig. 9. Relations between epicortical field potentials (DC) and PDS of a PT cell (MP) with single convulsive discharges (interictal recording). The tracings in *a* and *c* represent original registrations. In the line drawings in *b* and *d*, respectively, five consecutive records are superimposed in either case

dissimilarities can already be shown to occur with single seizure discharges. A typical finding is presented in Fig. 9. In this experiment the epicortical field potential and the MP of a PT cell were recorded simultaneously during repetitive neuronal PDS separated by intervals of at least 10 seconds. Tracings *a* and *c* represent original records denoting two types of PDS which originate either abruptly

from a preceding silent period (*a*) or after a gradual increase of
discharge frequency (*c*). In Figs. 9 *b* and *d*, respectively, five consecutive records of the MP and DC changes are superimposed in

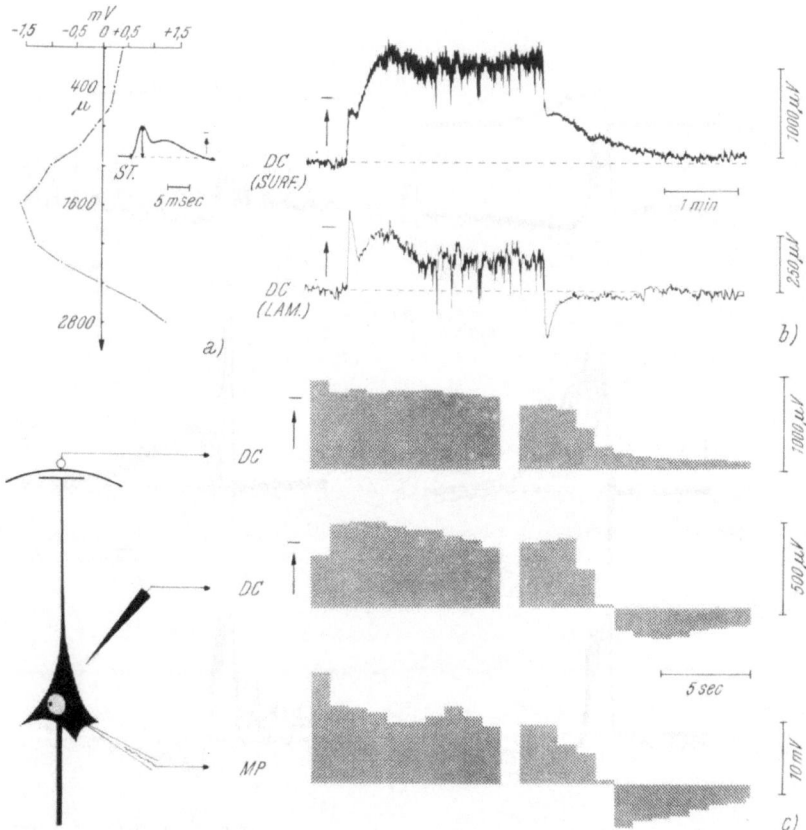

Fig. 10. Relations between laminar DC shifts and MP changes of a PT neuron
during series of convulsive discharges. *a*) Determination of the position of the
DC semi-microelectrode in the depth of the cortex by testing the field distribution
of the antidromic response to a pyramidal tract volley. *b*) Simultaneous DC
recordings from the surface and the fifth layer of the cortex during a series of
seizure discharges. *c*) Amplitude and polarity of MP and DC recordings during
seizure activity evaluated in 1 second intervals

either case. The curves indicate that each single PDS of a PT neuron
turns over to a steep *after-hyperpolarization*. This increase in MP
coincides with a slow surface negative wave which may depend, in

part, on the potential shift in the deeper cortical layer. However, the after-negativity in the DC record cannot be due entirely to such a volume conduction. Fig. 3, for instance, shows that the negative DC shifts decline gradually to the preceding level, while the MP of a PT neuron passes to a rapid and shorter lasting increase as soon as seizure activity stops. This considerable dissociation in time course may be due, in principle, to a variety of mechanisms. The present findings, however, suggest that different generators contribute to the DC displacement at the cortical surface. This assumption was studied by simultaneous recordings of the MP and of the DC component from the surface as well as from various depths of the cerebral cortex. A typical example of these experiments is presented in Fig. 10. The schematic drawing in Fig. 10 c indicates the location of the electrodes used in this investigation. The proper position of the DC semi-microelectrode in the depth of the cortex was tested by means of the antidromic response to a pyramidal tract volley. This procedure based upon studies by HUMPHREY (1968) is illustrated in Fig. 10 a. The traces in Fig. 10 b show two typical original DC recordings from the surface and from the fifth layer of the cortex during seizure activity of approximately 3 minutes duration. They indicate a clear-cut difference between the slow potential shifts at the cessation of the convulsive discharge. In contrast to the surface lead, the depth recording is characterized by a steep positive deflection, the time course of which corresponds to the after-hyperpolarization of pyramidal cells. This statement is supported by the diagram in Fig. 10 c. The columns represent the average amplitude, polarity and time course of the various potential shifts calculated from single convulsive attacks. They show that the DC shifts occurring in the fifth cortical layer are almost congruent with the MP changes of PT neurons. These findings confirm, at first, that the generation of DC deviations during seizure discharges can be explained, in principle, by a summation of slow membrane transients at cortical neurons. Furthermore they allow the conclusion that generator structures other than deeper cell bodies participate in producing the negative DC displacements encountered at the cortical surface during series of seizure discharges. This statement will be discussed in the following section.

### Discussion

The experiments aimed at a further clarification of the origin of cortical DC shifts known to occur during seizure activity. In this

context the study was focused on the relationship between DC deviations and MP changes of single PT cells in various convulsive states.
A gross evaluation of simultaneous recordings revealed at first a rather close correlation between the mean epicortical DC displacement and the mean depolarization of cortical neurons in the course of repetitive seizure discharges. Similar observations have been reported by Sugaya et al. (1964), Glötzner and Grüsser (1968) and by

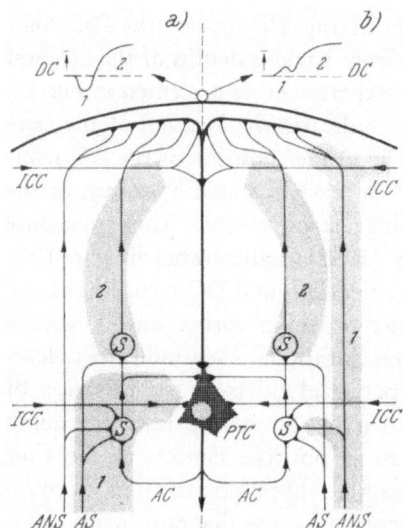

Fig. 11. Schematic presentation of neuronal circuits presumably involved in the generation of slow potential changes at the cortical surface during seizure activity. AS: specific afferents; ANS: non-specific afferents; AC: axon collaterals; ICC: intracortical connections; PTC: pyramidal tract cell; S: stellate cell

Ayala et al. (1970). A more detailed study of the records also showed, however, some systematic differences between DC and MP changes encountered especially at the onset and at the end of seizure activity. As a whole, these findings suggest that some of the faster waves in the surface leads reflect more rapid membrane transients in the deeper cellular layer, whereas the sustained displacement of the DC component depends, preferentially, on a direct activation of superficially located generator structures such as dendrites and/or glial cells. This interpretation will be discussed in greater detail in the schematic presentation of Fig. 11.

Previous studies of various authors have shown that neuronal PDS are usually initiated by a synaptic drive. If a presynaptic volley ascending, for instance, in a specific afferent pathway impinges upon a PT cell, the evoked PDS induces a source at the dendritic arborizations, before a passive electrotonic spread of the cellular depolarization takes place. In this way the rapid surface-positive deflection associated with the rise of the first neuronal PDS may be explained (*cf.*, channel 1 in Fig. 11 *a*). This interpretation would fit very well with observations by CREUTZFELDT *et al.* (1966 a, b). These authors found that the amplitude of the positive wave is related to the slope of the neuronal depolarization shift. It increases with the steepness of the PDS, and vice versa.

The afferent volleys conducted to pyramidal neurons are mediated either directly or by the way of axon collaterals also to stellate cells, the propagated discharges of which impinge in part on apical dendrites (*cf.*, channel 2 in Fig. 11 *a*). The secondary activation of these elements may be responsible for the delayed onset of the sustained negative DC shift. According to computations performed by RALL *et al.* (1967 a, b), dendritic EPSP's seem to differ from those of the soma membrane by a longer time constant. These conditions would actually favor the development of a sink at the cortical surface during continued seizure discharges. Afferent volleys ascending, for instance, in non-specific pathways or originating in the cortex itself may also be projected directly on dendrites in the outer cortical layer. The immediate activation of these superficial elements may account for the occurrence of negative-positive deflections as well as for the occasional absence of a positive wave at the onset of seizure activity. Such a mechanism is illustrated by channels 1 and ICC in Fig. 11 *b*.

The assumption that the sustained surface-negative DC shift originates mainly in the outer cortical layer is supported, furthermore, by the differences between MP and DC changes encountered at the termination of convulsive activity. It has been shown that the last neuronal PDS is always followed by a steep after-hyperpolarization. With single seizure discharges, this increase of MP coincides with a slowly decaying surface-negative wave which may depend, in part, on the positive potential shift in the deeper cortical layer. However, the after-negativity in the DC recordings cannot be attributed entirely to such a passive potential lead, since the time-courses of the DC and MP changes become completely different with repetitive seizure discharges. The findings point strongly to an enduring depolarization of generator structures located beneath the cortical surface. This conclusion is substantiated, moreover, by depth recordings of the DC component. In accordance with observations described by FERGUSON and JASPER (1971) DC tracings from the fifth cortical layer always show a steep positive shift as soon as seizure activity stops. In contrast to the surface leads the time course of these deeper DC deviations is closely related to the after-hyperpolarization in PT cells.

As has already been stressed, the structures responsible for the generation of DC displacements in superficial cortical layers may be dendrites, the known properties of which might well account for the observed potential shifts. In recent times there is, however, increasing evidence that so-called unresponsive cells which are presumed to be glial elements contribute to the generation of slow potential changes in the brain (Kuffler *et al.* 1966, Orkand *et al.* 1966, Grossman and Hampton 1968, Somjen 1970, Castellucci and Goldring 1970). The present study does not yield additional arguments as to this problem. In one of their papers Kuffler *et al.* (1966) pointed out that glial cells, by registering the changes of $K^+$ concentration in their environment, reflect activity changes of neuronal elements. It seems possible that also in this case such a linkage between neuronal and glial activity represents a complex generator.

## Summary

The relations between epicortical DC shifts and membrane potential changes (MP) of pyramidal tract (PT) cells during seizure discharges were studied in cats with the following results:

1. Generalized convulsive activity evoked by systemic application of pentylenetetrazol is accompanied by a surface-negative DC deviation. This effect coincides with paroxysmal depolarization shifts (PDS) in PT cells. During repetitive series of seizure discharges the mean epicortical DC displacement is closely related to the mean depolarization of the neuron.

2. At the onset of seizure activity, the first neuronal PDS is usually associated with a rapid positive or negative-positive deflection in the surface leads. The sustained negative DC displacement arises approximately at the top of the PDS showing a time-lag of 20–50 msec.

3. With the cessation of a seizure discharge, PT cells always develop a steep after-hyperpolarization, whereas the surface-negative DC shift declines gradually to the preceding level at a different time constant.

4. The findings suggest that some of the faster waves in the surface leads reflect more rapid membrane transients in the deeper cellular

layer. The sustained displacement of the DC component, on the other hand, can be attributed mainly to a direct activation of generator structures beneath the cortical surface. In this context the involvement of dendrites and glial cells is discussed.

## References

AYALA, G. F., H. MATSUMOTO, and R. J. GUMNIT: Excitability changes and inhibitory mechanisms in neocortical neurons during seizures. J. Neurophysiol. *33*, 73—85 (1970).

CASPERS, H.: Über die Beziehungen zwischen Dendritenpotential und Gleichspannung an der Hirnrinde. Pflügers Arch. ges. Physiol. *269*, 157—181 (1959).

— and W. SIMMICH: Cortical DC shifts associated with seizure activity. Proceedings of the international Symposium on comparative and cellular pathophysiology of epilepsy. Excerpta Medica International Congress Series, No. *124*, 151—162 (1966).

— and E.-J. SPECKMANN: DC potential shifts in paroxysmal states. In: JASPER, H. H., A. A. WARD, and A. POPE (eds.), Basic mechanisms of the epilepsies, pp. 375—388. Boston: Little, Brown & Co. 1969.

CASTELLUCCI, V. F., and S. GOLDRING: Contribution to steady potential shifts of slow depolarization in cells presumed to be glia. Electroenceph. clin. Neurophysiol. *28*, 109—118 (1970).

CREUTZFELDT, O., H. D. LUX, and S. WATANABE: Relations between EEG phenomena and potentials of single cortical cells. I. Evoked responses after thalamic and epicortical stimulation. Electroenceph. clin. Neurophysiol. *20*, 1—18 (1966 a).

— — — Relations between EEG phenomena and potentials of single cortical cells. II. Spontaneous and convulsoid activity. Electroenceph. clin. Neurophysiol. *20*, 19—37 (1966 b).

DICHTER, M., and W. A. SPENCER: Penicillin-induced interictal discharges from the cat hippocampus. I. Characteristics and topographical features. J. Neurophysiol. *32*, 649—662 (1969 a).

— — Penicillin-induced interictal discharges from the cat hippocampus. II. Mechanisms underlying origin and restriction. J. Neurophysiol. *32*, 663—687 (1969 b).

FERGUSON, J. H., and H. H. JASPER: Laminar DC studies of acetylcholine-activated epileptiform discharge in cerebral cortex. Electroenceph. clin. Neurophysiol. *30*, 377—390 (1971).

GLÖTZNER, F., und O.-J. GRÜSSER: Membranpotential und Entladungsfolgen corticaler Zellen, EEG und corticales Bestandpotential bei generalisierten Krampfanfällen. Arch. Psychiat. Nervenkr. *210*, 313—339 (1968).

GROSSMAN, R. G., and T. HAMPTON: Depolarization of cortical glial cells during electrocortical activity. Brain Res. *11*, 316—324 (1968).

GUMNIT, R. J., H. MATSUMOTO, and C. VASCONETTO: DC activity in the depth of an experimental epileptic focus. Electroenceph. clin. Neurophysiol. *28*, 333—339 (1970).

110    E.-J. Speckmann, H. Caspers, and R. W. Janzen

Humphrey, D. R.: Re-analysis of the antidromic cortical response. I. Potentials evoked by stimulation of the isolated pyramidal tract. Electroenceph. clin. Neurophysiol. 24, 116—129 (1968).

Klee, M. R.: Different effects on the membrane potential of motor cortex units after thalamic and reticular stimulation. In: Purpura, D. P., and M. D. Yahr (eds.), The Thalamus, pp. 287—317. New York-London: Columbia University Press. 1966.

— K. Offenloch, and J. Tigges: Cross-correlation analysis of electroencephalographic potentials and slow membrane transients. Science 147, 519—521 (1965).

Kuffler, S. W., J. G. Nicholls, and R. K. Orkand: Physiological properties of glial cells in the central nervous system of amphibia. J. Neurophysiol. 29, 768—787 (1966).

Loeschcke, H. H.: Über den Einfluß von $CO_2$ auf die Bestandpotentiale der Hirnhäute. Pflügers Arch. ges. Physiol. 262, 532—536 (1955/56).

— DC Potentials between CSF and Blood. In: Siesjö, B. K., and S. C. Sørensen (eds.), Ion homeostasis of the brain (Alfred Benzon Symposium III), pp. 77—96. Copenhagen: Munksgaard. 1971.

Matsumoto, H., and C. Ajmone Marsan: Cortical cellular phenomena in experimental epilepsy: interictal manifestations. Exptl. Neurol. 9, 286—304 (1964 a).

— — Cortical cellular phenomena in experimental epilepsy: ictal manifestations. Exptl. Neurol. 9, 305—326 (1964 b).

— G. F. Ayala, and R. J. Gumnit: Neuronal behavior and triggering mechanism in cortical epileptic focus. J. Neurophysiol. 32, 688—703 (1969).

O'Leary, J. L., and S. Goldring: DC potentials of the brain. Physiol. Rev. 44, 91—125 (1964).

Orkand, R. K., J. G. Nicholls, and S. W. Kuffler: Effect of nerve impulses on the membrane potential of glial cells in the central nervous system of amphibia. J. Neurophysiol. 29, 788—806 (1966).

Prince, D. A.: The depolarization shift in "epileptic" neurons. Exptl. Neurol. 21, 467—485 (1968).

— Electrophysiology of "epileptic" neurons: Spike generation. Electroenceph. clin. Neurophysiol. 26, 476—487 (1969).

— and K. J. Futamachi: Intracellular recordings from chronic epileptogenic foci in the monkey. Electroenceph. clin. Neurophysiol. 29, 496—510 (1970).

Purpura, D. P., J. G. McMurtry, C. F. Leonard, and A. Malliani: Evidence for dendritic origin of spikes without depolarizing prepotentials in hippocampal neurons during and after seizure. J. Neurophysiol. 29, 954—979 (1966).

Rall, W.: Distinguishing theoretical synaptic potentials computed for different somadendritic distributions of synaptic input. J. Neurophysiol. 30, 1138—1168 (1967 a).

— R. E. Burke, T. G. Smith, P. G. Nelson, and K. Frank: Dendritic location of synapses and possible mechanisms for the monosynaptic EPSP in motoneurons. J. Neurophysiol. 30, 1169—1193 (1967 b).

Somjen, G. G.: Evoked sustained focal potentials and membrane potential of neurons and of unresponsive cells of the spinal cord. J. Neurophysiol. 33, 562—582 (1970).

SPECKMANN, E.-J., H. CASPERS und R. W. JANZEN: Reaktionen spinaler Moto-neurone bei supraspinal induzierten Myoklonien. Z. EEG-EMG 2, 49—53 (1971).

SUGAYA, E., S. GOLDRING, and J. L. O'LEARY: Intracellular potentials associated with direct cortical response and seizure discharge in cat. Electroenceph. clin. Neurophysiol. 17, 661—669 (1964).

TSCHIRGI, R. D., and J. L. TAYLOR: Slowly changing bioelectric potentials associated with the blood-brain barrier. Amer. J. Physiol. 195, 7—22 (1958).

# Normal and Epileptic Synchronization
## at the Cortical Level in the Animal

J. SCHERRER and J. CALVET [1]

Unité de Recherches Neurophysiologiques de l'INSERM,
Faculté Pitié-Salpêtrière, Paris, France

Synchronization is a descriptive rather than an operative term. *It relates to the extent to which a population, or populations, of nervous elements are simultaneously active or silent* without prejudice to any particular mechanism.

The *concept of simultaneity* needs, however, to be more closely defined; it should be admitted that simultaneity in the strict sense of the word does not belong to biology, and what is to be expressed by simultaneity in the context of synchronization is a close timing of events during the phenomenon in question. Thus, two spikes delivered by two neurones can be considered to discharge simultaneously even if there is a time-lag of 0.1 msec. In the case of two cortical waves, on the other hand, which take 500 msec, a time-lag of 50 or 100 msec does not prevent us considering these two waves as synchronized.

This stresses the *operative aspect of the notion of synchronization.* In electrophysiology it largely depends on the technique used whether or not synchronization will be found; phenomena of synchronization which are not revealed by electrophysiological techniques may exist. Our present scope is limited as we do not intend to present a complete study of synchronization and its mechanisms

1 This paper sums up a fairly large number of data from the research work of several teams of the Laboratoire de Recherches Neurophysiologiques de la Pitié-Salpêtrière; data concerning the physiological aspects of synchronization by A. FOURMENT and M. THIEFFRY, ontogenesis by R. VERLEY and L. GARMA, convulsive activity by L. JAMI and isolated cortex by J. HIRSCH, J. F. HIRSCH and J. LANDAU.

as was attempted by BREMER (1958), FESSARD (1958) and more recently by ANDERSEN and ANDERSSON (1968). We shall only consider some electrophysiological aspects of wave-synchronization at the cerebrocortical level.

That synchronization phenomena exist in cortical activity has been more or less explicitly admitted since the discovery of the EEG; the finding that waves can be recorded at all, and that their amplitudes may increase, implies that in all probability large numbers of neurons are simultaneously active or silent. This tacit assumption underlies the widespread use of the terms "desynchronization" and "hypersynchronous".

There are now numerous observations which suggest that the EEG waves present the sum of postsynaptic potentials (CREUTZFELDT 1969, POLLEN 1969) and there are reasons to believe that the postsynaptic potentials which are responsible for the EEG waves are located in the pyramidal cells. These cells are oriented in such a way as to enable them to add their extra-cellular currents due to their membrane potential changes. Moreover, the amplitudes of EEG phenomena, when recorded extracellularly, have been proved to be maximal for radial derivations (SCHERRER 1965): the amplitude of wave-type phenomena when recorded by electrodes the upper tip of which is on the cortex, the lower at the lower limit of the cortex (or within the cortex, about 1000 $\mu$ apart from the surface electrode) is two to three times larger than when recorded from the commonly used surface electrodes. The potential gradients in the cortex are radially orientated.

These observations may be explained by the assumption that *cortical wave-synchronization is caused by a simultaneous or almost simultaneous occurrence of a relatively large number of postsynaptic potentials with the same electrical sign, in the pyramidal cells of the cortex.*

A priori it is probable that some relation between postsynaptic potentials and spikes may exist. That this is really the case will be shown later on. Therefore, synchronization is not only the sum of EPSPs and IPSPs but also includes changes in the rate of neuronal discharges.

Synchronization at the cortical level, whether representing either of these two phenomena as just mentioned can be considered from two different operative aspects: 1. when operative procedure is spatially limited, recording being done in one point of the cortex, an increased synchronization is revealed by an increase in amplitude of a given wave and/or by a simultaneous variation (increase or decrease) in neuronal discharges. 2. When the operative procedure is spatially extended (several cortical recordings more or less distant from one another) synchronization appears as simultaneous occurrence of an identical wave and/or of an identical variation in cell discharges in the different recording sites. However, single or multiple leads bring out the same phenomenon: an increase in the number of activated or silent neurons. The only difference is in the technique used.

In the following pages we shall first consider the relations between wave-type and single unit activity. Such a relation is necessary for brain-wave synchronization to be meaningful for the sending of messages. Afterwards, synchronization will be considered under normal physiological conditions and, finally, in seizures.

## Relations between Cortical Waves and Discharges under Normal Physiological Conditions

That there is some relation between cortical waves (*i.e.*, EEG) and discharges has been seen whenever electrocortical and unit activities were studied. Among the many studies on this subject we refer to Adrian and Moruzzi (1939), Andersen and Andersson (1968), Bishop and Clare (1952), Bremer (1958), Buchwald *et al.* (1966), Clare and Bishop (1955), Creutzfeldt *et al.* (1966), Eccles (1951), Enamoto and Ajmone Marsan (1959), Jasper and Stefanis (1965), Li *et al.* (1956), Petsche *et al.* (1970), Purpura *et al.* (1964), and Scherrer (1965).

Some authors, notably Creutzfeldt (1969), think that *no simple relation exists between cell discharges and electrocortical waves*. On the contrary, we consider it as proven for almost a decade ago (Calvet *et al.*, 1964) that there exists such a relation in a normal physiological state. Its existence was to be expected.

The use of partial radial cortical recordings with short interelectrode distances or transcortical recordings enabled us to consider EEG wave activity as being the result of three possible generators (Fig. 1 A, B and C). Generator A is located in the upper 500 $\mu$, *i.e.*, the superficial layers of the cortex. It underlies the surface-negative phenomena. When generator B is active, two partial dipoles can be observed; one is superficial and reaches down to approximately 1000 $\mu$, the other has the opposite sign and is found between 1000 and 2000 $\mu$ below the surface. The negativities of both the dipoles are at a depth of 1000 $\mu$. Since the potential difference produced

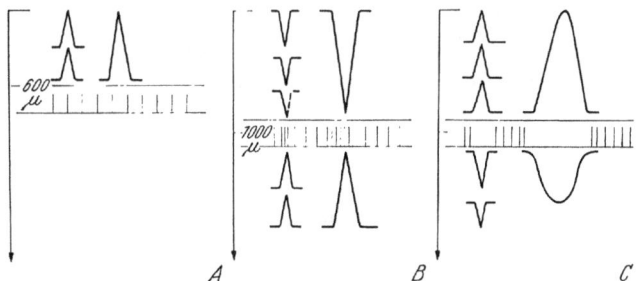

Fig. 1. The three types of cortical generators. The figure represents the intracortical level of each generator, the surface and deep dipoles of generators B and C and the relation of generators to cell discharge

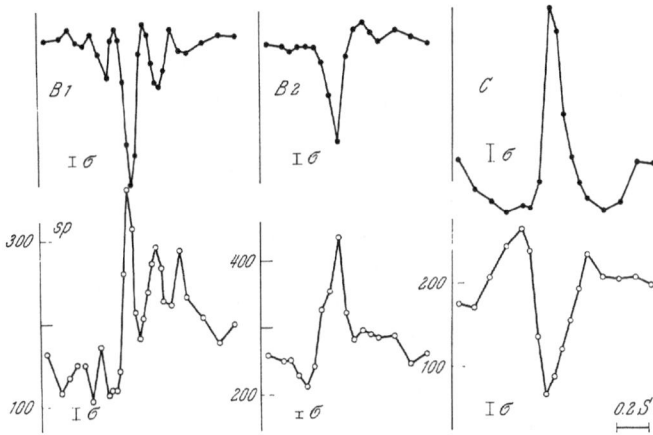

Fig. 2. Wave-spike relationship. *Upper traces*: Averaged spontaneous cortical waves. $B_1$: average of 186 surface positive waves recorded (during spindles). $B_2$: average of 191 surface positive "activating" waves. C: average of 103 surface negative sleep waves. *Lower traces*: Number of spike discharges computed during the $B_1$, $B_2$ and C waves respectively occuring at the same point as the wave. Cat: chronic preparation

by the deeper dipole is smaller than the one produced by the more superficially situated dipole, a transcortical derivation over the entire cortical depth which sums up the effects of both the dipoles will

8*

result in electropositive phenomena on the cortex whenever generator B is active. Generator C results in a positivity of the cortical
layers at 800–1000 $\mu$ with respect to both the surface and its deep
parts. In this case again (since the partial superficial dipole produces
higher potentials than the deep one) the total dipole resulting from
the addition of the two partial dipoles will result in surface-negative
events.

When generator A is active, there is no obvious interference with
any cellular discharges. In contrast, there is a remarkable relation between cellular discharges and waves caused by generators B
and C: with surface-positive waves (generator B) an increase of
discharges is found and with the activity of generator C, the opposite
phenomenon is observed. Fig. 2 demonstrates this by statistical
analysis. This analysis was performed on automatically selected
waves according to given polarity, amplitude and duration characteristics by means of a logical circuit. Time location of the peaks
of selected waves was recorded on magnetic memory. An averaging
of the amplitude of the waves was then performed for the peak time
and for samples 25, 50, 75 ... msec before and after peak time.
Another averaging after AD-conversion was performed for the same
time samples for cell discharges. The same intracortical electrode,
1 mm deep, was simultaneously used for recording cell discharges
and wave phenomena (Fig. 2). Spindle waves ($B_1$) and isolated
surface-positive waves ($B_2$), sometimes called "activating waves",
are accompanied by a considerable increase of cell discharges ($sp$)
whereas surface-negative waves reflecting the activity of generator C
are paralleled by a decrease of cell discharges. In Fig. 2, the surface-
negative waves are followed by, and even more distinctly preceded
by, some activity due to a B-type generator.

In view of the location of depolarization, the observations on
intracellular activity, and finally the increase in spike activity, one
could be tempted to assume that *generator B reflects a depolarization
of neuronal somata*, which means that the surface-positive wave may
be due to a summation of EPSPs. Similar arguments may reduce the
surface-negative waves which result from *generator C to a hyperpolarization of neuronal somata, i.e.,* a summation of IPSP's. Even
if these conclusions are not yet sufficiently substantiated, we take
them for granted enough in this context to speak about EPSP- and
IPSP-synchronization on the following pages.

EPSP-synchronization may be studied on surface-positive waves recorded in REM-sleep and in wakefulness. IPSP-synchronization, on the other hand, is studied on surface-negative waves recorded in the slow-wave sleep. Studies of the latter phenomenon are more advanced and therefore will be considered first. Physiological synchronization obviously has to be studied in animals and with

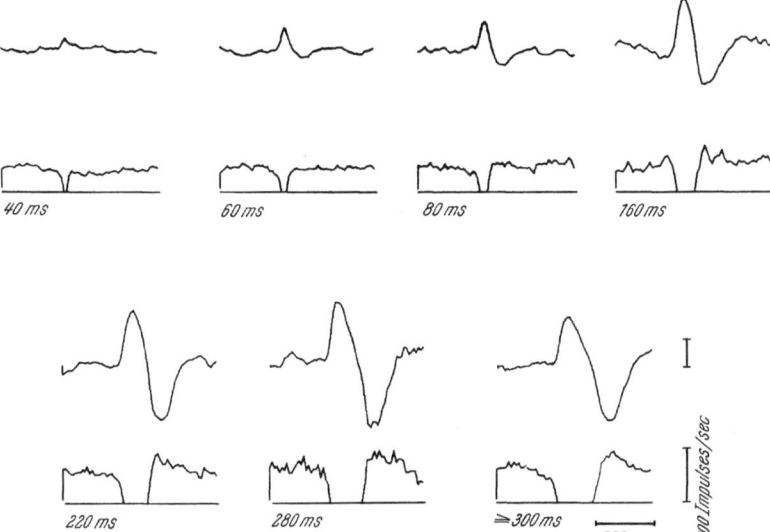

Fig. 3. Local IPSP synchronization. *Lower traces*: spike discharge. *Upper traces*: simultaneously recorded cortical waves. The successive figures are means of 227, 132, 84, 37, 44, 14, 69 averagings on Linc 8. Duration of spike suppression in msec. Cat: chronic preparation. In this and following wave recordings upward deflexion shows a negativity of surface electrode towards the deep electrode of the transcortical pair

implanted electrodes. The normal phases of wakefulness and sleep were the object of study.

## Synchronization of Cortical IPSPs

It was mentioned earlier that the *surface-negative waves caused by generator C,* most likely reflect a hyperpolarization of neuronal somata. They coincide with a decrease of the discharges of pyramidal cells. An increase of synchronization has, as already stated, a double consequence, namely an *increase of wave-amplitude in a given area and a larger diffusion of this wave to other cortical areas.* This is paralleled by *a local and distant decrease of discharges*. In Fig. 3, different degrees of synchronization are shown for a given recording point in the suprasylvian gyrus. When the synchronization is minimum, surface-negative amplitudes are low (not higher than

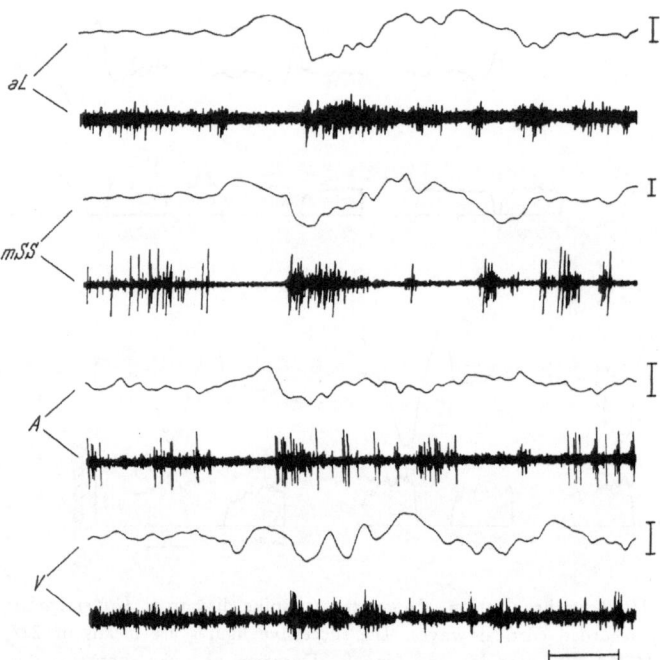

Fig. 4. Inhibitory surface negative wave recorded in different cortical regions. Electronical waves (upper traces) and related spike discharges (lower traces) are recorded simultaneously in anterior lateral (*aL*) and medial suprasylvian gyri (*mSS*); auditory (*A*) and visual (*V*) areas. In each record, a surface negative wave involving distinctly the four sites can be seen accompanied by a decrease in discharge. Cat: chronic preparation. Calibration: 500 $\mu$V. Time: 250 msec

100 $\mu$V) and have a short duration (about 40 msec); when their amplitude is large, they may reach up to 700 $\mu$V and last as much as 30 msec. The inhibition of cellular firing increases in parallel with the duration of the wave. In interpreting these data one has to take into account several factors among which two are inescapable: the increased hyperpolarization of a given neuron where an IPSP

is elicited, and the increased number of other neurons affected secondarily by these IPSPs. A third factor possibly plays an additional role, that is, some time dispersion of IPSPs, which could result in a shortened or prolonged duration of total inhibition.

The extension of the cortical area where simultaneous surface-negative waves can be recorded has been mapped by means of multiple implanted electrodes during sleep. Fig. 4 shows EEG activity and spike discharges recorded simultaneously in four different cortical

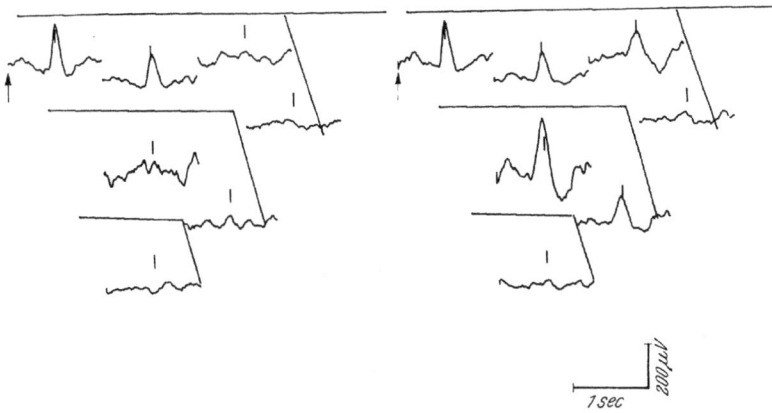

Fig. 5. Cortical extension of a surface negative wave during onset of sleep (left) and a later stage of slow wave sleep (right). Cat: chronic preparation (see text)

areas. Very definitely a surface-negative wave appears simultaneously in these four regions. Its inhibitory effect is more pronounced in the supra-sylvian gyrus.

The cortical extension of the surface negative wave is studied in Fig. 5. An averaging technique was used for automatically chosen spontaneous surface negative waves. Selection and averaging procedures were done on a Linc 8 computer during 90 second, first for posterior lateral derivation (arrow) according to given electrical polarity, amplitude and wave duration characteristics. Using the peak of the response (vertical bar) averaged responses for other areas are obtained during the same 90 seconds in middle and anterior lateral gyrus (top traces), suprasylvian gyrus (middle traces) and ectosylvian gyrus (lower traces). For each averaged responses the bar signals the peak of the response in the posterior lateral gyrus. A very definite difference in extension of the surface negative wave is seen between the averages performed during the onset of sleep and a later stage of sleep.

The various surface-negative waves recorded in different cortical areas do not rigorously coincide in time; there are frequently time differences of a few hundredths of a second which can be neglected when considering the duration of the wave. These time differences can be systematized for different regions (Calvet et al. 1971).

### Synchronization of Cortical EPSPs

EPSP-synchronization reflected by *surface-positive waves* (sometimes called "activating waves") which are related to an activation of type-B generators, can be studied in REM-sleep and during wakefulness (Calvet et al. 1965). As in the case of IPSP-, EPSP-synchronization is reflected by both an increased amplitude of the surface-positive deflection and a widespread extension of this wave over the cortex. The amplitude of this sort of waves during REM sleep may attain 300 $\mu$V, its duration 150–200 msec. The rate of cell discharges increases during the wave; it can double or even treble when the wave is large (Fig. 6).

The extent of the cortex involved in an activating wave during REM-sleep is fairly large: in cats it reaches far beyond the occipital area (Fig. 7) though its maximum amplitude is found in occipital derivations.

The surface-positive wave during wakefulness (Figs. 6 and 7) may be compared with the one during REM-sleep but the phenomenon is more definitly localized in the occipital areas. During wakefulness, activating waves seem closely related to eye movements.

Surface-positive waves during both wakefulness and sleep can be recognized by means of transcortical leads much more easily than by the commonly used mono- and bipolar surface-derivations.

When studied by means of several distant transcortical pairs of electrodes, the simultaneity of surface-positive waves is quite obvious. The time-lag between waves from different active areas does not exceed 10 to 20 msec which is within the resolution power of our system and can be neglected with respect to the wave duration.

### Synchronization during Ontogenesis

A follow-up study of the development of cortical synchronization during ontogenesis was made by Verley (1965). The appearance of spontaneous cortical waves and their increase in amplitude during ontogenesis is a well known fact; no doubt this increase is essentially related to the very maturation of the neurons whose volume increases

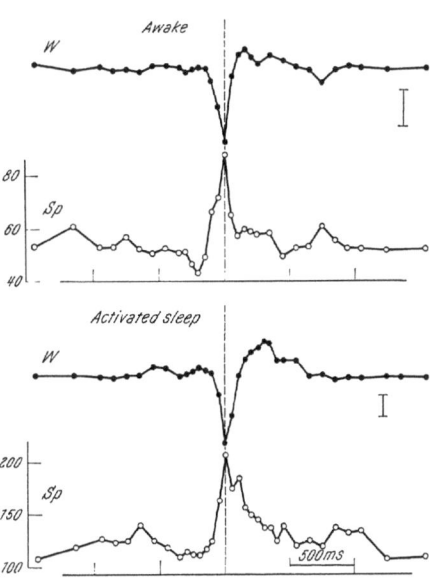

Fig. 6. Wave-spike relation during activating waves. Cellular discharges (Sp) accompanying cortical activating waves (W) during the waking state and REM sleep. Cat with implanted electrodes. Average of 191 activating waves during the waking state and 70 during REM sleep. Figures for the spikes, on the ordinate scale, indicate the average number of discharges per second. Calibration for the waves is 100 μV

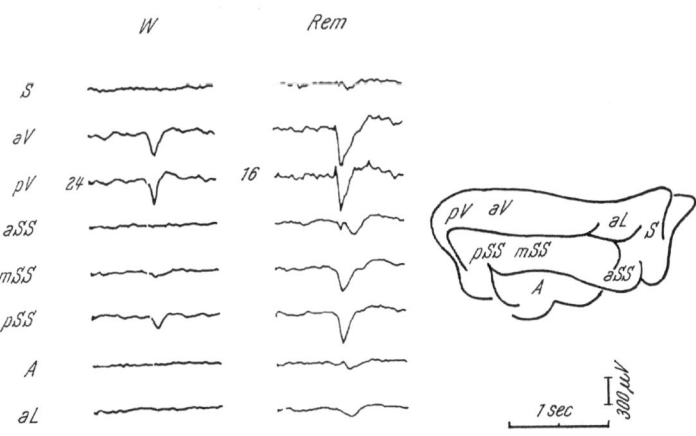

Fig. 7. Extension of activating waves. Average on Linc 8 computer of 24 activating waves during wakefulness compared with the average of 16 activating waves during REM (paradoxical) sleep. The extension of surface positive activating wave is larger during REM sleep. Cat: chronic preparation

and which become active. Already very early in ontogenetic development it can be observed that some cortical waves are accompanied either by an increase or a decrease of unit discharges (Fig. 8). Therefore, simultaneity, *i.e.*, synchronization of IPSPs or EPSPs at a

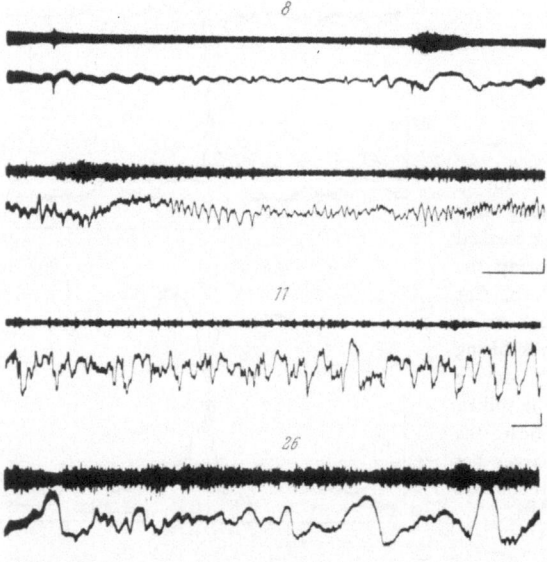

Fig. 8. Wave and spike relations during ontogenetic development. Rabbit of age 8, 11, and 26 days. At 8 days spike discharges may already coincide with isolated surface positive waves. Surface negative waves are progressively increasing in number and amplitude. At 3 weeks, they coincide systematically with discharge inhibition. Sweep speed: 500 msec; calibration: 50 $\mu$V for 8 days, 100 $\mu$V for 11 and 26 days. Cat: chronic preparation

given point of the cortex may be taken for granted (Garma and Verley 1967).

In rabbit the extension of synchronization was studied (Fig. 9) with a series of 4 to 6 closely aligned pairs of transcortical electrodes (mean distance between two pairs: 1 mm). Each pair of electrodes recorded the local activity only. This activity was integrated by measuring for 1 or 2 seconds the planimetric area delineated by positive and negative variations of the potential recorded. When

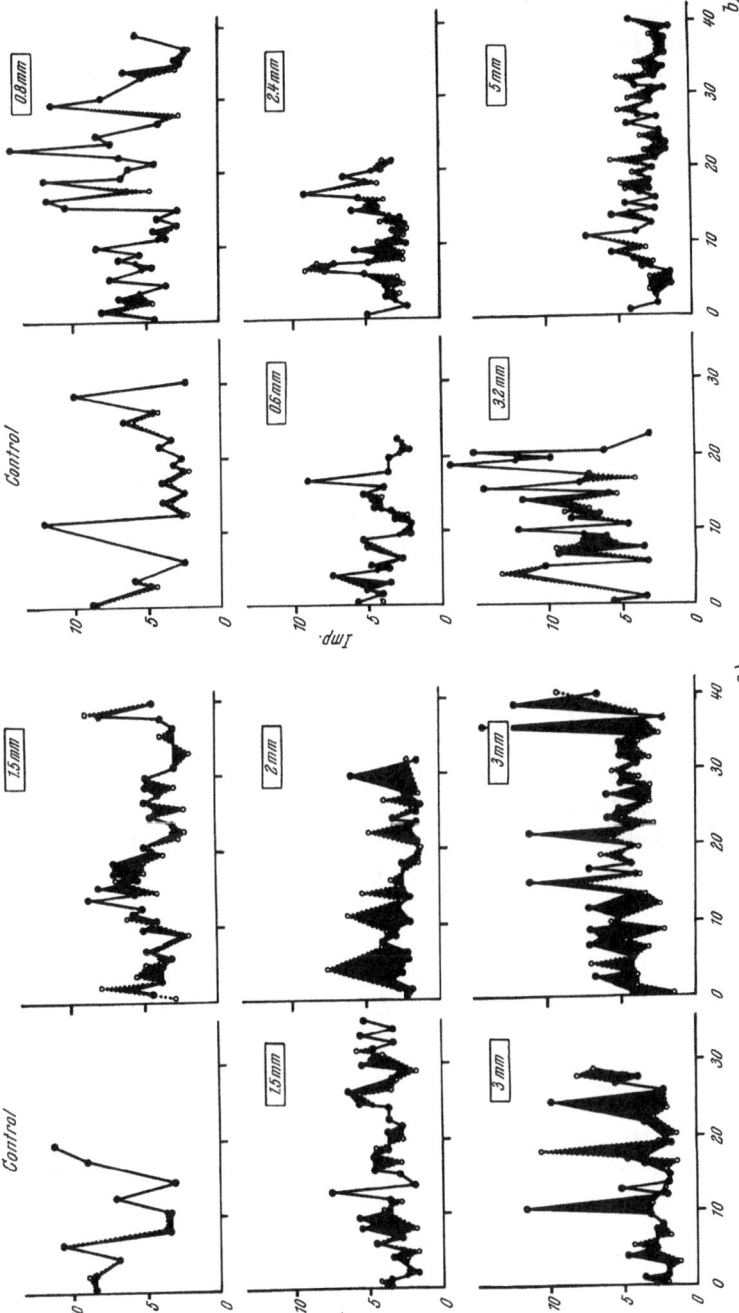

Fig. 9. Increase in synchronization during ontogenetic development (see text). Difference in integrated activity between two pairs of electrodes, situated at different distances (from 0.6 to 5 mm) in a 4 days (*a*) and a 30 days (*b*) old rabbit in acute preparation. Integrated activity is represented on ordinate in arbitrary units

the two levels do not coincide a shaded area represents the difference. It is obvious from the figure that in a 4 day old rabbit two pairs of transcortical electrodes 2 mm apart display activities quite different, whereas in a 30 day old rabbit, over a distance of 3, 2 to 5 mm there is a more similar activity.

## Synchronization in Convulsive States

Local synchronization in convulsive states can first be considered as an increase of electrocortical waves in amplitude. This is true of strychnine waves as well as of isolated convulsive waves between seizures and "clonic" phenomena at the end of seizures (these various convulsive phenomena are sometimes misnamed "spikes"). In contrast to surface-positive waves which, in physiological states where recorded by transcortical bipolar techniques, are not larger than 300–400 $\mu V$, and the surface negative waves which were hardly twice as high, convulsive waves may reach more than 4 mV peak-to-peak.

*The relations between these waves and the unit discharges have many similarities with those in physiological states* (JAMI 1972). With transcortical bipolar techniques, strychnine waves appear as either monophasic surface-positive or biphasic positive-negative waves; in both cases, however, the discharges occur mainly during positive deflections. A similar relation has been statistically proven for isolated interictal convulsive waves or those appearing at the onset of a seizure before the tonic stage. During the electrocortical convulsive clonic stage, the discharges begin simultaneously with the surface-positive wave, but the latter is distinctly shorter than the duration of the discharges.

On the other hand, it is not possible to find any time relationships between waves and cellular discharges in the tonic stage of electrocortical seizures. At these stage the potential fluctuations, being of lower voltage, are accompanied by a very high but almost constant rate of discharges.

The surface extension of waves seems highly variable according to the convulsive stage considered. Isolated convulsive waves may be restricted to a very limited zone in the cortex; this applies to strychnine waves which, in young animals (rabbits), involve a few mm² of cortex only. In adult animals, this area may be a little larger. Interictal convulsive waves do not generally spread very far over the cortex. On the contrary, during the clonic stage of epileptic seizures, the waves recorded can spread over large regions (Fig. 10). Since no statistical studies have been made on this subject, it is difficult to know whether this happens frequently.

When speaking of convulsive states, the word "synchronization" is

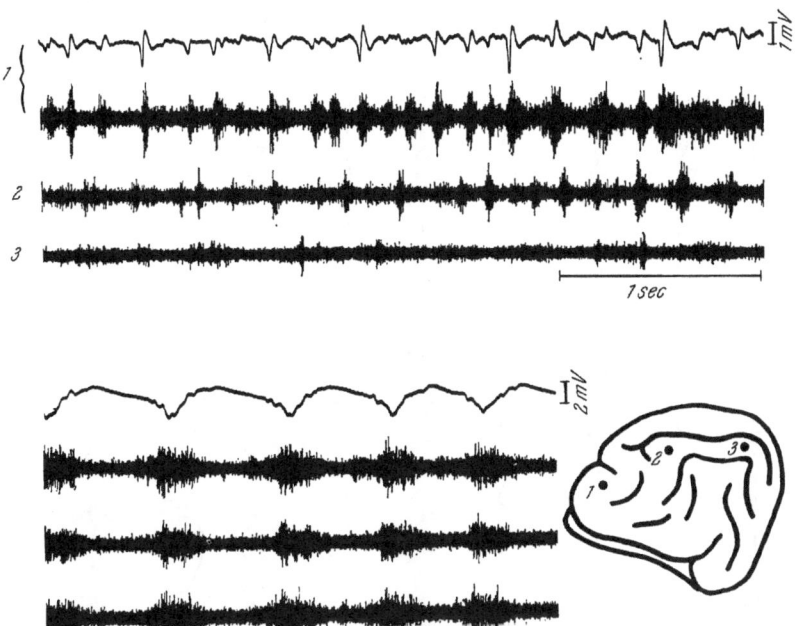

Fig. 10. Convulsive activity. Early phase (upper record) and clonic stage (lower record) during a cortical epileptic seizure. For area 1 the wave activity is compared with spike discharges. Definite relation between surface positive wave and spike discharge. Wave synchronization only during the clonic phase of the seizure for the 3 regions. Cat: acute preparation. Cardiazol seizure

often used to denominate an aspect different from the one dealt with so far, namely a simultaneous (or quasi simultaneous) occurrence of waves of a few hundredths or tenths of second duration. In convulsive episodes, synchronization indicates the fact that two or more cortical areas are at the same time affected by the same type of epileptic manifestations, be it clonic or tonic activities. Even a simultaneous occurrence of different convulsive phenomena in various cortical areas may be called "synchronization". In this case, the con-

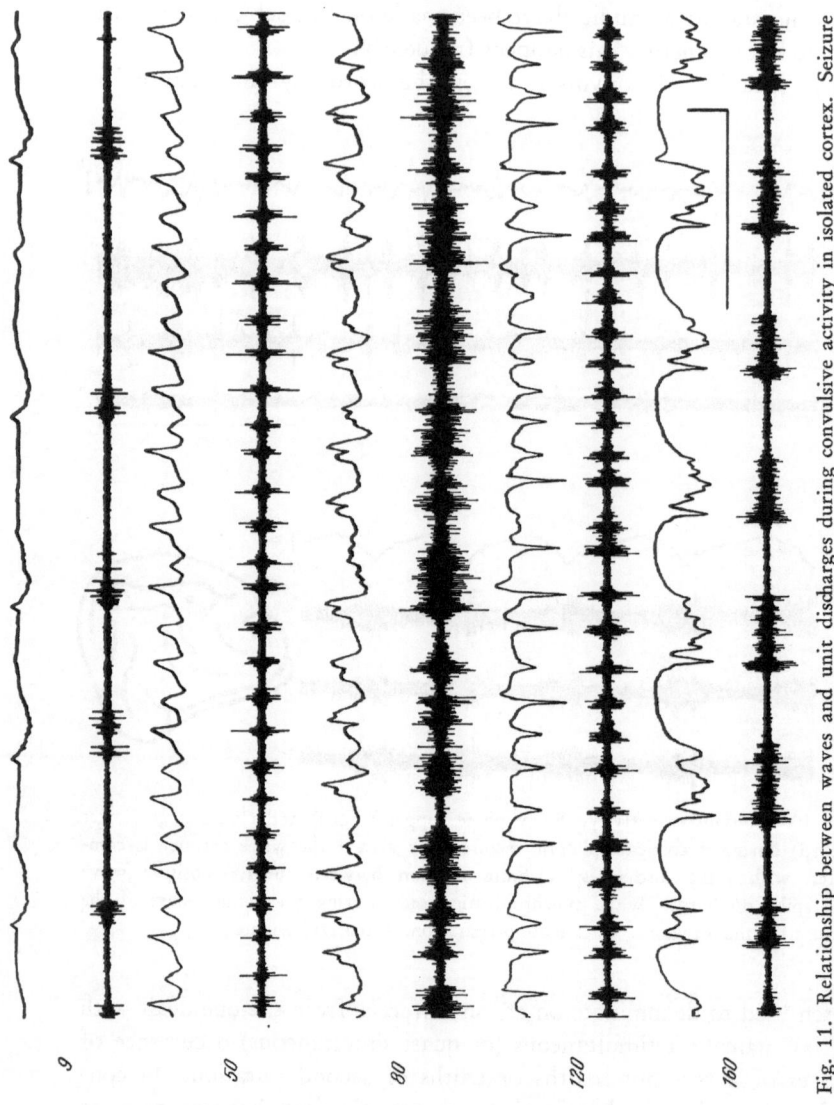

Fig. 11. Relationship between waves and unit discharges during convulsive activity in isolated cortex. Seizure recorded 9, 50, 80, 120, and 160 msec after electrical stimulation. ECoG: upper trace. Unit activity: lower trace. Definite relations between surface positive wave and unit discharges during clonic stage (120 and 160 sec). Cat. Chronically isolated suprasylvian gyrus. Calibration: 1 mV. Time: 500 msec

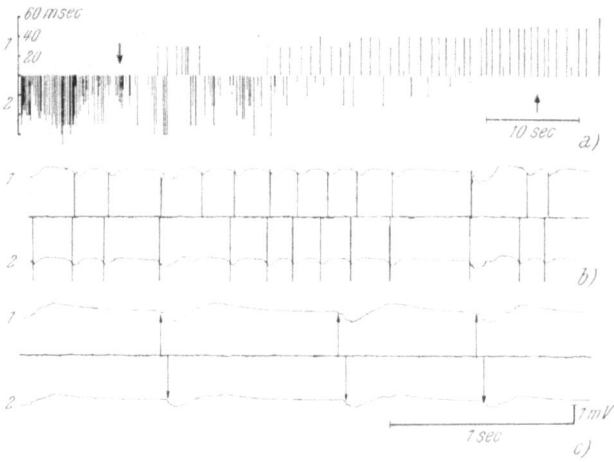

Fig. 12. Relative synchronization of paroxysmal waves in isolated cortex. End of clonic convulsive activity recorded by two pairs of transcortical electrodes located 15 mm apart. *b*) and *c*): recording of the ECoG. The beginning of each wave is indicated. *a*): time relation between wave onset. When electrode 2 leads vertical bar is down, up when it is lagging. Cat. Chronically isolated suprasylvian gyrus. Electrically induced seizure

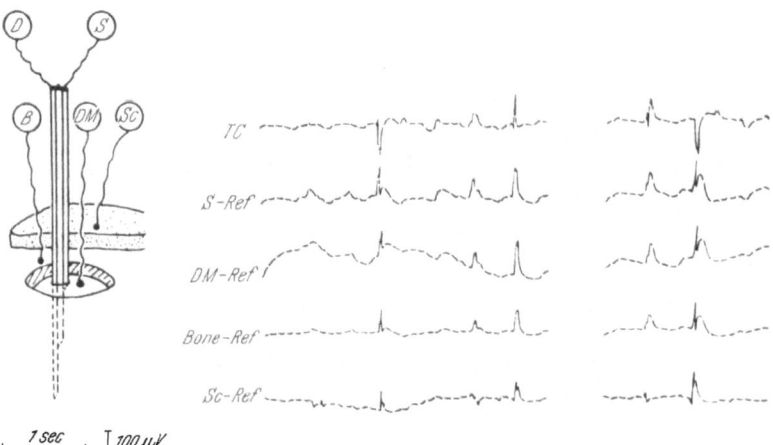

Fig. 13. Polarity distortion of surface positive wave. Recordings of surface positive and surface negative convulsive waves by a transcortical pair of electrodes and usual monopolar derivations. Electrical sign of the surface positive wave is inverted. Cat: acute preparation—convulsive waves induced by Bemegride

vulsive patterns observed in these areas may have time delays of up
to several seconds. The synchronization of spikes, on the other hand,
as already mentioned, takes place within the range of milli-seconds.
Therefore it would be adviseable to speak of *"wave synchronization"*,
*"stage synchronization"* and *"spike synchronization"*.
As has been known since the work of BURNS (1956) it is possible to
induce convulsive states in an isolated gyrus. This model may be
studied all the more easily since the abolition of the cortical connec-

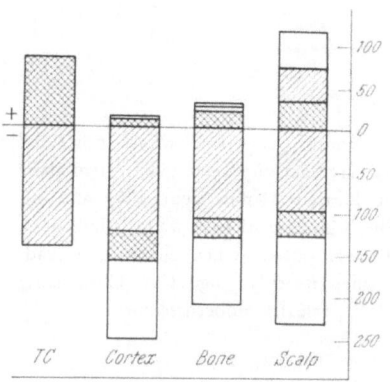

Fig. 14. Polarity distortion of convulsive
waves. Surface positive (black) and surface negative (stripped) waves detected
by transcortical leads as they appear in
monopolar recordings on cortex, bone
and scalp. During a 6 min recording
80 surface positive and 143 surface negative waves were registered. Black SP waves
maintain their sign on MP when they
are above the O line, reverse it when
below the O line. Stripped SN waves:
symmetrical convention. White: waves
recorded by MP without any transcortical
counterpart. Cat: acute preparation convulsive waves induced by Bemegride

tions to the deep structures of the brain does not alter the relation
between waves and cell discharges (FROST and GOL 1966, and
Fig. 11). When, in this type of preparation, isolated convulsive
waves are triggered, the total gyrus tends to be invaded by the
paroxysmal phenomenon. At first sight this invasion takes place
simultaneously for a given wave and in its accompanying unit
firing.
A careful study with several transcortical recordings at 3 mm distances shows, however, that there is a slight time delay between these
phenomena recorded from these different points, as if a low rate
of conduction were present (Fig. 12). Nevertheless, if the term synchronization is applied to phenomena with an approximate coincidence, it is adequate.
It is important to mention that when convulsive waves are recorded
with the common mono- and bipolar techniques, their polarity is
often reversed. This is shown in Fig. 13: the convulsive waves are
surface-positive, when transcortically recorded, but they are as a

rule surface-negative with monopolar derivation wherever the active electrode is located (cortex, dura, bone or scalp).

This was confirmed by a statistical study on 70 waves (Fig. 14). More generally, the commonly used derivations often distort the polarity of the cortical wave; activating waves can almost be suppressed. The *importance of both sign and amplitude distortions caused by recording technique* was emphasized by CALVET and SCHERRER (1961). The same was found in isolated gyrus (HIRSCH 1966). A descriptive study (FOURMENT et al. 1965) and an attempt to explain this phenomenon (JAMI et al. 1968) lead to the assumption that the distortion of the electrical sign of the waves may be explained by the curvature of the dipole-surface in which the activity is located. With this knowledge one is no longer surprised to see how many attempts at relating the EEG to cortical neuronal discharges have failed (CREUTZFELDT 1969).

The present paper is an attempt to classify and present the notion of electrical synchronization within the cortex and also to elucidate the conditions for it. As was said at the beginning, the term synchronization is only descriptive and without prejudice to the mechanisms underlying the simultaneous occurrence of brain waves. Yet these mechanisms are certainly different according to the type of synchronization in question (see ANDERSEN et al. 1962; BENNETT 1966; PURPURA et al. 1966). The mechanisms by which two spikes are synchronized are certainly different from that by which two waves or two convulsive states are synchronized. Conduction and transmission phenomena should also be implied in these different phenomena, but the modalities are likely to be different. Numerous synchronization phenomena may occur in a homogeneous neuronal aggregate, as may be concluded from the observations on the convulsive activities of isolated gyrus. Yet in other cases it seems that the connections between two distinct aggregates (*e.g.*, cortex and thalamus) favour the maintenance of synchronization.

Finally, the relation between synchronization and rhythmicity is close but undoubtedly complex. To sum up, there is still room for many investigations in this field.

## Summary

1. With respect to cortical activity the word "synchronization" may apply to the simultaneous or almost simultaneous occurrence of

spikes (action potentials), waves (based on EPSPs or IPSPs) or complex and long-lasting stages of activity. The present paper deals with wave-type synchronization in the cerebral cortex.

2. From an operational point of view, wave synchronization may be detected in one point of the cortex and is characterized by an increase in wave potential; when recording from different points at the same time, synchronization is characterized by the simultaneous occurrence of waves of similar shape.

3. Massive physiological EPSP-synchronization underlies activating waves seen during REM sleep and wakefulness. They are associated with an increase of neuronal discharges. Massive IPSP-synchronization underlies slow-wave sleep. During these waves nerve cells are inhibited.

4. Neuronal synchronization phenomena appear early in ontogenetic development.

5. Synchronization in convulsive states results in a maximum summation of PSPs in a given point of the cortex. This event may or may not involve a large cortical area.

6. Cortical waves and neuronal discharges are closely related even in physiological states and also in some convulsive states. The recordings of cortical waves may be considerably distorted by the surface derivations commonly used.

## References

Adrian, E. D., and G. Moruzzi: Impulses in the pyramidal tract. J. Physiol. (London) *97*, 153 (1939).

Andersen, P., and S. A. Andersson: Physiological basis of the alpha rhythm. New York: Appleton-Century-Crofts. 1968.

— and J. C. Eccles: Inhibitory phasing of neuronal discharge. Nature *196*, 645—647 (1962).

Bennett, M. V. C.: A comparative study of neuronal synchronization. In: Purpura, D. P., and M. D. Yahr (eds.), The Thalamus, pp. 173—181. New York: Columbia University Press. 1966.

Bishop, G. H., and M. H. Clare: Sites of origin of electric potentials in striate cortex. J. Neurophysiol. *15*, 201 (1952).

Bremer, F.: Considération sur l'origine et la nature des ondes cérébrales. Electroenceph. clin. Neurophysiol. *1*, 177—193 (1949).

— Cerebral and cerebellar potentials. Physiol. Rev. *38*, 357—388 (1958).

Buchwald, J. S., E. S. Halas, and S. Schramm: Relationship of neuronal spike populations and EEG activity in chronic cats. Electroenceph. clin. Neurophysiol. *21*, 227—238 (1966).

Burns, B. D.: The production of after-bursts in isolated unanaesthetized cerebral cortex. J. Physiol. *127*, 168—188 (1955).

Calvet, J., M. C. Calvet et J. Scherrer: Etude stratigraphique corticale de l'activité électroencephalographique spontanée. Electroenceph. clin. Neurophysiol. *17*, 109—125 (1964).

— — and J. Langlois: Diffuse cortical activation waves during so-called desynchronized EEG patterns. J. Neurophysiol. *28*, 893—907 (1965).

— A. Fourment et M. Thieffry: Activité électrique des zones corticales de projection et d'association pendant le sommeil (1972). In preparation.

— et J. Scherrer: Relations des décharges unitaires avec les ondes cérébrales spontanées et la polarisation corticale. C. R. Acad. Sci. *252*, 2297—2299 (1961).

Clare, M. H., and G. H. Bishop: Properties of dendrites; apical dendrites of the cat cortex. Electroenceph. clin. Neurophysiol. *7*, 85—98 (1955).

Creutzfeldt, O. D.: Neuronal mechanisms underlying the EEG. Basic mechanisms of the epilepsies, pp. 397—410. Boston: Little, Brown & Co. 1969.

— S. Watanabe, and H. D. Lux: Relations between EEG phenomena and potentials of single cortical cells. II. Spontaneous and convulsoid activity. Electroenceph. clin. Neurophysiol. *20*, 19—37 (1966).

Eccles, J. C.: Interpretation of action potentials evoked in the cerebral cortex. Electroenceph. clin. Neurophysiol. *3*, 449—464 (1951).

Enomoto, T. F., and C. Ajmone Marsan: Epileptic activation of single cortical neurons and their relationship with electroencephalographic discharges. Electroenceph. clin. Neurophysiol. *11*, 199—218 (1959).

Fessard, A.: Les mécanismes de synchronisation interneuronique et leur intervention dans la crise épileptique. In: Bases physiologiques et aspects cliniques de l'épilepsie, pp. 37—60. Paris: Masson. 1958.

Fourment, A., L. Jami, J. Calvet et J. Scherrer: Comparaison de l'EEG recueilli sur le scalp avec l'activité élémentaire des dipôles corticaux radiaires. Electroenceph. clin. Neurophysiol. *19*, 217—229 (1965).

Frost, J. D., Jr., and A. Gol: Computer determination of relationships between EEG activity and single unit discharges in isolated cerebral cortex. Exptl. Neurol. *14*, 506—519 (1966).

Garma, L., et R. Verley: Activités cellulaires corticales étudiées par électrodes implantées chez le lapin nouveau-né. J. Physiol. Paris *59*, 357—376 (1967).

Hirsch, J. C.: Variations selon les modalités d'enregistrement de l'électro-corticogramme d'un gyrus cortical isolé. C. R. Acad. Sci. Paris *263*, 778—780 (1966).

— J. Landau-Ferey, J.-F. Hirsch, and J. Scherrer: Electrocorticogramme et activités unitaires d'un gyrus cortical isolé en préparation chronique. J. Physiol. Paris *61*, 387—402 (1969).

Jami, L.: Patterns of cortical population discharges during metrazol induced seizures in cats. Electroenceph. clin. Neurophysiol. (1972), in press.

— A. Fourment, J. Calvet et M. Thieffry: Etude sur modèle des méthodes de détection EEG. Electroenceph. clin. Neurophysiol. *24*, 130—145 (1968).

Jasper, H., and G. Stefanis: Intracellular oscillatory rhythms in pyramidal tract neurones in the cat. Electroenceph. clin. Neurophysiol. *18*, 541—553 (1965).

Li, C. L., C. Cullen, and H. H. Jasper: Laminar microelectrode analysis of cortical unspecific recruiting responses and spontaneous rhythms. J. Neurophysiol. *19*, 131—143 (1956).

Petsche, H., P. Rappelsberger, and R. Trappl: Properties of cortical seizure potential fields. Electroenceph. clin. Neurophysiol. *29*, 567—578 (1970).

Pollen, D. A.: Discussion on the generation of neocortical potentials. Basic mechanisms of the epilepsies, pp. 411—419. Boston: Little, Brown & Co. 1969.

Purpura, D. P., T. L. Frigyesi, J. G. McMurtry, and T. Scarff: Synaptic mechanisms in thalamic regulation of cerebello-cortical projection activity. In: Purpura, D. P., and M. D. Yahr (eds.), The Thalamus, pp. 153—172. New York: Columbia University Press. 1966.

— R. J. Shofer, and F. S. Musgrave: Cortical intracellular potential during augmenting and recruiting responses. II. Patterns of synaptic activities in pyramidal and non-pyramidal tract neurons. J. Neurophysiol. *27*, 131—151 (1964).

Scherrer, J.: Analyse de l'activité électro-corticale spontanée. Actualités Neurophysiologiques. Sixième série, pp. 201—221. Paris: Masson. 1965.

Verley, R.: Recherches sur le développement des activités électro-corticales avec des électrodes corticales radiaires. J. Physiol. Paris *57*, 407—436 (1965).

# Discussion to the Papers
## of Raabe et al., Elul, Kuhnt, Speckmann et al., and Scherrer et al.

LEHMANN: I have a question to Dr. SCHERRER: In your results on different latencies between the waves recorded from isolated slices of cortex, where was the reference electrode, and how did you record the different traces?

SCHERRER: In all our transcortical recordings we measured the potential difference between a surface electrode situated on the cortex and a deep one located vertically below the superficial electrode. Inter-electrode distance was 1000 or 2000 $\mu$. In the latter case we registered radially the whole cortex, in the first one only the superficial dipole. As already said, the results are identical: for B and C generators the amplitude of the event is larger when recording with inter-electrode distances between 0 and 1000 $\mu$. In all our figures downward deflection reflects a surface-positive wave (superficial electrode becoming positive with respect to the deep one), and upward deflection a surface negative wave.

In the isolated suprasylvian gyrus we made several transcortical derivations (up to six). Each derivation is from a pair of electrodes as described above. There is no common reference. The distance between two transcortical derivations is usually 3 mm (15 mm from the anterior to the posterior pair of transcortical electrodes).

GLOOR: The location of the reference electrode is very important. It is very difficult to find a suitable reference point which is uncontaminated by EEG. Transcortical recordings would be preferable for such experiments.

VERZEANO: Dr. ELUL, you mentioned, in your presentation, some events which you called synaptic and other events which you called non-synaptic. What kind of events did you designate as non-synaptic?

ELUL: A non-synaptic mechanism for production of wave activity in cortical neurons may involve intrinsic oscillations of the membrane potential. Such oscillations, which are predicted from the Hodgkin-

Huxley equations (HODGKIN and HUXLEY 1952) may be due to lowering of the resting potential, or lowering of the extracellular calcium concentration (HUXLEY 1959). Although initially these mechanisms have been discussed with relation to the squid giant axon, the effect of calcium in eliciting changes in cortical electrical activity has been recently demonstrated in the cat (ADEY, personal communication), and sustained oscillations due to steady depolarization are exhibited by Aplysia neurons (ELUL and ADEY 1972). Depolarization, in particular, may be common in the fine dendrite arborizations, and may be enhanced by EPSP's, thus providing a point of juncture between synaptic and non-synaptic sources of unitary wave activity.

VERZEANO: How did you estimate the percentage of cells in the network which correspond to a particular category?

ELUL: I do not have direct measurements of the number of cells which must be synchronized to produce a detectable gross cortical potential. However, a comparison with evoked potentials yields relevant information: in the unanesthetized cat, unpatterned light flashes elicit gross cortical potentials of approximately the same magnitude as the spontaneous EEG recorded from the same surface electrode. Now exploration by a micropipette reveals that only 10–20% of the impaled cells exhibit responses to the light flash (CYRULNIK et al. 1972). The responses in individual impaled cells are synchronized quite closely. This suggests that spontaneous cortical potentials also may require synchronization of 10–20% of the total cell population.

VERZEANO: Dr. SCHERRER, from the data you have, would you say that the *action* potentials are related to the generation of the EEG? Or would you implicate some other events, slower and graded events, such as postsynaptic potentials which *precede* and determine the development of action potentials?

SCHERRER: There are only statistical relationships between waves and spikes in our records. This is true of both increase with surface positive waves and decrease during surface negative waves. This does not exclude, however, that some of the cortical neurons behave in a different way.

PETSCHE: There is an additional point worth considering: even with transcortical recordings we are not able to register all cortical slow-frequency events. There are numerous potentials which do not extend over areas wider than a few hundred micron in all directions.

Even patterns of self-sustained activities with such properties have been observed by us. The EEG is still more complicated than we think.

SCHERRER: To study intracortical waves we used pairs of electrodes with inter-electrode distances of 200 $\mu$; with such a rather small inter-electrode distance inversely oriented cortical dipoles were able to be located. Of course it would also be interesting to use smaller inter-electrode distances, but we have not done this.

As far as the question of tangentially orientated dipoles is concerned we did not find any when using pairs of horizontally arranged electrodes with an inter-electrode distance of 200, 1000, or 2000 $\mu$; more precisely: the potential differences found were many times smaller than the differences found with the same inter-electrode-distance when vertically oriented.

A tangential potential difference should also appear if two radially orientated generators evoke wave-phenomena of different shape.

LUX: If the expression "generator" refers to distinct synaptic or other activities of nerve cells in the cortical network, I do not see a way to specify such a generator from EEG recordings. PURPURA and McMURTRY [J. Neurophysiol. 28, 166 (1965)] have presented a perplexing example of how poor a correlation between a cortical surface potential and observable postsynaptic potentials can be. As for example, the augmenting potential after thalamic stimulation is especially liable to vary drastically and even to change polarity with relatively weak transcortical polarization. Although the surface responses, on most nerve cells, usually go apparently parallel with the sequence of EPSP's and IPSP's, the postsynaptic activity is only minimally affected by transcortical currents. This suggests that there are generators of the cortical surface response which usually escape detection at the single cell level. On the other hand, a considerable postsynaptic activity seems to exist which does not directly contribute to the generation of the recorded surface potential, in spite of a possibly large influence on the state of excitability of the intracortical environment. An example of this is the observation of intralaminar, synaptically generated current flow around a single cell with a greater current density horizontally than perpendicularly to the cortical surface. Synaptic activity produces local impedance changes which can be tested. Intralaminar impedance measurements made simultaneously with potential recordings of brain potentials are of considerable value for the location of the sources which may generate intracortical and surface potentials.

GLOOR: Dr. LUX, let me ask you a question to make a point quite clear: did you find a greater current in the tangential direction?

LUX: Yes, this was observed in the immediate surrounding of single nerve cells in the cat's motor cortex in certain stimulus situations.

GLOOR: What kind of synaptic system may be involved in generating this current?

LUX: We applied single submaximal stimuli to specific thalamic nuclei and repetitive stimulation at a rate which produces cortical augmenting responses. Cortical stimulation also often produces comparable potential field distributions.

ELUL: In an analysis of cortical dipole orientation which has been made using 3 extracellular microelectrodes, spaced about 50 $\mu$ apart and introduced simultaneously into the cortex, some data were obtained on the question of propagation of activity in the cortex (ELUL 1962). Briefly, in the lightly anesthetized cat, the tangential dipoles are of the same order of magnitude as those perpendicular to the cortical surface. In contrast, in animals showing marked Nembutal spindles, there is practically no activity tangential to the cortical surface, whereas large dipoles are discernable perpendicular to the surface.

LUX: We did not evaluate barbiturate spindle activity. Although our electrode arrangements may look similar to that in your experiments there are methodical differences. With respect to a distant reference electrode your observation may hold good. With an intralaminar reference point, which is essential if one wishes to know the potential distribution around single cells, however, the situation may be entirely different. In this situation, a strong field orientated tangentially may show up which is cancelled otherwise.

TÖMBÖL: There is a less prominent horizontal fibre system in the fifth layer too, which consists of the axons of a few pyramidal cells of the same layer, but mainly of the collaterals of other pyramidal cells. The vertical organization in the cortex may be formed by the so-called small neurons which establish vertical connections.

NAQUET: I have three questions, one to Dr. CASPERS and two to Dr. SCHERRER: Dr. CASPERS, do you find any difference in the variation of DC-shifts between seizures induced by metrazol and seizures following the application of penicillin?

CASPERS: The relations between cortical DC-shifts and neuronal PDS presented in our slides were studied during generalized seizure discharges evoked by systemic administration of pentylenetetrazol. Additional investigations were performed after local application of penicillin to the cerebral cortex. The comparative measurements did not reveal significant differences between the two experimental conditions, provided that both the DC component and the unit activity were recorded from the immediate area of the cortical focus. With a stepwise displacement of either of the active recording electrodes from the focal area, however, increasing dissociations between DC and single unit responses were observed.

NAQUET: Dr. SCHERRER, did you find any differences between the beginning of a seizure after the application of penicillin and a seizure induced by metrazol *i.v.*? I saw in one picture that comes from Mme. JAMI some giant depolarization shifts analogous to the ones described after application of penicillin. To me it seems to be the same event.

SCHERRER: There is a difference in seizures induced by metrazol or by locally applied penicillin. In the latter case convulsive activity is rapidly generalized, but this does not mean that there is necessarily a coincidence in time of EEG activity in the entire cortex; one cortical area may already display a tonic activity whereas others produce irregular convulsive waves.

NAQUET: A second question to Dr. SCHERRER: In another slide from Mme. JAMI you showed that after metrazol there was no synchronization at the beginning of the seizure; I think this is against the idea of a single pacemaker. One may admit that the seizure starts in one point and progressively spreads and I would like to know if you have some idea about the mechanisms of this spreading. This type of propagation reminds me of the work of GREEN and co-workers in the hippocampus who showed that the propagation may be not synaptic but "en tâche d'huile".

PETSCHE: I can confirm Dr. SCHERRER's findings about the starting of seizure patterns at different times and at different places of the cortex. This cannot be caused by a single subcortical pacemaker since we found the same when the hemisphere in question was neuronally isolated before. We also observed the prolonged duration of seizure patterns in the kind of experiments Dr. SCHERRER told us about. I remember one very regular spike-and-wave pattern that went on for a period of 9 minutes.

SCHERRER: What were the connections between the cortex and the deep areas of the brain in your isolation experiments?

PETSCHE: The capsula interna and the corpus callosum were separated, the cortical blood supply was left intact.

NAQUET: My last question is directed to Dr. KUHNT: You showed some modification of the pattern of the discharge by changing the frequency. MENINI and ROSTAIN in my laboratory [J. Physiol. *62*, 3: 414—415 (1970)] found the same with baboon at 25 flashes per second. Why did you stop at 15 c/s? Your work was done under Nembutal anesthesia in cat; are you sure that this may be related to the specific frequency of light stimulation of photosensitive patients?

VERZEANO: Dr. SCHERRER's findings are in agreement with ours. We have shown, some years ago, that there is a relation between the gross waves and the frequency of the neuronal discharge when they are both recorded by the same microelectrode at the same point. This does not mean that the action potentials themselves summate to generate gross waves. What may be happening is that the frequency of neuronal discharge determines the magnitude of the postsynaptic or of other graded potentials whose summation results in the development of gross waves.

LUX: It would be advantageous to know more about how and where the gross waves are generated.

BRAITENBERG: When you find only a small proportion of neurons in synchrony, is it your impression that there is any spatial correlation between them; I mean: are they lumped together, or spaced at equal intervals or any such thing?

ELUL: The groups of cortical nerve cells synchronized by subcortical influences according to the model proposed in my presentation, may be viewed as functional columns or functional aggregates of neurons. It is significant to note that even in the primary visual cortex only 20 per cent of the cells respond to unpatterned visual stimulus, so that even in this area there are cells which subserve more complex functions and not simply respond to light and dark (CYRULNIK et al. 1972). It should however be kept in mind that these groupings of neurons are only *temporary* combinations from the large pool of available nerve cells.

PETSCHE: The question of the functional significance of the columnar structure may better be discussed in connection with the neuronanatomists' papers. Dr. ELUL, you said that the rate of the cells active at one time is about 20%. What do you mean by "active at one time"? Do you mean all cells firing almost simultaneously or do you consider synaptic potentials too? When we were studying the hippocampus by microelectrodes, we were often surprised by the relatively small amount of cellular firing compared with the vivid slow wave activity.

ELUL: I would like to clarify my earlier statement. My usage of the adjective "active" is somewhat misleading. All nerve cells are active all the time, in terms of subthreshold wave activity as well as in terms of spike discharge. In viewing the gross EEG, however, we are only offered a glimpse of those particular cells which happen to be synchronized at that time. Activity is not necessarily related to firing. In sleep the firing rate often decreases, and yet the amplitude of the EEG shows a significant increase.

NAQUET: Dr. KUHNT, you showed some modification of the pattern of the discharge by changing the frequency. MENINI and ROSTAIN in my laboratory [J. Physiol. 62, 414—415 (1970)] found the same with baboon at 25 flashes per second. Why did you stop at 16/sec? Your work was done under Nembutal anesthesia in cat, are you sure that this may be related to the specific frequency of light stimulation of photosensitive patients?

KUHNT: You are right Dr. NAQUET. In my talk I referred only the behavior of the compound IPSP and the way it changes with different repetition rates of the light stimuli. With a repetition rate of 16/sec none of the cells I recorded from developed a compound IPSP. The IPSP always disappeared between 8 and 16/sec stimulation. This was partly dependent on the light intensity. Cells responding in a stimulus-locked manner have been stimulated with higher frequencies as well. But they are not of interest regarding the IPSP. The highest repetition rate of light stimuli used in this study was 50/sec. Only 2 cells were able to follow this frequency regularly.

Regarding the second question let me say something concerning the barbiturate anesthesia. The animals were given a dosage of nembutal (30 mg/kg bodyweight i.p.) at the beginning of the experiment; after this only a mixture of flaxedil, Ringer-solution ans laevulose was given. During the first 6 hours the animal was "recovering" and no records were made. The whole experiment lasted up to 48 hours and significant changes in the IPSP were never seen. That is, it was not possible to separate cells according to the time after the barbiturate anesthesia.

The other part of the question is more difficult to answer. It is always dangerous to relate findings from animal experiments to humans. However, I only like to point out that regarding the EEG or the VEP under these stimulating conditions there exist some good similarities between both species. If similarly the above mentioned unitary findings are applicable to humans, it would imply that under this stimulation (16/sec) the inhibition is missing, that is, it is missing under normal conditions. From this we have to conclude that the missing inhibition cannot be the main cause of the photosensitive epilepsy, as I mentioned in the paper.

# Changes of Focal Potentials by Iontophoretic Application of Glutamic Acid and Gamma-Amino-Butyric Acid

A. Herz and W. Zieglgänsberger

Max-Planck-Institut für Psychiatrie, München, Germany

The current belief is that EEG-waves are built up predominantly by summating postsynaptic transients evoked synchroneously in a large population of neurones (Creutzfeldt et al. 1966 a, b; Humphrey et al. 1968). Such synchronized activity of nerve cells can evoke potentials of considerable magnitude as seen for example in epileptic seizures. The number of cells underlying such potentials is yet unknown.

Our intention was to correlate electrical phenomena evoked in various structures of the central nervous system with the number of nerve cells involved. Excitatory and inhibitory amino acids were applied microelectrophoretically and the resultant "micro-foci" were analysed with respect to their generating field, by means of twin-multibarrelled micro-pipettes.

Generally, the microelectrophoretic method is used to apply small amounts of drugs close to a single neurone and to study changes in firing rate or the membrane properties of that cell. This method has been employed in relatively few investigations for studying focal potentials which are generated from the summated membrane transients of a restricted number of cells (Curtis et al. 1959, 1960, 1961, 1962, Krnjević et al. 1966 b, Biscoe and Straughan 1966, Phillis et al. 1967). This may be due to the fact that the interpretation of such studies is difficult. For example, the amplitude of focal potentials in the spinal cord which were induced by ventral and dorsal root stimulation, was reduced either by an excitatory or by an inhibitory amino acid (Curtis et al. 1959, 1960). The present communication concerns detailed data about the parameters of such focal potentials obtained and analysed with a special technique.

The first part of this study describes the action of glutamic acid and GABA on slow waves evoked in the caudate nucleus of rabbits by thalamic stimulation. The second part deals with the analysis of the field in which these slow waves are generated. The caudate nucleus proved to be especially appropriate for this purpose since

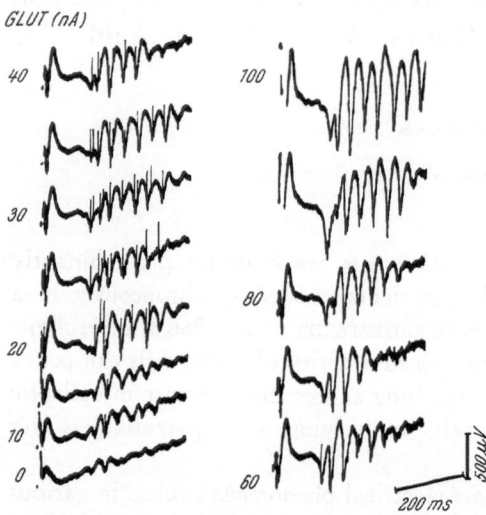

Fig. 1. Changes of focal potentials in the caudate nucleus by microelectrophoretic application of GLUT. Electrical stimulation of the medial thalamus at a frequency of 1/s. The columns should be read from bottom to top. Downward deflection indicates positivity. The numbers give the phoretic current in nA($= 10^{-9}$ A)

under the experimental conditions slow wave responses were rather small in amplitude. Thus, changes induced by application of an excitatory amino acid can easily be seen. After all surgery was done under ether anesthesia, the animals were immobilized with gallamine (FLAXEDIL). The head of the caudate nucleus was penetrated with a four barrelled micropipette; one barrel contained glutamic acid, another GABA. The two remaining barrels were filled with NaCl and were used to record focal potentials, and to perform current controls. Release of the drug from a pipette is proportional to the current applied and therefore the dosage will be given in nanoamperes ($= 10^{-9}$ A). Bipolar stimulating electrodes were placed in the midline areas of the thalamus. Stimulation frequency was 1/s (for experimental details see HERZ et al. 1970).

Fig. 1 shows the effect of glutamic acid (GLUT) on potentials evoked in the caudate nucleus by thalamic stimulation. Before GLUT application the evoked response is rather small in amplitude and not well differentiated. During GLUT application the potentials increase con-

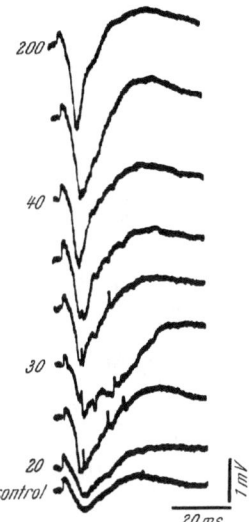

Fig. 2. Increase of the primary response evoked in the caudate nucleus by electrical stimulation of the medial thalamus under the action of increasing doses of GLUT. A similar response as in Fig. 1 is reproduced on a shorter time base. For further explanation see Fig. 1

siderably and become readily separable into three components: a positive/negative primary response, a negative/positive wave with an intermediate isoelectric part during which no neuronal discharges can generally be seen (discharge-free interval), and finally, spindle-like waves.

A primary response on a shorter time base is shown in Fig. 2. Prior to GLUT application the wave is small in amplitude. With small doses of GLUT, however, the amplitude increases and a single neurone can be observed to discharge on the ascending part of the wave. The amplitude of the wave continues to increase as the GLUT dose is increased. The discharges of the single neurone vanish, probably due to a depolarization block.

The effect of an increase of stimulus intensity upon the discharge-free interval can be seen from Fig. 3. Small doses of GLUT (25 nA) were applied to evoke some background activity. A discharge-free interval becomes evident at increased intensities of thalamic stimulation. Other experiments (Krnjević et al. 1966 a, b) indicate that this interval is probably the equivalent of synchronized inhibitory postsynaptic transients. This discharge-free interval is followed by spindle waves.

Fig. 1 shows the increase of such spindle waves with increasing doses of GLUT. A neurone (or neurones) begins to discharge at a low

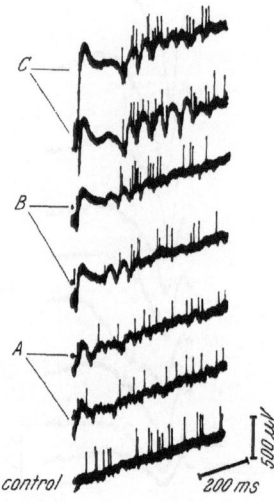

Fig. 3. Focal potentials evoked in the caudate nucleus by increasing intensities of thalamic stimulation (A stimulus intensity = 3 V; B = 8 V; C = 30 V). A constant GLUT dose of 25 nA was applied throughout. For further explanation see Fig. 1

dose predominantly at the sharp positive deflection of the spindle wave. This common finding is in contrast to most other spike/wave relations described up to now (cf. Curtis et al. 1959, 1960, Creutzfeldt et al. 1966 a, b). The neuronal discharges in these recordings are generally associated with the negative-going phase. The spindles, characterized by their arcade shape, undergo a further increase in amplitude with larger doses of GLUT whereas the neurone which is recorded from simultaneously appears inactivated at a certain dose level.

Microelectrophoretic application of the depressant gamma-aminobutyric acid (GABA) evoked the mirror effect. Waves which were enlarged with GLUT were reduced or even abolished with simultaneous application of GABA. The effect of GABA on the primary response and spindle waves respectively, can be seen from Fig. 4. Increasing doses of GLUT are applied (b, c, d). In e, GABA (80 nA) was applied concomitantly and the response is markedly reduced. In microelectrophoretic studies, the phoretic current may seriously

Fig. 4. Decrease of focal potentials by GABA. Left: *a–d* increasing doses of GLUT (0, 20, 80, 200 nA); *e*: GLUT 80 nA, GABA 80 nA; *f*: GLUT 80 nA + cationic current 80 nA. Right: *a–c*: increasing doses of GLUT (0, 20, 60 nA); *d*: GLUT 60 nA + GABA 60 nA; *e*: GLUT 60 nA + GABA 100 nA; stimulation intensity 5 V. Note the shortening of the inhibitory slow wave with increasing doses of GLUT (*c*) in this experiment

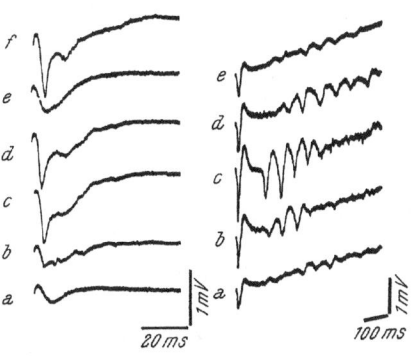

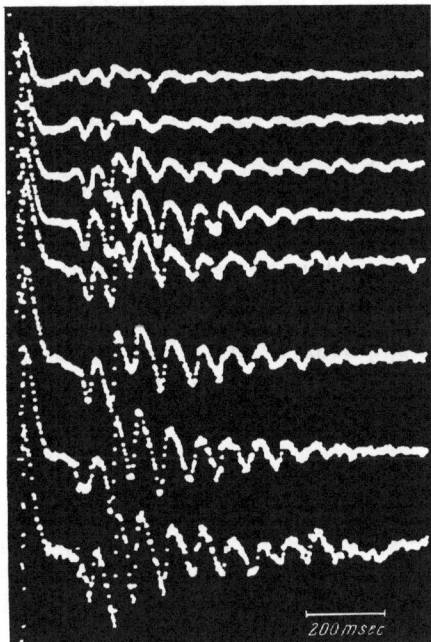

Fig. 5. Phase constancy of focal potentials in the caudate nucleus by electrical stimulation of the medial thalamus during increasing doses of GLUT. The records refer to GLUT doses between 0 nA (above) and 100 nA (below). Each record curves averaged from 40 sweeps

interfere with the effects of the drug. For this reason, extensive controls were performed to exclude current effects (*f*). The right part of this figure illustrates the effect of GABA on spindle waves. In *b*, *c*, GLUT was continually increased and the response increased

10  Synchronization

concomitantly. The frequency and phase of the spindles did not change. The inhibitory slow wave was shortened markedly with GLUT and was relengthened with simultaneous GABA application (c, d). In a few experiments where well developed focal potentials were already present prior to GLUT application, GLUT caused only

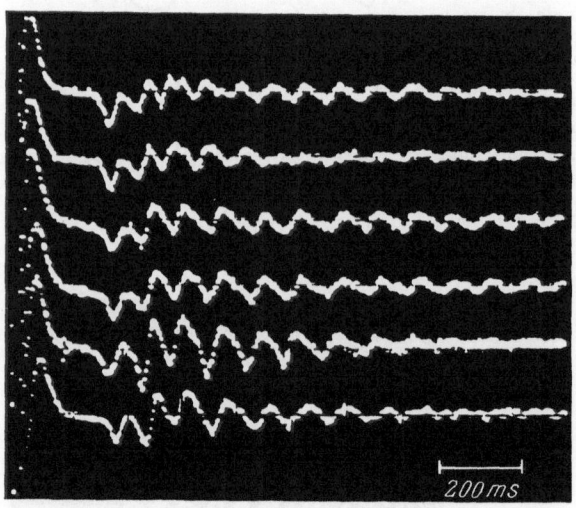

Fig. 6. Phase constancy of focal potentials evoked by electrical stimulation of the medial thalamus at different depths in the head of the caudate nucleus. (Above: record just after entrance into the caudate nucleus; below: about 2 mm deeper.) 50 nA of GLUT were applied throughout. Each record represents curves averaged from 40 sweeps

small or even no observable change. Control experiments were performed in the cortex where quite distinct focal potentials could be generally evoked prior to GLUT application. In the cortex, the effect of GLUT application was slight. Thus, it is evident that the effectiveness of GLUT and GABA depends largely on the initial magnitude of the response.

The phase correlation of the response remained constant with GLUT application. This is illustrated in Fig. 5 which shows curves each averaged from 40 sweeps corresponding to different GLUT dosages. Such phase constancy was also obvious when the head of the caudate nucleus was penetrated stepwise from above to the bottom. This can be seen from Fig. 6 which shows the potentials evoked in different depths. It can be concluded from these studies that the mechanism involved in the generation of complex responses must be remote from the site of application because GLUT application does not apparently interfere with the generation of the pattern. This is clearly evident from the phase constancy of the slow wave response

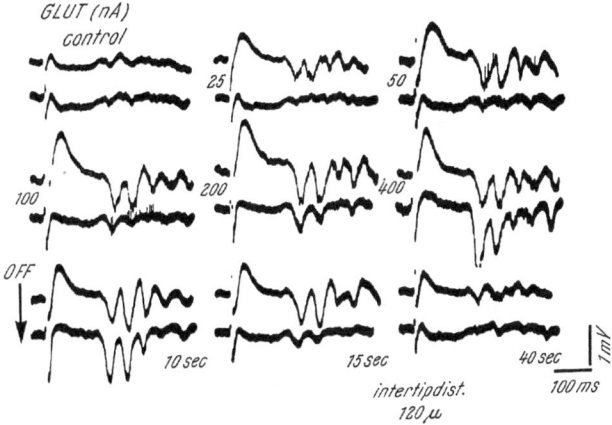

Fig. 7. Focal potentials recorded from both tips of a double multibarrelled electrode in the caudate nucleus after thalamic stimulation. The intertip distance between both tips was 120 μ. [Upper record: potential at the tip from where GLUT is released; lower record: potential picked up at the electrode tip 120 μ away. The numbers indicate the dosage of GLUT (nA). Lower trace: Reversal of the GLUT effect at different intervals after cessation of GLUT application (400 nA)]

during GLUT application (see also Buchwald *et al.* 1961, Heuser *et al.* 1961).

The constancy of the responses makes it possible to study some parameters of the generating field. For this purpose, twin electrodes were employed (see Herz *et al.* 1970). Two electrodes of the type described above were fixed together at certain distances (70 up to 240 μ). An experiment performed with this type of electrode is illustrated in Fig. 7. The intertip distance in this case was 120 μ. Prior to GLUT application thalamic stimulation evokes only a small response at both tips. Application of 25 nA of GLUT resulted in a large increase of amplitude of the focal potential recorded at this tip (upper trace). No change is observed at the remote electrode

120 $\mu$ away. A small increase in amplitude becomes detectable at the remote electrode at a dose level of 50 to 100 nA. With this dose, single neurone activity is recorded simultaneously at the remote electrode indicating that the concentration in this region is sufficient to cause the neurones to discharge.

An increase of dose is followed (200 nA) by an increase in amplitude at both electrodes. It should be noted that at a dose of 400 nA the

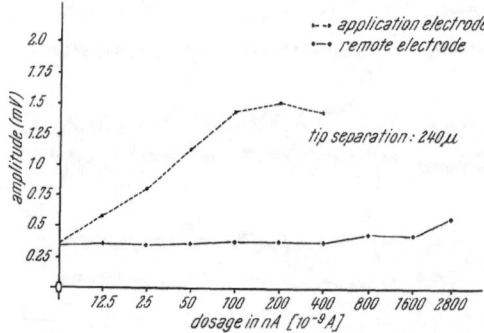

Fig. 8. Evaluation of an experiment in which increasing doses of GLUT were released from one tip of a double multibarrelled electrode (intertip distance 240 $\mu$). Artefacts prevented the evaluation of the responses at the application electrode when GLUT dosages beyond 400 nA were applied (see text)

amplitude of the focal potential recorded at the application electrode becomes smaller. This decrease in amplitude at high doses was typical and might be due to a depolarization block of neurones in the close vicinity of the application electrode. The third group of recordings shows the reversion of the effect at different times after cessation of GLUT application. It can be seen that the reversion required more time at the application electrode than at the remote electrode. This can be explained by a pool of GLUT persisting at the application electrode. 40 seconds after cessation of GLUT application the control level is reached once again. The amplitude of the potential declines steadily after cessation of GLUT application and does not show a step after the current is "switched off"; thus, it can be seen that current effects do not play an important role.

A quantitative evaluation of such an experiment is shown in Fig. 8.

The intertip distance in this case was 240 $\mu$. A considerable increase in amplitude was observed with large amounts of GLUT. The focal potentials at the remote electrode did not change appreciably even at an excessive dose. When the GLUT dose was increased up to

Fig. 9. Schematic drawing illustrating the evaluation of field parameters. $D$ = intertip distance. The change of the potentials does not go beyond $D/2$. For further explanation see text

Table 1. *Parameters of Fields at Different Radii*

|  | Radius of potential fields | | |
|---|---|---|---|
|  | $\sim 40\ \mu$ | $\sim 65\ \mu$ | $\sim 120\ \mu$ |
| Increase in amplitude by glutamic acid (mV) | 0.3 – 0.8 | 0.7 – 1.5 | 0.8 – 1.6 |
| Field gradient | $\sim 0.5\ mV/40\ \mu$ | $\sim 1.0\ mV/65\ \mu$ | $\sim 1.2\ mV/110\ \mu$ |
| Numbers of neurones maximally involved | 24 | 102 | 490 |

100 to 200 nA, the amplitude of the focal potential recorded at the application electrode reaches a plateau and thereupon decreases with higher doses. Further increase of the phoretic current causes many electrical artefacts such that clear records cannot be obtained during drug application. In the above experiments, the first responses to appear after cessation of GLUT application were used for evaluation.

The gradients and other parameters of the generating fields can be estimated from such data. Fig. 9 illustrates schematically the principle of measurement. When the GLUT-induced increase of amplitude of the focal potential is recorded exclusively at the application electrode, one can assume that the generating field does not reach beyond the half distance between the tips (D/2). Three classes of intertip-distances were distinguished: 80 $\mu$, 130 $\mu$, 220 $\mu$ (Table 1). It can be seen that the generating fields show very steep gradients. Using the assumption that the generating fields are spherical, the maximal number of neurones contributing to these fields were estimated. The calculations are based on histological measurements of the density of neurones in the caudate nucleus ($9 \times 10^4$ neurones/mm$^3$). From these results, it appears that a rather small number of neurones is capable of building up a complex potential field of considerable magnitude.

It is interesting to compare these data of the GLUT-induced focal potentials with the extent of diffusional spread of GLUT. In earlier studies, the diffusion of this substance within the cortex and the caudate nucleus was investigated by means of twin electrodes similar to those used in the present investigation (see Herz et al. 1970). A comparison of the results of the two investigations shows that GLUT spreads over an area of the same range as the area in which glutamate induces the increase of focal potentials.

The basic mechanism underlying focal potentials has been elucidated by Humphrey et al. (1968) who showed that spike activity of groups of cells cannot summate. On the other hand, considerable evidence exists that the evoked slow responses are built up by inhibitory and excitatory membrane transients (Creutzfeldt et al. 1966 a, b). Our findings support this contention. Intracellular studies in the spinal cord from this laboratory indicate that at a low dose level of GLUT, a depolarization occurs without a detectable conductance change (Bernardi et al. 1972). At the same time, hyperpolarizing transients increase and depolarizing transients decrease. The ionic conductance of the membrane is increased at higher dosages thereby reducing depolarizing and hyperpolarizing transients. Such inactivation of the cell might be the reason for the often observed decrease in amplitude of the focal potential following large doses of GLUT. The increase of inhibitory transients by the action of lower doses of GLUT might therefore contribute a great deal to the development of the slow wave amplitude. The contribution of excitatory transients to the increase of the slow waves cannot easily be estimated. This might be the subject of a more detailed study about the mechanism underlying the generation of this complex response.

Finally, the question arises about the relevance of these studies to the topic of this symposium dealing with synchronization of neuronal populations. In contrast to the cortex the caudate nucleus proved to be very suitable for measuring such potential fields and to correlate electrical and histological data. Differences in these structures might be partly due to the well known "low excitability" of the caudate nucleus and the different anatomical organization: no obvious orientation in the caudate nucleus while lamination is present in the cerebral cortex. A more detailed analysis of differing susceptibility of various CNS structures to different drugs might be a reasonable approach to investigating EEG phenomena associated with synchronization in general and seizures in particular. It should be noted that with glutamic acid application no self-sustaining process could be evoked either in the cortex or in the caudate nucleus. But on the other hand it is conceivable that the well known convulsants applied in this way might evoke them and therefore this method may prove to be useful also in this field.

## Summary

Changes of focal potentials induced by microelectrophoretic application of an excitatory (glutamic acid) and an inhibitory (GABA) amino acid were investigated in the caudate nucleus of non-anesthetized rabbits. The size of these fields was determined by means of twin-multibarrelled electrodes with intertip distances between 70–240 $\mu$.

The responses evoked by electrical stimulation of midline structures of the thalamus were very small before glutamate application; the response increased greatly after application of glutamic acid and three different components could then be distinguished: a positive/negative primary wave, a flatter negative/positive wave during which no discharge activity of single neurones was observed (discharge-free interval) and finally, spindle waves. Microelectrophoretic application of GABA greatly inhibited the glutamate-induced potentials. The experiments with twin-multibarrelled electrodes showed that the size of these potentials is rather small and corresponds fairly well to the spread of glutamate in tissue. Potentials of more than 1 mV are generated in fields with a radius in the range of 100 $\mu$. Using histological studies the maximal number of neurones contributing to focal potentials could be calculated. This complex response appears to be built up by a relatively small number of neurones.

# References

Bernardi, G., W. Zieglgänsberger, A. Herz, and E. Puil: Intracellular studies concerning the action of L-glutamic acid on spinal cord neurones of the cat. Brain Res. (in press, 1972).

Biscoe, T. J., and D. W. Straughan: Microelectrophoretic studies of neurones in the hippocampus. J. Physiol. (London) *183*, 341—359 (1966).

Buchwald, N. A., E. J. Wyers, T. Okuma, and G. Heuser: The caudate spindle. I. Electrophysiological properties. Electroenceph. clin. Neurophysiol. *13*, 509—518 (1961).

Creutzfeldt, O. D., S. Watanabe, and H. D. Lux: Relations between EEG phenomena and potentials of single cortical cells. I. Evoked potentials after thalamic and epicortical stimulation. Electroenceph. clin. Neurophysiol. *20*, 1—18 (1966 a).

— — — Relations between EEG phenomena and potentials of single cortical cells. II. Spontaneous and convulsoid activity. Electroenceph. clin. Neurophysiol. *20*, 19—37 (1966 b).

Curtis, D. R., and R. Davis: Pharmacological studies upon neurones of the lateral geniculate nucleus of the cat. Brit. J. Pharmacol. *18*, 217—246 (1962).

— J. W. Phillis, and J. C. Watkins: The depression of spinal neurones by y-amino-n-butyric acid and β-alanine. J. Physiol. (London) *146*, 185—203 (1959).

— — — The chemical excitation of spinal neurones by certain acidic amino acids. J. Physiol. (London) *150*, 656—682 (1960).

— — — Actions of amino acids on the isolated hemisected spinal cord of the toad. Brit. J. Pharmacol. *16*, 262—283 (1961).

Herz, A., W. Zieglgänsberger, and G. Färber: Microelectrophoretic studies concerning the spread of amino acids in brain tissue. Exp. Brain Res. *9*, 221—235 (1969).

Heuser, G., N. A. Buchwald, and E. J. Wyers: The caudate spindle. II. Facilitatory and inhibitory caudate-cortical pathways. Electroenceph. clin. Neurophysiol. *13*, 519—524 (1961).

Humphrey, D. R.: Re-analysis of the antidromic cortical response. II. On the contribution of cell discharge and PSPs to the evoked potentials. Electroenceph. clin. Neurophysiol. *25*, 421—442 (1968).

Krnjević, K., M. Randić, and D. W. Straughan: An inhibitory process in the cerebral cortex. J. Physiol. (London) *184*, 16—48 (1966 a).

— — — Nature of a cortical inhibitory process. J. Physiol. (London) *184*, 49—77 (1966 b).

Phillis, J. W., A. K. Tebecis, and D. H. York: A study of cholinoceptive cells in the lateral geniculate nucleus. J. Physiol. (London) *192*, 695—713 (1967).

## Discussion

VERZEANO: Did you find changes in excitability in other regions of the brain in which spindles develop?

HERZ: In most of our experiments in the cortex glutamate was not very effective in increasing the focal potentials. This may be due to the fact that rather high potentials were observed in this region even before glutamate application. Often a decrease of the potentials was observed there. Probably the neurones are inactivated by the high glutamate dosages.

VERZEANO: Do you think that glutamate has a role as a synaptic transmitter?

HERZ: This question has not been answered as yet but there is good evidence that it is a transmitter in some regions of the brain.

SCHERRER: If you increase the amount of substance do you find more cells active close by the point of electrophoretical application?

HERZ: Initially there is probably an increase in excitability of the neurones in the immediate neighbourhood of the glutamate releasing tip. With higher doses more remote cells are involved as well. At the same time, the neurones in the immediate vicinity of the application electrode become inactivated by over-depolarization.

SPECKMANN: Can you distinguish, in this particular case, between the potential field and the drug field?

HERZ: This is an important question. In other studies we investigated the diffusion of glutamate and found that the area reached by the substance is rather similar to the field in which changes of the potentials are observed. From this one may conclude that the neurones are activated by glutamate itself and not via neuronal pathways *i.e.,* by synaptic activation.

In this context, the "gradient" given in the table should only describe the drop of potential within a certain distance.

SPECKMANN: If you apply a drug with excitatory or inhibitory actions, you will get alterations of neuronal activity with quite different responses of the $pO_2$ and $pCO_2$. This in turn will evoke alterations in blood flow, which influence the spatial distribution of the drug. Therefore the area affected directly by the drug may vary considerably in each single case.

HERZ: If the glutamate effects were not due to changes in neuronal activity, the antagonistic effects of GABA would be difficult to explain.

# Pacemakers, Synchronization, and Epilepsy[1]

M. Verzeano

Department of Psychobiology, University of California, Irvine, California

## I. The Nature of Synchronization

Few words in the vocabulary of neurophysiology have been more misused than "synchronization". This term originated in the mid-thirties, when it was believed that brain waves were generated by the synchronous discharge and the summation of action potentials produced by cortical neurons. A few years later (Renshaw et al. 1940; Li and Jasper 1953) it was shown that synchronous discharge of neurons could not be found, and could not be related to the development of brain waves. In the mid-fifties (Verzeano and Calma 1954; Verzeano 1955, 1956) it was shown that the neuronal discharge which accompanies the development of periodic gross waves, in the cortex and in the thalamus, is sequential rather than synchronous, i.e., several groups of neurons discharge in a regular sequence in relation to each gross wave. Around the same epoch, new hypotheses were presented involving graded potentials, synaptic and dendritic, which would not have to occur synchronously in order to summate and generate brain waves. From that time on, the term "synchronization" was reserved for the development, in the electroencephalogram or the electrocorticogram, of rhythmic waves such as those which occur in the alpha rhythm or in sleep spindles. It is still used in this sense at the present time, even though there is sufficient evidence to indicate that nothing is truly synchronous in any aspect of brain wave development. However, since, so far, neurophysiologists and electroencephalographers have not agreed on a better term, "synchronization" will be used in this article to denote the development of rhythmic waves, wherever they may occur, in cortical or subcortical structures.

1 Aided by grant NS-07145 from the National Institutes of Health.

Since neurons do not discharge synchronously but sequentially, and since their action potentials do not seem to summate, a question arises about what the neurons do and which elementary potentials summate, to result in the formation of rhythmic waves.

More recent investigations based on the use of recordings obtained, simultaneously, with several electrodes or microelectrodes (Verzeano 1956; Verzeano and Negishi 1960, 1961, Mescherskii 1961, Verzeano 1963, Verzeano et al. 1965, 1970, Andersen and Andersson 1968, Petsche and Šterc 1968, Petsche and Rappelsberger 1970) have indicated that synchronization is a dynamic process characterized by a highly organized circulation of activity through the neuronal networks of the cortex and of the thalamus. The detailed analysis of this circulation, conducted by Verzeano and his collaborators in recent years, has shown that 1) the degree of rhythmicity (or the level of synchronization) of the gross waves corresponds to the degree of rhythmicity in the circulation of neuronal impulses, to the velocity of circulation, and to the number and level of activity of the neurons involved in it, and 2) that each passage of circulating activity through the network causes the development of a fringe of inhibition around the pathway of circulation and in its wake, whose duration increases with the degree of synchronization.

Since the circulating activity follows along curved lines and shows frequent reversals of direction (Fig. 2 b, c, d) and progressive shifts in the location of the groups of neurons involved in it (Verzeano 1956; Verzeano and Negishi 1960), it has been concluded that the pathway of circulation extends along a series of loops distributed through the neuronal networks.

These findings are summarized in Fig. 1, which shows a diagrammatic representation of the relations between the degree of synchronization of the gross waves and the parameters of the circulating activity, Fig. 2, which shows corresponding examples of actual recordings from the brain, and Fig. 3, which shows the incorporation of neurons and the increase of activity within the pathway of circulation.

The velocity of circulation of neuronal activity through the networks, estimated by the time which it requires to cover the distance between two successive microelectrode tips, varies from 0.5 to 8 mm/sec (Verzeano 1956; Verzeano and Negishi 1959), a very low figure, which indicates that many synapses are traversed in the process. The explanation for the wide range of velocities is found when recordings are obtained with arrays of microelectrodes whose tips are separated by distances greater than 200 $\mu$ (Figs. 4 and 5). Under such conditions it can be seen, in some cases, that the activity approaches from one side of the array, appears at one or two tips at one extremity of the line along which the tips are displayed ($E_1$, $E_2$ and loop a' in Fig. 4), abandons this region to advance towards the tips which are in the center of the array (loop b') and, finally, abandons this region to appear at the tips located at the other extremity of the line ($E_3$ $E_4$ and loop c'). This sequence of events suggests a pathway of circu-

---

Fig. 1. Relations between neuronal discharge, circulation of neuronal activity and synchronization of the gross waves, as they appear when recorded simultaneously with four microelectrodes with tips separated by 100 to 150 $\mu$. Left: diagrammatic representation of oscilloscope tracings, showing the progressive changes which take place from the desynchronized to the hypersynchronized state. Right: diagrammatic two-dimensional representation of hypothetical neuronal networks, showing the neuronal activity corresponding to each successive state. $E_1$, $E_2$, $E_3$, $E_4$, represent the microelectrodes through which oscilloscope tracings $E_1$, $E_2$, $E_3$, $E_4$, (at left) would be obtained; the circles represent neurons in the two-dimensional networks; the degree of darkness in each circle represents the degree of excitation of that particular neuron; the arrow represents the direction of the circulation of neuronal activity. a) Oscilloscope: infrequent clustering of the neuronal action potentials, no circulation of neuronal activity, no synchronization of the gross waves. Neuronal network: scattered, sporadic neuronal activity, at low level of excitation; no circulation of neuronal activity. b) Oscilloscope: increased clustering of action potentials; occurrence of neuronal activity in regular succession at each one of the tips of the microelectrodes, indicating circulation through the neuronal network; decreased activity in the interval (T) between successive passages of circulating activity through the network; gross waves slightly synchronized. Neuronal network: neuronal activity at higher level of excitation, concentrated mostly in the pathway of circulation (arrow); decreased activity outside this pathway. c) Oscilloscope: high degree of clustering of action potentials; increase in the velocity of circulating activity ($\Delta$ $t_1$ < $\Delta t$); activity abolished in an increased interval ($T_1$) between successive passages of circulating activity through the network; gross waves fully synchronized. Neuronal network: neuronal activity at high level of excitation in the pathway of circulation; no activity outside the pathway. d) Oscilloscope: extreme degree of clustering of action potentials; high velocity of circulation ($\Delta$ $t_2$ < $\Delta$ $t_1$); further increase in the interval ($T_2$) between successive passages of circulating activity through the network; neuronal activity in this interval abolished; hypersynchronized gross waves of high amplitude. Neuronal network: neuronal activity at very high level of excitation concentrated exclusively in an enlarged, multilane pathway of circulation; completely abolished outside this pathway.
G. W.: gross waves recorded within the same networks, by the same microelectrodes. From: Verzeano 1963

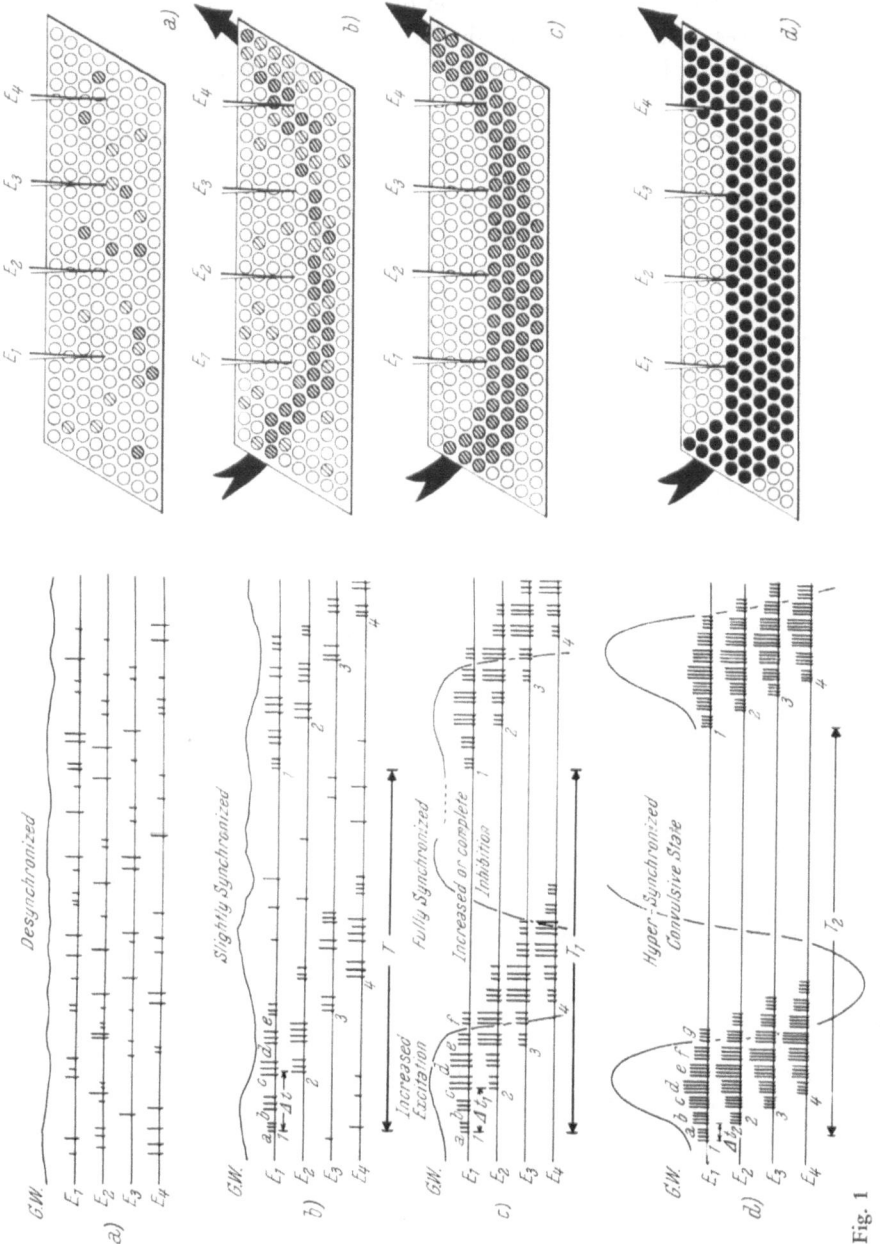

Fig. 1

lation which follows along a series of loops (a', b', c') whose "locus" advances progressively through the neuronal network. Fig. 5 shows actual recordings from the brain, illustrating the development of such a process.

When the tips of the microelectrodes are displayed along one loop of the pathway (as in the right side of Fig. 1) the circulating neuronal activity travels directly from one tip to the next and its apparent velocity is high; when the tips of the microelectrodes are displayed across several loops (as in Fig. 4), the circulating activity travels the complete length of these loops before moving from one tip to the next and its apparent velocity is low. Since the pathway of circulation at a given place and time cannot be known in advance,

---

Fig. 2. Relations between neuronal discharge, circulation of neuronal activity, and synchronization of gross waves, in thalamic neuronal networks, shown by recordings obtained, simultaneously, with four microelectrodes (a, b, c, d). Channels a, b and c show neuronal action potentials; channel d shows action potentials as well as gross waves.

a) Recordings obtained from the n. ventralis medialis of the waking cat under gallamine. b) Another section of the same recording as in A, in which some "synchronization" of the gross waves occurs in wakefulness (x to y) and in which clustering of neuronal action potentials and circulation of neuronal activity can be seen (at 1-2-3-4), in association with the gross waves. c) Recordings obtained from the same animal in the same experiment, with the same array of microelectrodes at the same thalamic location, in sleep induced by sodium pentobarbital; the increase in the synchronization of the gross waves is accompanied by an increase in the clustering of neuronal action potentials, an increase in the regularity of the circulation of neuronal activity (at 1-2-3-4), an increase in the number of neurons involved, and an increase in the duration of the period of silence ($T_1$) between successive passages of circulating activity through the network. d) Recordings obtained, in the "hypersynchronized" preconvulsive state, from another animal, under gallamine, 15 minutes after the administration of 0.9 mg/kg of picrotoxin. The clustering of neuronal action potentials is very marked, the circulation of neuronal activity is highly regular (1-2-3), a large number of neurons is involved in it, and the period of silence between successive passages of activity ($T_2$) through the network is much longer. These changes correspond to a great increase in the amplitude of the gross waves. The time line and amplitude calibration (0.5 mV) under c) apply to a), b), and c); the time line and amplitude calibration (0.5 mV) under d) apply only to d). Distances between the microelectrodes: in a), b), and c): ab = 120 $\mu$; bc = 100 $\mu$; cd = 155 $\mu$; in d): ab = 50 $\mu$; bc = 80 $\mu$; cd = 90 $\mu$; microelectrode d in d) shows only gross waves. Negative up in all tracings.

Note the frequent reversals in the direction of circulation (at 1, 2, 3, 4): in b) (from a-b-c-d to d-c-b-a); in c) and d) (from d-c-b-a to a-b-c-d)

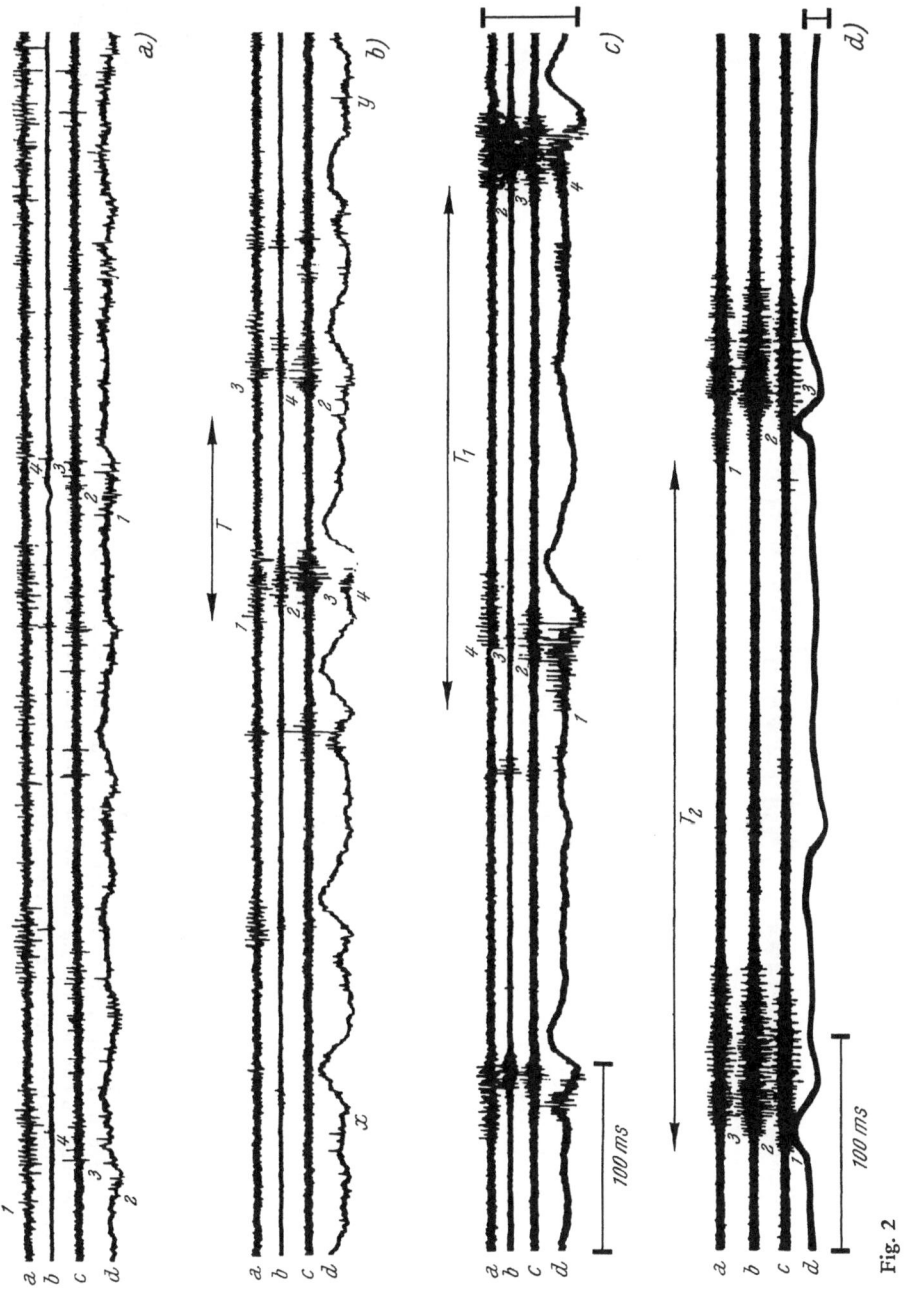

Fig. 2

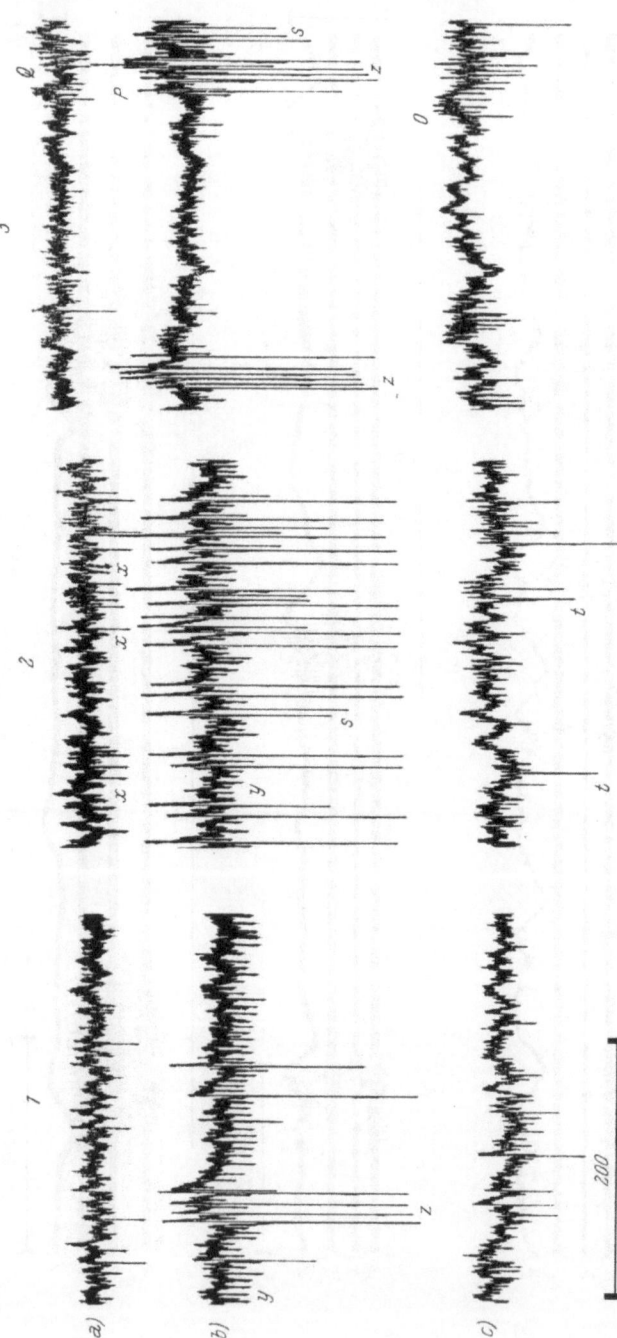

Fig. 3. Incorporation of neurons into the pathway of circulating activity, during synchronization. Recordings obtained, simultaneously, with three microelectrodes (*a, b, c*) displayed along an antero-posterior line, in the n. anterior ventralis of the cat, showing activity during "interspindle" intervals in 1 and 2 and during the development of a thalamic gross wave spindle in 3. The action potentials of neurons z and s, which were already clustered in 1 and 2, become more densely clustered in 3. The action potentials of neurons x, y, and t, which were not clustered in 1 and 2, are incorporated, in 3, into the circulation of neuronal activity, at O-P-Q. Animal under gallamine; recordings obtained 26 minutes after the injection of 10 μg of 5 HT, directly at the recording point. Distances between microelectrodes: AB = 194 μ; BC = 246 μ. Time in msec. Negativity down in all tracings. (From Verzeano and Mahnke, in press)

the position of the microelectrode tips along loops or across loops is a matter of chance, so that some recordings will show the patterns of Figs. 1 and 2 and others the patterns of Figs. 4 and 5.

Further studies of the pathway of circulation of neuronal activity by means of arrays of microelectrodes whose tips are arranged in such a way that they form a triangle and, thus, provide a tridimensional view of the process (Fig. 6), have shown that the circulation is not limited to one plane but shifts continually from one plane to

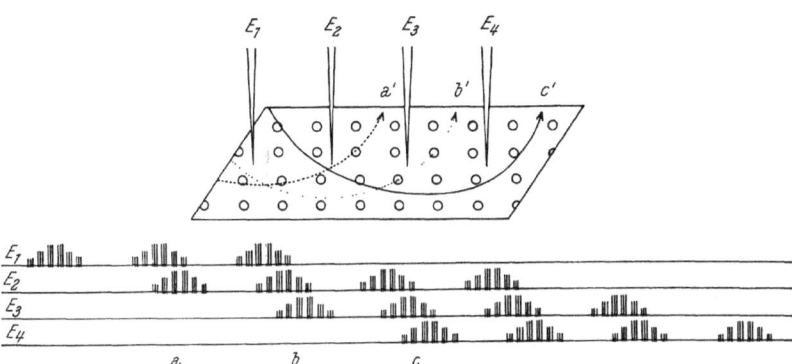

Fig. 4. Two-dimensional diagrammatic representation of the configuration of the pathway of circulation of neuronal activity as determined with four microelectrodes ($E_1$, $E_2$, $E_3$, $E_4$) with tips displayed along a straight line. Activity appears at one extremity of the line ($E_1$) and proceeds by successive steps (a, b, c, etc.) towards the other extremity ($E_4$), suggesting a pathway of circulation which follows along a series of loops (a', b', c') whose "locus" advances progressively through the neuronal network. (From VERZEANO and NEGISHI 1961)

another. If the series of loops followed by the pathway of circulating activity in one plane (Fig. 4) are extended to three-dimensional space, the three-dimensional pathway might conform to a helix.

## II. Concepts Which Require Reexamination

In relation to the nature of synchronization three widely accepted concepts require reexamination: One is that the gross waves of the electroencephalogram result exclusively from the summation of postsynaptic potentials; the second, that the reversal in the polarity of the waves observed by means of an electrode which penetrates from the surface to the depth of the cortex, is due to changes in the position of the electrode tip with respect to dipoles generated by sinks and sources of somata and dendrites of pyramidal cells; and the third, that in the process of synchronization, the thalamic nuclei "drive" the cortex in a unidirectional mode, without feedback information coming back from the cortex to the thalamus.

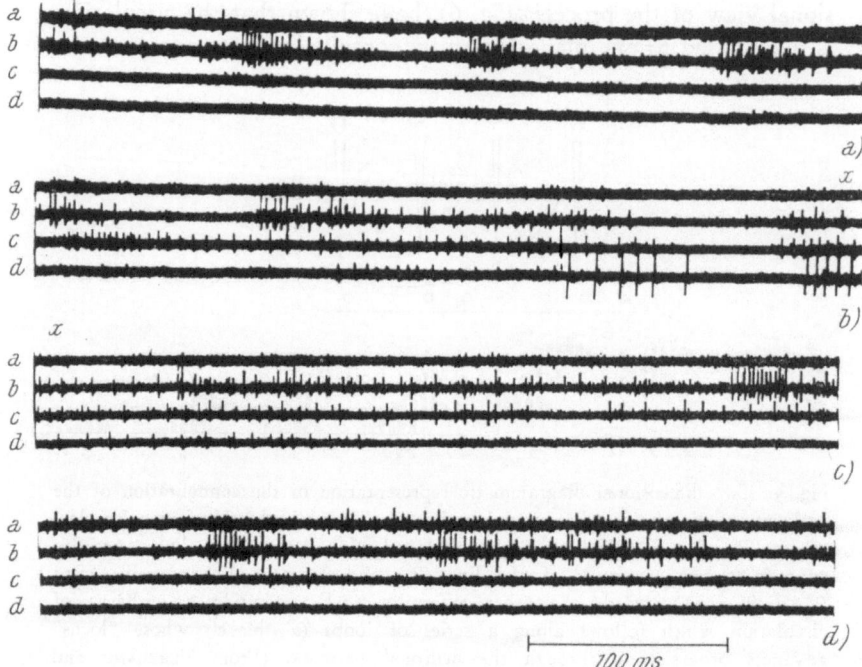

Fig. 5. Actual circulation of neuronal activity showing the patterns diagrammatically represented in Fig. 4. Recordings obtained from the n. reticularis thalami of the cat, with four microelectrodes with tips (a, b, c, d) displayed along a straight line. a) and b) are continuous and show activity circulating in successive steps from tip a to tip d; c) and d), also continuous, recorded 16 seconds after a–b, show activity circulating in successive steps from tip d to tip a. Anesthesia: sodium pentobarbital 6 mg/kg. Negativity up. (From VERZEANO and NEGISHI 1961)

## A. The Nature of the Summating Potentials

A large body of evidence supports the view that postsynaptic potentials participate in a great measure to the generation of cortical gross waves (PURPURA 1959; PURPURA and SHOFER 1964; CREUTZFELDT et al. 1966 a, 1966 b). However, the summation of postsynaptic potentials alone does not explain the development of synchronized gross waves in fiber tracts, such as have been found in the internal capsule and the pyramidal tract (VERZEANO 1955), or in the optic

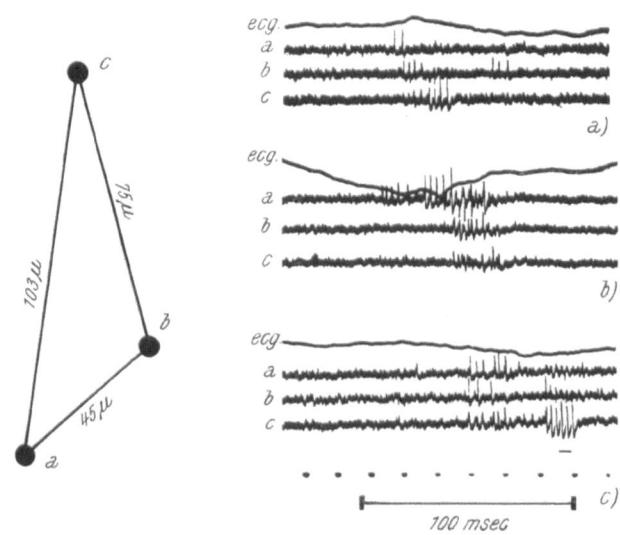

Fig. 6. Pathway of circulating neuronal activity in a tri-dimensional network. Simultaneous tracings obtained out with three microelectrodes whose tips (a, b, c) were arranged as shown in the diagram at the left; location, n. lateralis posterior of the cat. In *a*), activity circulates from a to b then to c and back to b. In *b*), activity arrives first at a, then progresses toward the center of the triangle (all three microelectrodes show action potentials simultaneously). In *c*), activity appears first near the center of the triangle (all three microelectrodes show action potentials simultaneously) then moves toward the vicinity of c (high amplitude action potentials on tracing). Anesthesia: sodium pentobarbital 6 mg/kg. Negativity up. (From VERZEANO and NEGISHI 1960)

nerve and optic tract (DOTY and KIMURA 1963; LAUFER and VERZEANO 1967).

Attempts to explain these findings have been made by stating that the gross waves found in white matter are due either to electrotonic spread to subcortical fibers of potentials which develop in the cortex, or to field effects resulting from cortical activity.

11*

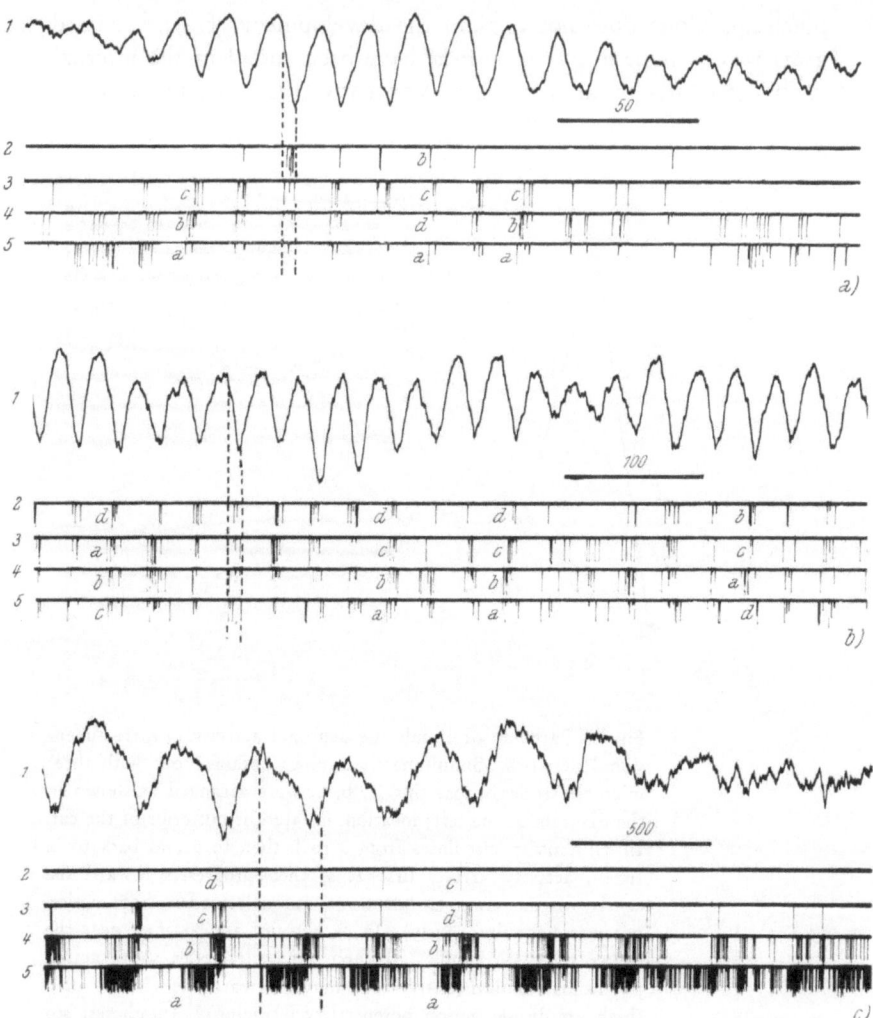

Fig. 7. Phase relations between gross waves and axonal action potentials in the optic tract. *a*): Tracing 1 shows oscillations at a frequency of 55/sec, recorded by means of a gross electrode in the optic tract of the cat, under maintained illumination. Tracings 2, 3, 4, and 5 show the pulses corresponding to the axonal action potentials, simultaneously recorded from the same location and separated, by an amplitude discriminator, into four amplitude ranges decreasing in equal steps, from 2 to 5. Thus, action potentials in tracing 2 occurred near the tip of the electrode, action potentials in tracing 5 occurred far from it. One vertical dotted line coincides with the first action potential of a group of action potentials, the other with the last action potential of the same group; this illustrates

Further investigation shows that these explanations are not satisfactory, for the following reasons:

1. Electrotonic effects along nerve fibers decrease exponentially with distance from the point of origin, according to the space constants of the fibers. Recordings obtained from subcortical regions show that the amplitude of subcortical gross waves does not change with the distance from the cortex. It remains the same at various distances from the cortex, as far as the pyramidal decussation (VERZEANO 1955). Similarly, the amplitude of slow oscillations which occur in the optic nerve and optic tract in relation to the discharge of ganglion cells in the retina, does not depend on the distance from the retina; these oscillations cannot be explained by the development of field effects since, around the optic nerve and optic tract, there are no structures from which these effects can be transmitted.

2. Gross waves which occur in fiber tracts (Fig. 7) are closely and consistently associated in time and in phase with fiber action potentials (VERZEANO 1955; LAUFER and VERZEANO 1967). This suggests that the two phenomena are generated at the recording point and are not electrotonically transmitted along the axons from their respective somata.

3. Since rhythmic gross waves of considerable amplitudes occur in fiber tracts and since the cerebral cortex contains a large number of fiber bundles going into it and out of it, and interconnecting many cells, the possibility has to be considered that these waves participate in an important way in the generation of the electroencephalogram and the development of synchronization.

## B. The Existence of Dipoles

When evoked cortical activity is studied by means of a penetrating microelectrode which detects the activity of several neurons simultaneously, it is found that the time and phase relations between the

---

the phase relations between the action potentials and oscillations: the action potentials always occur during the initial part of the positive phase. The letters a–d illustrate, for some cases, the successive appearance of the groups of action potentials in the different amplitude ranges, indicating activity approaching or moving away from the tip of the electrode. b): Similar relations, recorded after cessation of illumination, when the frequency of the oscillations has decreased to 35/sec. c): Similar relations, recorded a few minutes later, still in darkness, when the frequency of the oscillations has decreased to 3–6/sec. Time in msec. (From LAUFER and VERZEANO 1967)

gross and the neuronal responses remain the same whatever the cortical depth at which they are recorded (VERZEANO 1970; VERZEANO et al. 1970 b). This is illustrated in Fig. 8 which shows a study of these relations, analyzed by means of analog and digital computers. a is the average gross response to a brief flash of light, recorded by a microelectrode penetrating through the marginal gyrus of the

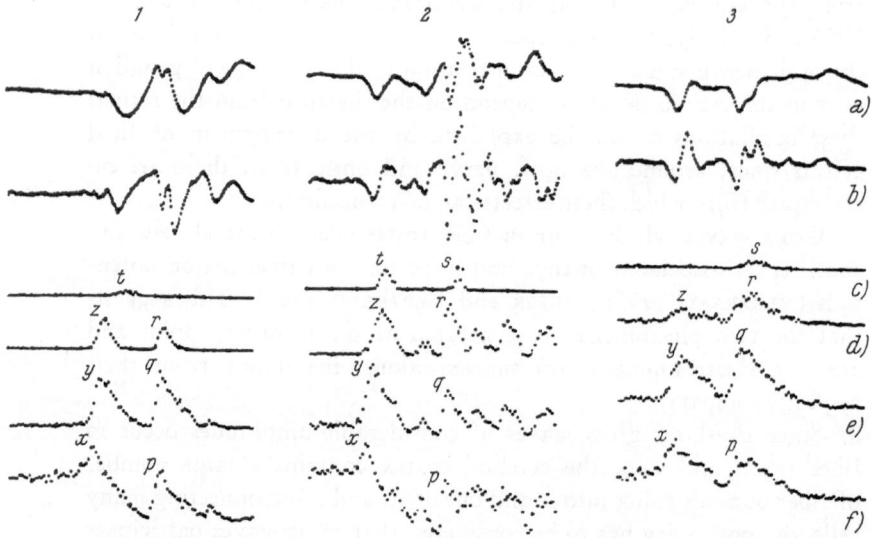

Fig. 8. Relations between gross response and neuronal discharge, at various depths, in the visual cortex of the cat under gallamine. a): average gross response; b): its average first derivative; f), e), d), c): time histograms of neuronal action potentials in the amplitude ranges 30 to 60, 60 to 100, 100 to 140 and over 280 μV, respectively, show highest probabilities of discharge (x, y, z, t and p, q, r, s) occurring at progressively later times in progressively higher amplitude ranges; data were obtained by means of a penetrating microelectrode whose tip recorded gross and neuronal responses simultaneously, at each successive depth: 0.75, 1.5, and 2.25 mm, corresponding to columns 1, 2, and 3, respectively. Note that, whatever the depth and whatever the waveform and polarity of the gross response, the relations between the highest probabilities of neuronal discharge and the phase of the gross response remain the same. Stimulus: brief flash of light, triggering the sweep. Number of sweeps: column 1, 300; column 2, 370; column 3, 290. Duration of sweep: 125 msec. (Modified from VERZEANO 1970)

cat; b its average first derivative with respect to time (which gives an indication of the slope of the gross response at any moment); c, d, e, and f are the time histograms of action potentials (probability of discharge with respect to time) previously separated by a pulse-height discriminator into four amplitude ranges corresponding to four different groups of neurons. Columns 1, 2 and 3 show the

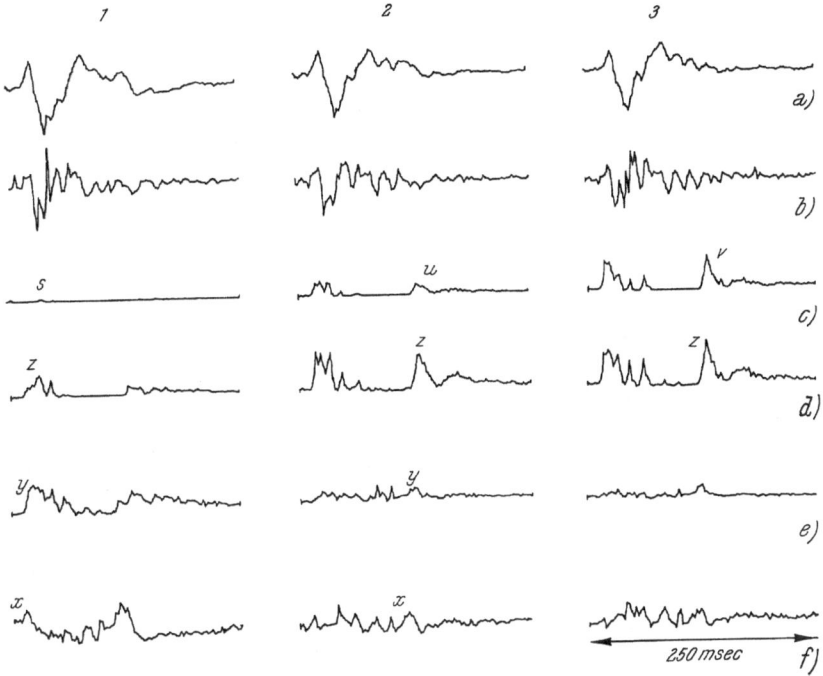

Fig. 9. The action of metrazol on gross and neuronal responses. *a*): Average gross response, computed from 200 responses evoked by visual stimulation with brief flashes of light, in the lateral geniculate body of the cat. *b*): First derivative of the gross response with respect to time. *f*): Time histogram of neuronal activity in the amplitude range 20–50; *e*: 50–80; *d*: 80–130; *c*: over 130 $\mu$V. Column 1: control. Columns 2 and 3: 22 and 36 minutes, respectively, after injection of metrazol directly into the lateral geniculate. x–y–z–s and x–y–z–u: high probabilities of neuronal discharge appearing at successively later times in successively higher amplitude ranges. Note that metrazol enhances the neuronal discharge in the higher amplitude ranges (z, u, v), and depresses the neuronal discharge in the lower amplitude ranges (x, y), which shows that discharge in the different ranges represents the activity of different elements. (From VERZEANO *et al.* 1970 b)

results obtained as the microelectrode penetrates to 0.75, 1.5, and 2.25 mm within the cortex. It can be seen that, at any depth and whatever the waveform or the polarity of the gross response, the peaks of the time histograms occur progressively later and are of progressively smaller magnitude from the lower to the higher amplitude ranges so that the high probabilities of discharge of the action potentials of low amplitudes correspond to negative slopes, while those of high amplitudes correspond to positive slopes or to peaks of the gross response; furthermore, each progression (x, y, z, t and p, q, r, s) corresponds to a major deflection of the gross response. This suggests that each deflection corresponds to a convergence of activity over several sets of neurons and interneurons the activity of different sets corresponding to different phases of gross waves. Similar relations in time and in phase, between gross and neuronal evoked responses, which maintain their consistency at any depth at which they are recorded and whatever the waveform of the gross response, are found in the lateral geniculate body (VERZEANO 1970; VERZEANO et al. 1970 b).

Separation by means of pharmacological agents provides additional evidence to indicate that neuronal responses in the different amplitude ranges are due to different elements. This is illustrated in Fig. 9 which shows the average gross response to a brief flash of light, recorded from the lateral geniculate body of the cat (a), its average first derivative with respect to time (b), and four time histograms of neuronal activity, showing the probability of neuronal discharge in the amplitude ranges 20–50 (f), 50–80 (e), 80–130 (d), and above 130 $\mu$V (c). Column 1 shows the control recordings, columns 2 and 3 show the changes which develop in these parameters 22 and 36 minutes, respectively, after the injection of metrazol directly into the lateral geniculate body. It can be seen that, after the injection, the waveform of the gross response changes, with the appearance of additional oscillations, the discharge of action potentials of low amplitude decreases (e and f, columns 2 and 3), while the discharge of action potentials of high amplitudes increases greatly (c and d, columns 2 and 3). Furthermore, in the highest amplitude range, above 130 $\mu$V, a new group of action potentials appears about 130 msec after the stimulus (u in column 2, c), and the probability of discharge of this late group is greatly increased as more time elapses after the administration of metrazol (v in column 3, c). Thus metrazol depresses the activity of elements responsible for the low amplitude discharge and enhances the activity of elements responsible for the high amplitude discharge.

The remarkable consistency of these phenomena, at every cortical level as well as in thalamic relay nuclei, cannot be explained on the basis of currents flowing along dipoles. It can only be explained by the convergence of activity over sets of neuronal and interneuronal elements interconnected in a specific way. It is the nature of the responding elements and the sequence in which they are activated

Fig. 10. Intracellular recordings from medial thalamic neurons (MeT, lower traces) during 8/sec stimulation of the internal capsule (IC) (at open triangles) at the level of the rostral limit of the ipsilateral head of the caudate nucleus. Upper traces are surface recordings from the motor cortex (MC). *a*): IC stimulus coincides with the declining phase of a spontaneous action potential discharge in a MeT neuron.

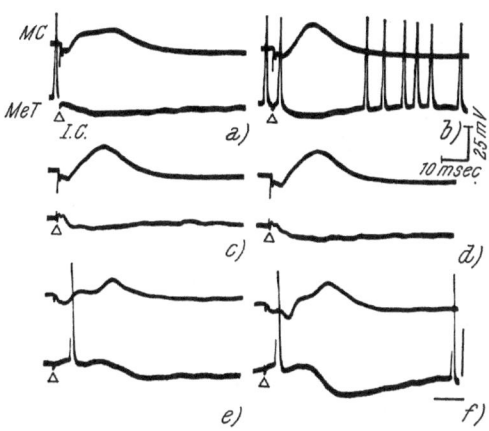

The IC evoked prolonged IPSP exhibits the shortest latency observed in this study. *b*): same cell, IC stimulus elicits a short latency EPSP with a superimposed single action potential discharge followed by an IPSP. *c*) and *d*): another cell, IC stimuli elicit 1.5–2 msec latency EPSPs and succeeding prolonged 30–70 msec IPSPs. Small field potential is seen prior to the evoked EPSP. *e*)–*f*), a third MeT neuron, IC stimuli elicit prolonged EPSPs compounded of several components indicative of polysynaptic excitation of the MeT neuron. In *e*), 10 msec after the IC stimulus, abrupt development of a second major component of the IC evoked depolarization is demonstrable. In *f*), prominent late component of IC evoked EPSP is seen. IC evoked EPSPs, in *e*) and *f*), are succeeded by prolonged (50–80 msec) IPSPs. IC stimuli, in *a*)–*f*), elicited typical anti- and orthodromic responses in the motor cortex. Voltage calibrations are for the intracellular traces. Negativity upwards in the surface evoked responses and downward in the intracellular records. (From FRIGYESI and SCHWARTZ 1972, by permission)

at a particular cortical or subcortical location that determines the waveform, the successive deflections, and the polarity of the gross waves.

## C. Cortical Feedback Loops

The concept that, in the generation of synchronized gross waves, the thalamus drives the cortex, without information being transmitted from the cortex to the thalamus, is based on a series of studies in which it was shown that the removal of large cortical areas did not suppress the thalamic rhythmic activity (Adrian 1941, Morison and Bassett 1945). However, recent morphological studies (Scheibel and Scheibel 1966) indicate that large cortical areas possess reciprocal connections with a number of thalamic nuclei, while electrophysiological studies (Frigyesi and Schwartz 1972) show that stimulation of the internal capsule has orthodromic excitatory and inhibitory effects, by means of mono- and polysynaptic pathways, on neurons located in the rostral part of the ventro-lateral and in the medial thalamic regions (Fig. 10). These effects may in turn influence other thalamic nuclei and trigger the development of cortical spindles (Frigyesi 1971; Frigyesi and Schwartz 1972). This suggests that cortico-thalamic feedback loops, which conduct impulses to the thalamic nuclei, may be important in the coordination of thalamo-cortical synchronization. The extensive removal of the cortex may not suppress the rythmic activity of the thalamus but it does not follow that, under normal conditions, when synchronization is highly coordinated over large thalamic and cortical regions, the influence of a cortico-thalamic feedback activity for which there is strong evidence, has to be deemed impossible or nonexistent.

## III. Pacemaker Mechanisms

The concept of a pacemaker mechanism located in the non-specific nuclei of the thalamus, was originated by Morison and Dempsey who demonstrated that electrical stimulation of these nuclei causes

Fig. 11. Basic pace-making mechanism in the n. ventralis anterior of the thalamus. Neuronal and wave activity in the n. ventralis anterior of the thalamus of the Macaque, in light barbiturate anesthesia. Recordings obtained, simultaneously, with four microelectrodes (a, b, c, d) whose tips were located along a straight line; micro-electrode a recorded action potentials as well as slow waves; microelectrodes b, c, and d recorded neuronal action potentials only. The tracings are continuous, from A to B. The appearance of activity in succession, at the tips of microelectrodes, indicating circulation of neuronal activity through the neuronal network, is shown at 1, 2, 3, 4. The period of progressive development of activity extends from A to x; the period of maximum activity extends from y to z and the slow waves which develop during this period can be seen on tracing a; the alternations of groups of high amplitude and low amplitude action potentials are designated s–x; the period of silence extends from z to B (see text). Distances between the tips of the microelectrodes: 140 $\mu$. Negativity up. (From Verzeano et al. 1970 a)

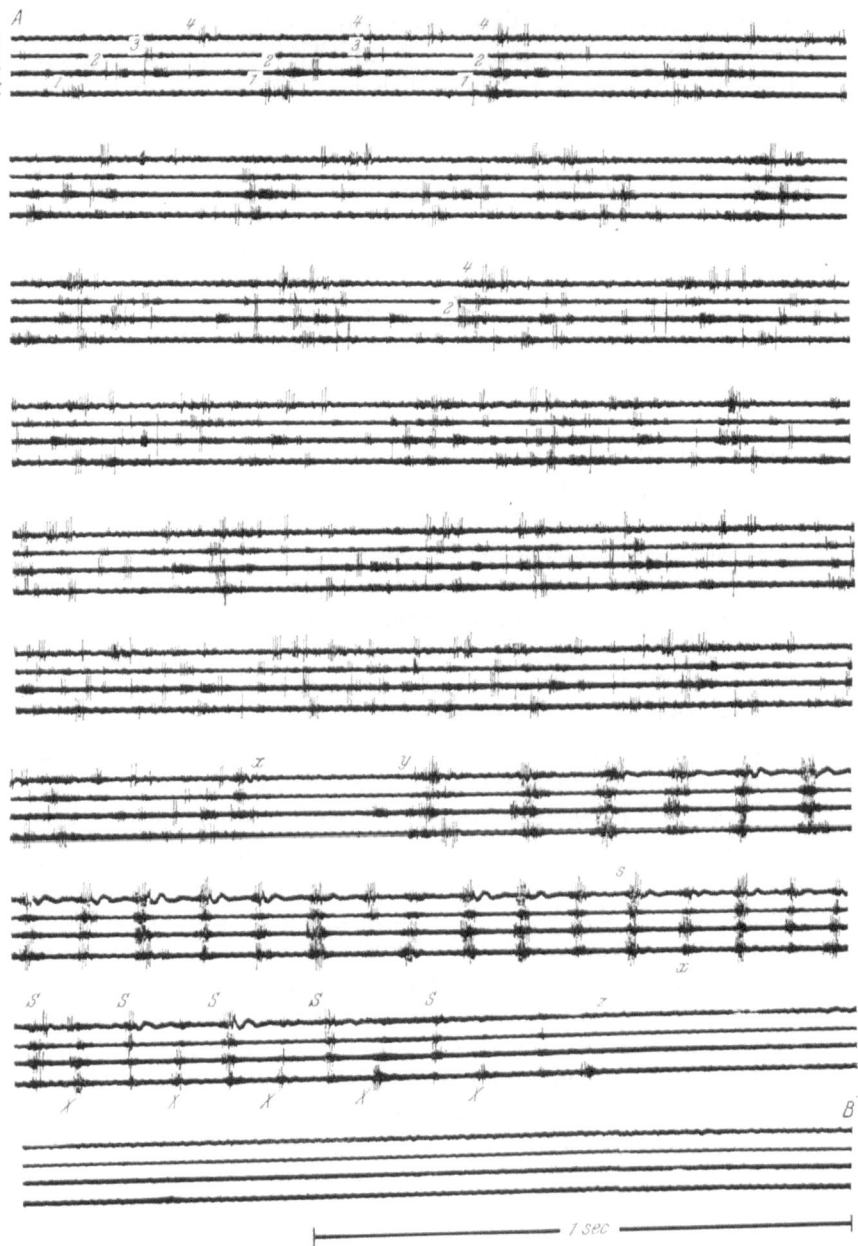

Fig. 11

the development of rhythmic waves over very wide areas of the cortex (Morison and Dempsey 1942). This concept was further developed by Jasper and his collaborators (Hanbery and Jasper 1953, Jasper 1954) who showed that several nonspecific nuclei, more particularly the ventralis anterior and the reticularis, participate in the distribution of rhythmic activity to the cortex.

More recent investigations (Verzeano and Negishi 1960, 1961; Verzeano et al. 1965, 1970 a) have shown that the interaction between the circulating neuronal activities of several thalamic nuclei forms the cellular basis of the pacemaker mechanism. This is particularly evident in the thalamus of the monkey, and is illustrated in Fig. 11, 12 and 13.

Fig. 11 shows recordings of neuronal activity and gross waves obtained, simultaneously, with four microelectrodes (a, b, c, d) from the n. ventralis anterior of the macaque monkey under light barbiturate anaesthesia. Microelectrodes b, c and d record only neuronal action potentials; microelectrode a records both neuronal action potentials and gross waves. Recordings begin in the upper left hand corner of the picture (A) and continue, uninterrupted, to the lower right hand corner (B). At the beginning, the neuronal spikes are clustered and activity appears, in succession, at the tips of the four microelectrodes (as in the example 1, 2, 3, 4); as time goes on, the frequency of discharge and the clustering of the action potentials of each neuron increase, the delays between the appearance of activity at the successive microelectrode tips become shorter, new neurons are activated, and a larger number of neurons is progressively implicated in this process. At the end of this period, which lasts for 8 to 12 seconds, the frequency of discharge, the degree of clustering of the action potentials, and the number of neurons involved reach a maximum. During this period of maximal activity, which lasts from 2 to 4 seconds (y to z), the delays between the appearance of activity at consecutive microelectrode tips are very short, the neuronal discharge is highly periodic, groups of high amplitude action-potentials alternate regularly with groups of low amplitude action potentials, one group of amplitudes being recorded by one set of microelectrodes, the other group by another set of microelectrodes (S, X). There follows a period of complete silence, which lasts from 3 to 5 seconds, after which the whole sequence of events (from A to B) repeats itself in the same order, indefinitely, as long as the animal remains at a level of relaxed wakefulness, light sleep or

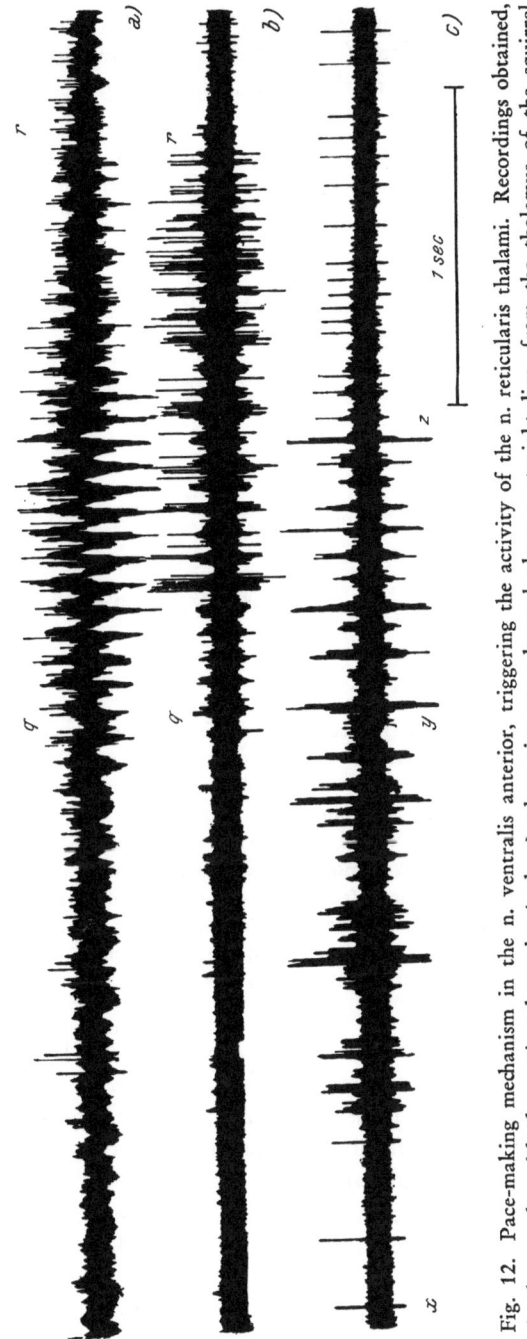

Fig. 12. Pace-making mechanism in the n. ventralis anterior, triggering the activity of the n. reticularis thalami. Recordings obtained, simultaneously, with three microelectrodes (a, b, c) whose tips were located along a straight line, from the thalamus of the squirrel monkey, in the waking state. The tips of microelectrodes a and b were in the n. reticularis, the tip of microelectrode c was in the n. ventralis anterior. Tracing c shows a period of progressive development of activity (x–y), followed by a period of maximum activity (y–z). The different periodicities of the groups of action potentials of different amplitudes which occur during the period x–y became coordinated (locked-in) during the period of maximum activity y–z. The period of maximum activity in the n. ventralis anterior (y–z) triggers the circulation of activity and the synchronization of the gross waves in the n. reticularis (q–r). Negativity up. (From Verzeano et al. 1970 a)

anesthesia. It is during the period of maximal activity (y to z) that the gross waves are synchronized (as can be seen on tracing a) and that the process is transmitted to the n. reticularis (Fig. 12) and to the cortex (Fig. 13).

In addition to its influences on the nucleus reticularis, the activity of the nucleus ventralis anterior may affect the activity of the

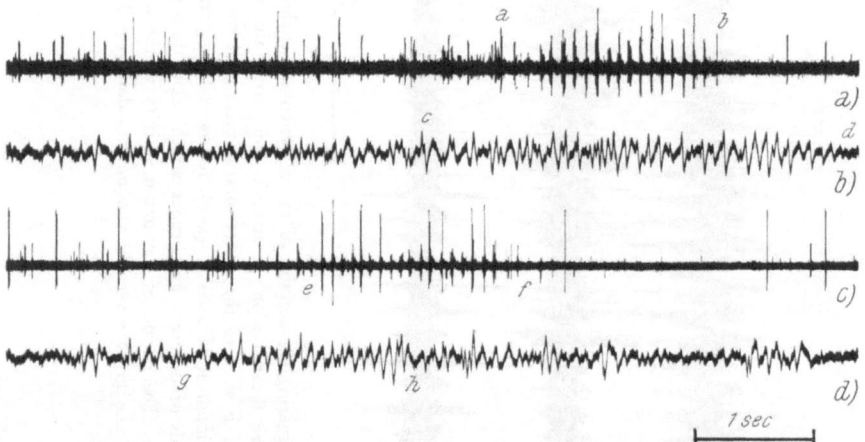

Fig. 13. Relations between thalamic and cortical activities. Recordings obtained, simultaneously, with two microelectrodes (a and c) and two macroelectrodes (b and d), from the thalamus and the cortex of the Macaque, in light barbiturate anesthesia. Microelectrodes a and c were in the n. ventralis anterior, right and left, respectively. Macroelectrodes b and d were on the surface of the right and left parietal cortex, respectively. A period of maximum neuronal activity is shown in each of the thalamic tracings (a–b and e–f). A train of cortical slow waves is shown in each of the cortical tracings (c–d and g–h). The cortical spindles c–d and g–h develop during the periods of maximum activity a–b and e–f, in the n. ventralis anterior. Negativity up. (From VERZEANO et al. 1970 a)

nucleus ventralis medialis (VERZEANO et al. 1965, 1970 a). Reciprocally (FRIGYESI 1971), impulses from the mid-line nuclei may affect the activity of the nucleus reticularis. These reciprocal actions suggest the existence of intrathalamic feedback loops along which the circulation of neuronal activity is maintained.

Rhythmic phenomena, at the cellular and at the gross wave level occur not only in the non-specific nuclei, but in other thalamic nuclei as well (VERZEANO 1955; VERZEANO and NEGISHI 1961; NEGISHI et al. 1962; ANDERSEN and ECCLES 1962; ANDERSEN and

SEARS 1964) and the question arises whether the pacemaking mechanism is exclusively located in the non-specific system. According to ANDERSEN and ANDERSSON (1968) every thalamic nucleus, whether specific or non-specific, has a pacemaking mechanism of its own, and controls the rhythmicity of a definite region of the cortex. However,

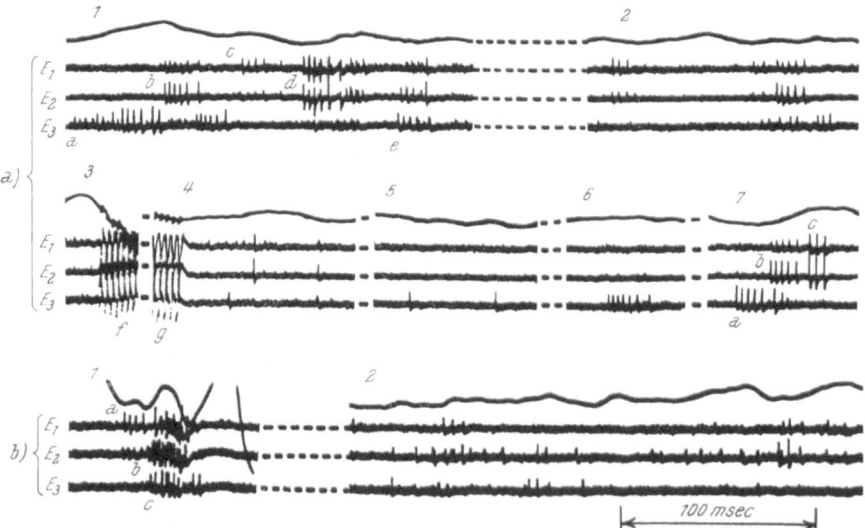

Fig. 14. Disruption of circulation of neuronal activity in the thalamus, by stimulation of the mid-brain reticular formation. a): Recordings obtained, simultaneously, with three microelectrodes ($E_1$, $E_2$, and $E_3$), from n. ventralis lateralis of cat in light sleep (sodium pentobarbital 7 mg/kg), showing the effects of stimulation of reticular formation. Strips 1 and 2, before stimulation, circulation of activity along the direction $E_3 E_2 E_1$, returning along $E_1 E_2 E_3$ can be seen at a, b, c, d, e. Strips 3, 4, and 5, stimulation (f to g) and its effects. Strips 6 and 7, return to original condition, 10 seconds after stimulation. Circulation of activity along direction $E_3 E_2 E_1$, can, again, be seen at a, b, c. b): Recordings obtained in the same experiment as in a, above, after administration of an additional dose of sodium pentobarbital (7 mg/kg). Strip 1, before stimulation; circulation of activity along the direction $E_1 E_2 E_3$ can be seen at a, b, c. Strip 2, after stimulation. Stimulation occurred during the period marked with dotted lines. In all tracings of this figure, the dotted lines represent sections where a continuous record has been cut, and the line showing slow oscillations is the simultaneous recording of the electrocorticogram. Negativity up. (From VERZEANO and NEGISHI 1960)

it is only in the non-specific nuclei that the time and phase relations between neuronal activity and gross waves are fully consistent

(VERZEANO 1955; VERZEANO and NEGISHI 1961), and it is only in the n. ventralis anterior that can be found the highly organized sequences of progressive development of circulating activity, followed by maximal synchronization and by the triggering of rhythmic processes in the n. reticularis, so closely associated with the appearance of cortical spindles, as shown in Figs. 11, 12, and 13. This suggests that the non-specific nuclei perform a function different from that of other thalamic nuclei.

It is possible that under conditions of moderate synchronization, when considerable irregularity and variability are found in the distribution of rhythmicity in the thalamus and in the cortex, each thalamic nucleus may serve as its own pacemaker and may control the appropriate region of the cortex. When rhythmicity is generalized and greatly coordinated, as in the highly synchronized states, such as "slow wave" sleep (VERZEANO and NEGISHI 1961), the process may be under the control of the non-specific nuclei. In all cases, however, even though the pacemaking mechanism may be based on the activity of thalamic nuclei, it is probable that its stability and its coordination are greatly influenced by cortico-thalamic feedback loops.

## IV. Synchronization, Desynchronization, and Epilepsy

From the considerations mentioned above, it may be concluded that the synchronized state, which characterizes relaxed wakefulness and "slow-wave" sleep, is based on the activity of a neuronal system distributed through all thalamic nuclei, communicating with the cortex, and coordinated by the non-specific nuclei. The coordinating mechanism is the highly organized circulation of neuronal activity through cortical and thalamic networks and through cortico-thalamic feedback loops.

---

Fig. 15. The action of metrazol on patterns of neuronal discharge and of circulation of neuronal activity in the thalamus. Recordings obtained from the n. centralis lateralis of the cat, with three microelectrodes (a $= 2 \mu$; b $= 3 \mu$; c $= 4 \mu$) displayed along a straight line. Distances between tips: a to b $= 120 \mu$; b to c $= 130 \mu$. a) control record before the administration of metrazol; b) after metrazol 4 mg/kg; c) and d): after additional doses of 8 mg/kg. Increased rates of neuronal discharge can be seen at u, v, w, x, y and z; increased rate at which neurons are activated can be seen at v, w, y and z; increased regularity in circulation of neuronal activity can be seen (at 1–2–3) in c) and d). Negativity up. (From NEGISHI, BRAVO, and VERZEANO 1963)

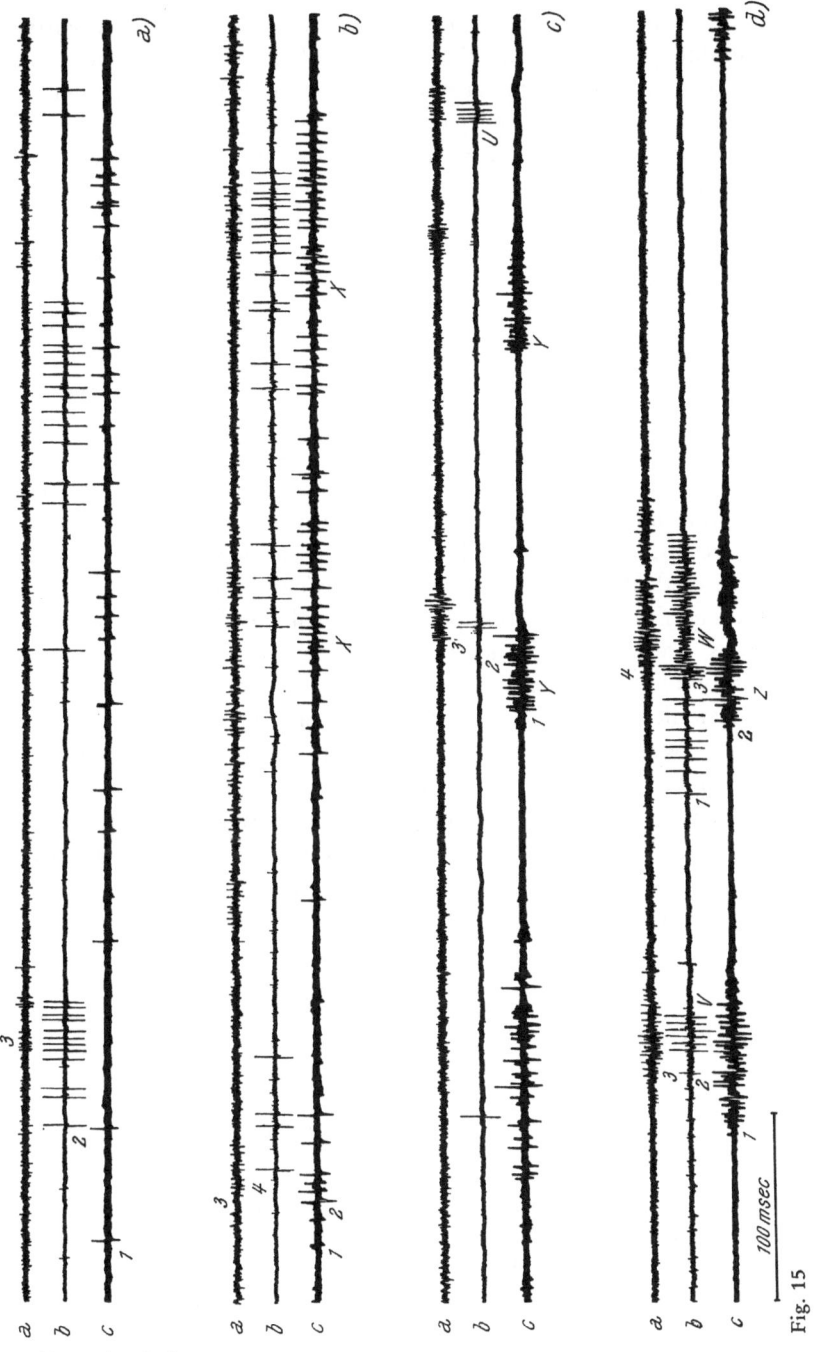

Fig. 15

The desynchronized state, which characterizes alertness and attentive wakefulness, is based on the activity of the ascending reticular activating system (Moruzzi and Magoun 1949) whose action on the brain is mediated, in part, by relay stations located in the thalamic nuclei (Starzl *et al.* 1951). When these relay stations are activated by reticular stimulation, the synchronized state and the circulation of neuronal activity are disrupted (Fig. 14). Thus, two overlapping neuronal systems operate in the thalamus: one concerned with circulation of neuronal activity and synchronization, another concerned with the disruption of circulation of neuronal activity and desynchronization.

When either one of these systems is overdriven by means of pharmacological agents, the development of epileptiform discharge may result. This is illustrated in Figs. 15, 16, and 17. Fig. 15 a) to c) shows the action of metrazol on the neuronal activity of the n. centralis lateralis. As the convulsant agent is injected and the gross wave record evolves toward the hypersynchronized state, the rate of neuronal discharge (action potentials/unit time) increases, the clustering of neuronal action potentials is enhanced, the rate at which neurons are activated (neurons activated/unit time), within the pathway of circulating neuronal activity, increases, the regularity and periodicity of the circulation of neuronal activity become more marked. When the preconvulsive state is reached (d), the passages of circulating activity through the network are very rhythmic, very rapid, and involve a large number of neurons discharging at a very high rate. The progressive increase in the rate of neuronal discharge under the influence of metrazol can be seen with particular clarity by following the evolution of the pattern of discharge of the single neuron shown on tracing b, from a) to d) Fig. 16 shows the devel-

---

Fig. 16. The action of picrotoxin on gross waves and neuronal discharge in the cortex. Recordings obtained from the cingulate gyrus of the cat, with four microelectrodes (a = b = c = d = 2 $\mu$); tips of b, c and d were displayed along a straight line at distances b to c = 80 $\mu$, c to d = 90 $\mu$; tip of a was located at a distance of 50 $\mu$ from b on a line forming an angle of 100° with the line going through b, c and d. Thus, microelectrode d was recording the development of gross waves within the same neuronal network from which microelectrodes a, b and c were recording patterns of neuronal discharge and of circulation of neuronal activity. a) control; b) after the administration of picrotoxin 0.6 mg/kg; c) and d): after the administration of additional doses of 0.6 mg/kg. The amplitude scale refers to channel d only. Negativity up. (From Negishi, Bravo, and Verzeano 1963)

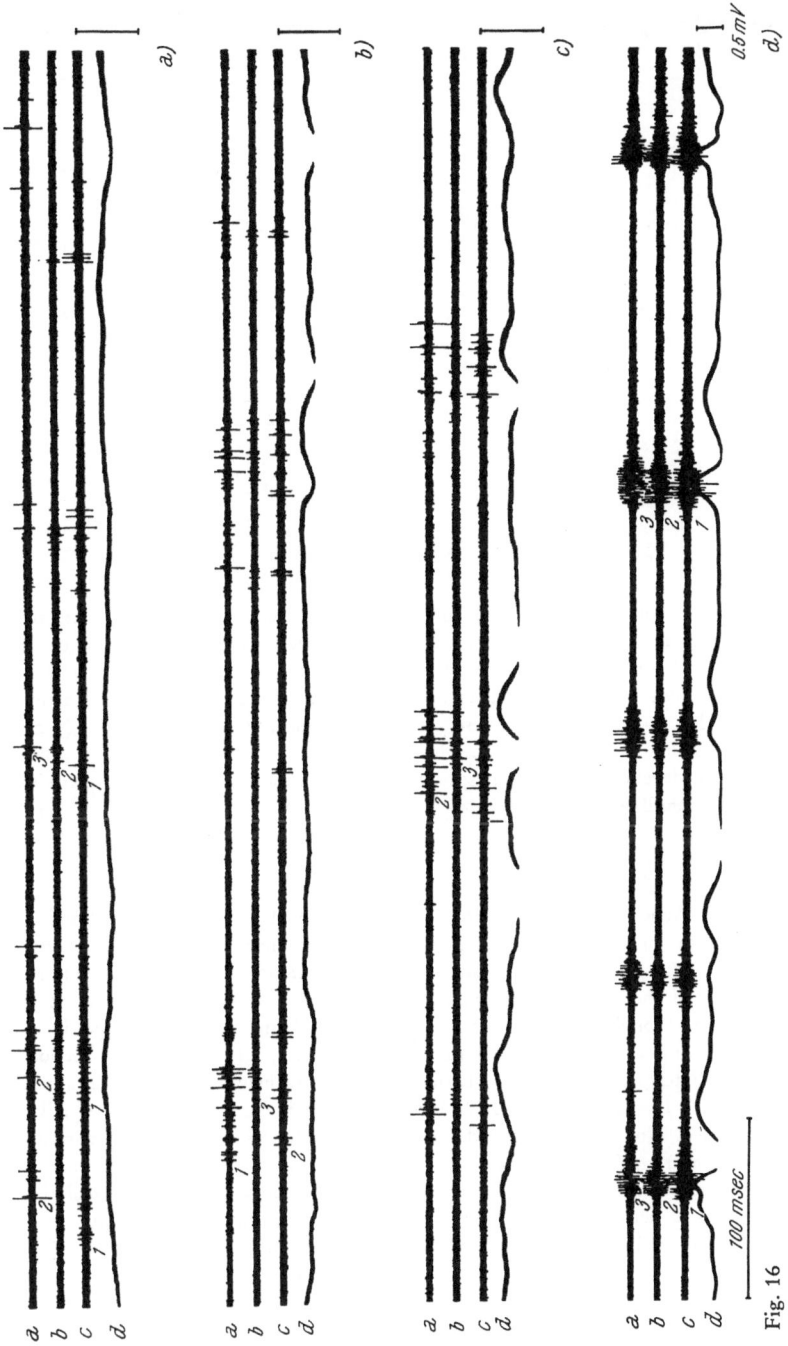

Fig. 16

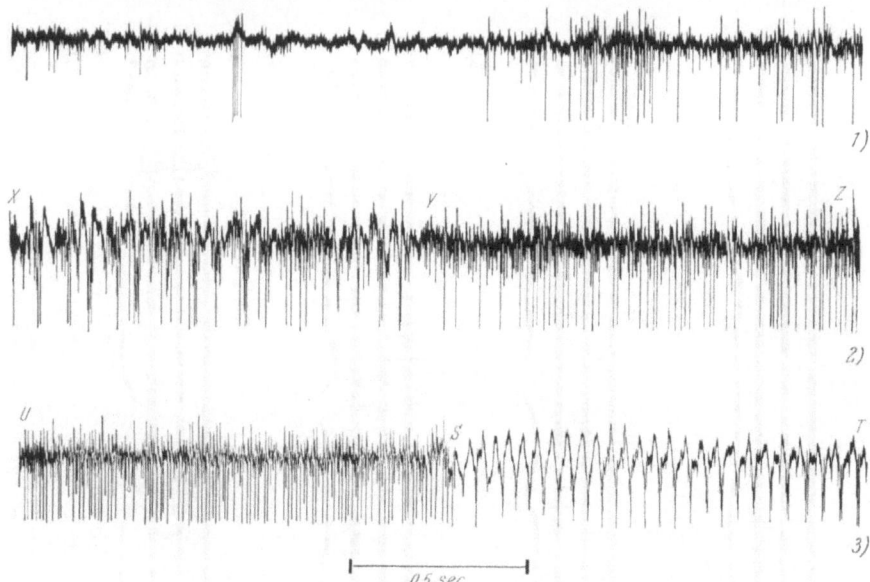

Fig. 17. Action of carbamylcholine on the electrical activity of n. lateralis dorsalis of the cat. 1: control recording; 2: 7 minutes after the local injection of 20 μg of carbamylcholine: after a brief period during which there is an increase in the amplitude of the gross waves accompanied by clustering of neuronal action potentials (x–y), a period of marked desynchronization follows (y–z); 3: 11 minutes after injection, the desynchronization continues (u–s), after which a seizure develops (s–t), which spreads rapidly to the whole cortex. Negativity down. (From Babb *et al.* 1972)

---

Fig. 18. Action of 5 HT on circulation of neuronal activity and synchronization in the thalamus. Recordings obtained, simultaneously, with three microelectrodes (a, b, c) displayed along an antero-posterior line in the n. anterior ventralis of the cat under gallamine. Upper section: during the development of gross wave "spindles". 1: control; 2, 3, and 4: 12, 37, and 50 minutes, respectively, after the injection, directly into the nucleus, of 10 μg of 5 HT. Lower section: during "interspindle intervals". Each one of the tracings in 5, 6, 7, and 8 was obtained a few seconds after 1, 2, 3, and 4, respectively, just after the "spindle" had passed. Note an increase in the density of the clustering and the incorporation of new neurons (z) into the clusters, both in 1 to 4 and in 5 to 8; the increase in the duration of the silent period from x–y in 1, to x–y in 4; the circulation of neuronal activity in the direction c–b–a, at OPQ. d: scalp recording from the homolateral occipital region, showing no changes after the injection of 5 HT. Distances between microelectrode tips same as in Fig. 3. Negativity down. Time in msec. (From Verzeano and Mahnke, in press)

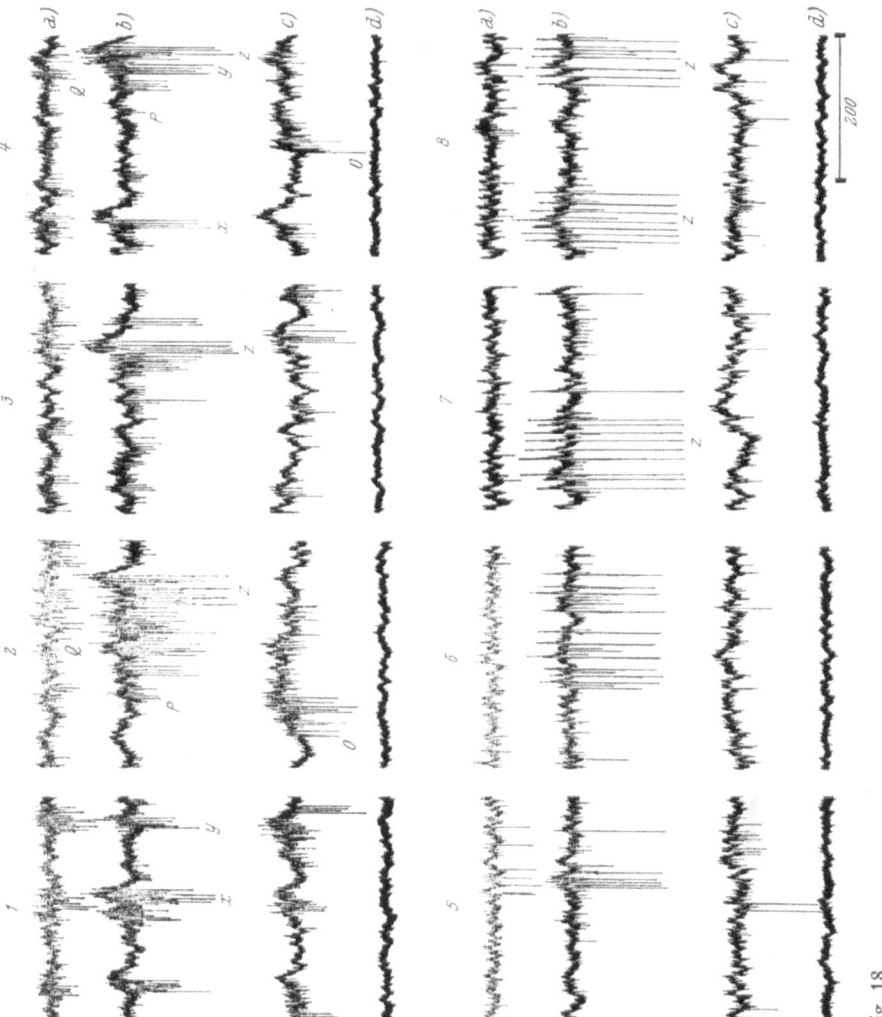

Fig. 18

182    M. VERZEANO

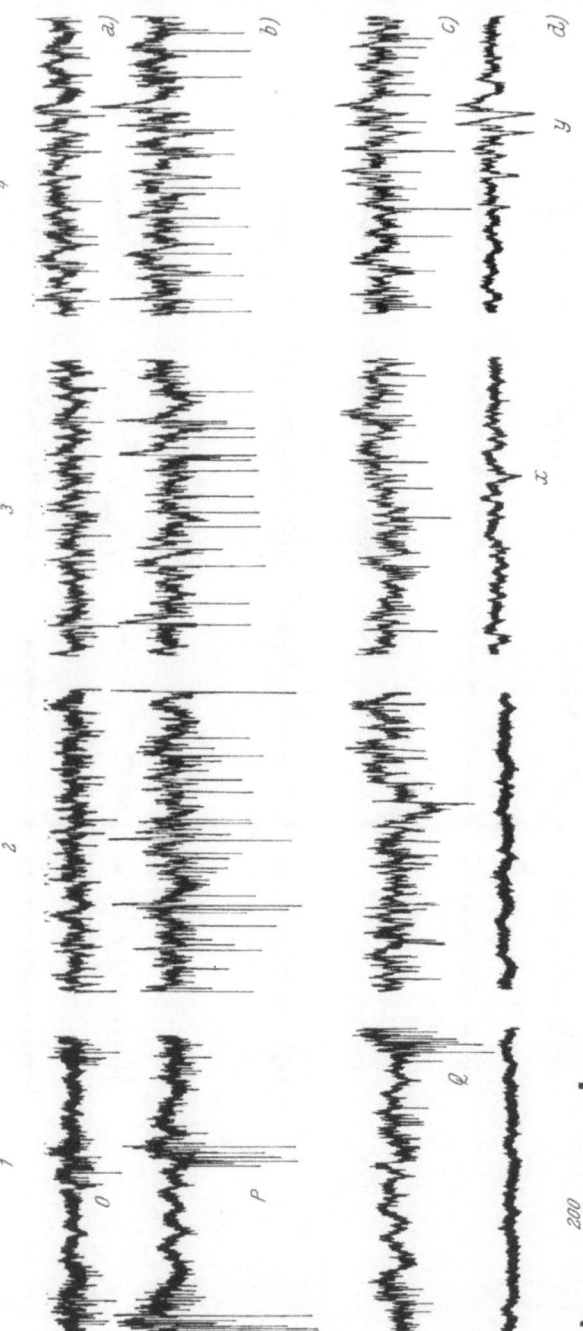

Fig. 19. Disruption of circulation of neuronal activity and of synchronization by carbamyldholine. Recordings obtained, simultaneously, with three microelectrodes (a, b, c) displayed along an antero-posterior line, in the n. anterior ventralis of the cat under flaxedil. 1: high degree of synchronization with circulation of neuronal activity in the direction a-b-c (at OPQ), reached 50 minutes after the injection, directly into the nucleus, of 10 μg of 5 HT. 3 minutes after the recordings in 1 were taken, 10 μg of carbamyldcholine were injected at the same location. 2: 2 minutes after the injection of carbamyldcholine desynchronization begins; 3: 6 minutes after injection, desynchronization continues and epileptiform waves appear in the electroencephalogram (x in tracing d); 4: 11 minutes after injection, desynchronization is complete and epileptiform waves (y) are larger. Distances between microelectrodes, same as in Fig. 3. Time in msec. Negativity down. (From VERZEANO and MAHNKE, in press)

opment of similar changes leading to the convulsive state in the cortex under the influence of picrotoxin. In both examples the convulsive state is reached by overdriving the synchronizing system. Fig. 17 shows the development of a convulsive state reached by overdriving the desynchronizing system by injecting, directly into the thalamus, the cholinergic agent carbamylcholine, in a cat maintained under gallamine. Injections of cholinergic agents into the thalamus of unrestrained animals similarly result in the development of generalized seizures (BABB et al. 1972).

Recent investigations have provided evidence which indicates that different synaptic transmitters may be operating in the two neuronal systems: serotonin in the synchronizing system and acetylcholine in the desynchronizing system. By means of anatomical and histochemical methods, a system of neurons containing 5 HT has been found to extend rostrally from the raphe complex of the reticular formation (JOUVET 1969, FUXE et al. 1970). By means of cholinesterase staining, a system of cholinergic and cholinoceptive neurons has been found to extend from the mid-brain reticular formation to the thalamus, more particularly to the non-specific nuclei (SHUTE and LEWIS 1967, OLIVIER et al. 1970, LYNCH and MOSCOW 1971).

The findings obtained by electrophysiological methods corroborate those obtained by histochemical methods. The injection of 5 HT directly into the non-specific nuclei of the thalamus (VERZEANO and MAHNKE 1972) causes an increase in the clustering of neuronal action potentials, an increased inhibitory action between passages of circulating neuronal activity through the networks and an increase in the amplitude of the gross waves (Figs. 3, 18, and 19). As mentioned above, the injection of cholinergic agents directly into the same nuclei (BABB et al. 1972) causes a brief period of increased clustering of action potentials, followed by prolonged periods of increased discharge rates and desynchronization. When the dose of cholinergic agent injected is small, the activity returns to normal from 40 to 60 minutes after injection. When the dose is higher, the desynchronization is followed by epileptiform discharges and generalized epileptic seizures (Fig. 17). Furthermore, after synchronization has been achieved by local injection of 5 HT (Fig. 19), the injection of a very small dose of carbamylcholine disrupts the synchronizing process and leads to the development of epileptiform discharge (VERZEANO and MAHNKE 1972).

Normally, the activities of the synchronizing and desynchronizing systems are in balance. An increased influence of the first leads to a state of relaxation, drowsiness or sleep. An increased influence of the second leads to arousal, alertness and wakefulness. The possibility has to be considered that, under pathological conditions, one or the other system may be overdriven, either by an excessive action of its own synaptic transmitter or by a diminished action of its antagonist, leading to the development of epileptic seizures.

## Summary

1. The study of the electrical activity of the cortex and of the thalamus, by means of arrays of microelectrodes, reveals that "synchronization" is based on the activity of a neuronal system distributed through all thalamic nuclei, communicating with the cortex and coordinated by pace-making mechanisms originating in the non-specific nuclei. The coordinating mechanism is the highly organized circulation of neuronal activity over intrathalamic, intracortical and cortico-thalamic feedback loops.

2. The analysis of time and phase relations between axonal potentials and rhythmic waves recorded from fiber tracts suggests that rhythmic activity may develop, locally, in such tracts, without electrotonic or field effects transmitted from other regions. Because of the numerous fiber tracts present in the cortex, it is possible that axonal oscillations may participate, with post-synaptic potentials, in the development of the electroencephalogram.

3. The analysis of time and phase relations between gross and neuronal evoked responses, in the cortex as well as in the thalamus, shows that each deflection of the gross response corresponds to a convergence of activity over several sets of neurons and interneurons, the activity of different sets corresponding to the different phases of the gross waves. These relations remain consistent, at different cortical depths and show that the waveform, the successive deflections and the polarity of the gross waves, depend on the nature of the responding neuronal elements and the sequence in which they are activated. These findings cannot be explained by the existence of dipoles and, for this reason, the concept of dipole may have to be abandoned.

4. Microinjection of 5 HT directly into the non-specific nuclei, increases synchronization; microinjection of cholinergic agents causes

desynchronization. This suggests that two over-lapping neuronal systems operate in the thalamus: one—possibly serotonergic—responsible for pacemaking and synchronization, the other—possibly cholinergic—responsible for desynchronization and arousal. Each system can be overdriven: the first by convulsant agents, such as metrazol; the second by cholinergic agents, such as neostigmine or carbamylcholine. Overdriving either system leads to the development of epileptic discharge.

It is proposed that, under normal conditions, the two systems are in balance. When, under pathological conditions, the balance is disrupted, epileptogenic processes may develop.

## References

ADRIAN, E. D.: Afferent discharges to the cerebral cortex from peripheral sense organs. J. Physiol. *100*, 159—191 (1941).

ANDERSEN, P., and S. A. ANDERSSON: Physiological basis of the alpha rhythm. New York: Appleton-Century-Crofts. 1968.

— and J. C. ECCLES: Inhibitory phasing of neuronal discharge. Nature *196*, 645—647 (1962).

— and T. A. SEARS: The role of inhibition in the phasing of spontaneous thalamo-cortical discharge. J. Physiol. *173*, 459—480 (1964).

BABB, T. L., M. BABB, J. H. MAHNKE, and M. VERZEANO: The action of cholinergic agents on the electrical activity of the non-specific nuclei of the thalamus. Int. J. Neurol. (in press, 1972).

CREUTZFELDT, O. D., S. WATANABE, and H. D. LUX: Relations between EEG phenomena and potentials of single cortical cells. I. Evoked responses after thalamic and epicortical stimulation. Electroenceph. clin. Neurophysiol. *20*, 1—18 (1966 a).

— — — Relations between EEG phenomena and potentials of single cortical cells. II. Spontaneous and convulsoid activity. Electroenceph. clin. Neurophysiol. *20*, 19—37 (1966 b).

DOTY, R. W., and D. S. KIMURA: Oscillatory potentials in the visual system of cats and monkeys. J. Physiol. *168*, 205—218 (1963).

FRIGYESI, T. L.: Intracellular studies of neurons in the thalamic reticular nucleus during caudate, capsular and thalamic stimulation. Fed. Proc. *30*, 324 (1971).

— and R. SCHWARTZ: Reciprocal innervation of ventrolateral and medial thalamic neurons by corticothalamic pathways. In: FRIGYESI, T. L., E. RINVIK, and M. D. YAHR (eds.), Corticothalamic Projections and Sensorimotor Activities. New York: Raven Press. (In press, 1972.)

FUXE, K., T. HÖKFELT, and U. UNGERSTEDT: Morphological and functional aspects of central monoamine neurons. Int. Rev. Neurobiol. *13*, 93—126 (1970).

HANBERY, J., and H. H. JASPER: Independence of diffuse thalamo-cortical projection system shown by specific nuclear destruction. J. Neurophysiol. *16*, 252—271 (1953).

Jasper, H. H.: Functional properties of the thalamic reticular system. In: Adrian, E. D., F. Bremer, and H. H. Jasper (eds.), Brain Mechanisms and Consciousness, pp. 374—401. Springfield: Charles C. Thomas. 1954.

Jouvet, M.: Biogenic amines and the states of sleep. Science 163, 32—41 (1969).

Laufer, M., and M. Verzeano: Periodic activity in the visual system of the cat. Vision Research 7, 215—229 (1967).

Li, C. L., and H. H. Jasper: Microelectrode studies of the electrical activity of the cerebral cortex in the cat. J. Physiol. 121, 117—140 (1953).

Lynch, G., and S. Moscow: Personal Communication (1971).

Mescherskii, R. M.: The vectorgraphical characteristic of spontaneous rabbit brain cortex activity. Sechenov Physiological Journal of the USSR (English Translation) 47, 419—426 (1961).

Morison, R. S., and D. L. Bassett: Electrical activity of the thalamus and basal ganglia in decorticate cats. J. Neurophysiol. 8, 309—314 (1945).

— and E. W. Dempsey: A study of thalamo-cortical relations. Amer. J. Physiol. 135, 281—292 (1942).

Moruzzi, G., and H. W. Magoun: Brain stem reticular formation and activation of the EEG. Electroenceph. clin. Neurophysiol. 1, 455—473 (1949).

Negishi, K., M. C. Bravo, and M. Verzeano: The action of convulsants on neuronal and gross wave activity in the thalamus and in the cortex. Electroenceph. clin. Neurophysiol. Suppl. 24, 90—96 (1963).

— E. S. Lu, and M. Verzeano: Neuronal activity in the lateral geniculate body and the nucleus reticularis of the thalamus. Vision Research 1, 343—353 (1962).

Olivier, A., A. Parent, and L. J. Poirier: Identification of the thalamic nuclei on the basis of their cholinesterase content in the monkey. J. Anat. 106, 37—50 (1970).

Petsche, H., and P. Rappelsberger: Influence of cortical incisions upon synchronization pattern and travelling waves. Electroenceph. clin. Neurophysiol. 28, 592—600 (1970).

— and J. Šterc: The significance of the cortex for the travelling phenomenon of brain waves. Electroenceph. clin. Neurophysiol. 25, 11—22 (1968).

Purpura, D. P.: Nature of electrocortical potentials and synaptic organizations in cerebral and cerebellar cortex. Int. Rev. Neurobiol. 1, 47—163 (1959).

— and R. J. Shofer: Cortical intracellular potentials during augmenting and recruiting responses. I. Effects of injected hyperpolarizing currents on evoked membrane potential changes. J. Neurophysiol. 27, 117—132 (1964).

Renshaw, B., A. Forbes, and B. R. Morison: Activity of isocortex and hippocampus: Electrical studies with micro-electrodes. J. Neurophysiol. 3, 74—105 (1940).

Scheibel, M. E., and A. B. Scheibel: Patterns of organization in specific and nonspecific thalamic fields. In: Purpura, D. P., and M. D. Yahr (eds.), The Thalamus, pp. 13—46. New York: Columbia University Press. 1966.

Shute, C. C. D., and P. R. Lewis: The ascending cholinergic reticular system: neocortical, olfactory and subcortical projections. Brain 90, 497—520 (1967).

Starzl, T. E., C. W. Taylor, and H. W. Magoun: Ascending conduction in reticular activating system, with special reference to the diencephalon. J. Neurophysiol. 41, 461—477 (1951).

VERZEANO, M.: Sequential activity of cerebral neurons. Arch. Internat. Physiol. Biochim. *63*, 458—476 (1955).
— Activity of cerebral neurons in the transition from wakefulness to sleep. Science *124*, 366—367 (1956).
— The synchronization of brain waves. Acta Neurol. Latinoamer. *9*, 297—307 (1963).
— Evoked responses and network dynamics. In: WHALEN, R. E., R. F. THOMPSON, M. VERZEANO, and N. M. WEINBERGER (eds.), The Neural Control of Behavior, pp. 27—54. New York: Academic Press. 1970.
— and I. CALMA: Unit-activity in spindle bursts. J. Neurophysiol. *17*, 417—428 (1954).
— R. DILL, G. NAVARRO, and E. VALLECALLE: The action of metrazol on spontaneous and evoked activity. Physiol. and Behav. *5*, 1099—1102 (1970 b).
— M. LAUFER, PHYLLIS SPEAR et SHARON MCDONALD: L'activité des reseaux neuroniques dans le thalamus du singe. Actualités Neurophysiologiques *6*, 223—252 (1965).
— — — and S. MCDONALD: The activity of neuronal networks in the thalamus of the monkey. In: PRIBRAM, K., and D. E. BROADBENT (eds.), Biology of Memory, pp. 239—271. New York: Academic Press. 1970 a.
— and J. H. MAHNKE: In Press.
— and K. NEGISHI: Neuronal activity and states of consciousness. Proc. Int. Cong. Physiol. Buenos Aires. August 1959.
— — Neuronal activity in wakefulness and in sleep. In: WOLSTENHOLME G. E. W., and M. O'CONNOR (eds.), The Nature of Sleep, pp. 108—130. (Ciba Symposium.) London: Churchill. 1961.

# Discussion

GLOOR: How can you be sure that convulsant drugs actually activate the synchronizing system?

VERZEANO: The changes caused by the administration of convulsant drugs are identical with the spontaneous activation of the synchronizing system: increased clustering of the action potentials, increased number of neurons involved in the pathway of circulation, increased velocity of circulation, increased period of silence between successive passages of circulating activity through the network (see Figs. 1, 2, 15, and 16).

GLOOR: How real are these systems? Perhaps there are two kinds of responses of the same cells?

VERZEANO: There are two possibilities: one, that the two kinds of activity correspond to two different neuronal systems; two, that they correspond to two different synaptic systems operating on the same kind of neurons. I am inclined to believe that we deal with two neuronal systems because, when we give metrazol, we see that different kinds of *action* potentials are simultaneously affected in opposite ways: in spontaneous activity, low amplitude discharge is depressed while high amplitude discharge is enhanced; similarly, in evoked activity, low amplitude *early* discharge is depressed while high amplitude *late* discharge is enhanced (Fig. 9). Since we deal with *action* potentials I don't see how this could be explained on the basis of a single neuronal system.

# Experimental Research on the Ontogenetic Development of the Mechanism of Epileptic Synchronization. Comparative Study on the Immature and Mature Brain

D. VOLANSCHI

Institute of Neurology and Psychiatry of the Academy of Medical Sciences, Bucharest, Romania

Experimental studies on the evolution of the paroxysmal synchronization of epileptic type during the ontogenetic development of the mammalian brain from birth to adulthood have been performed in the past 20 years by BISHOP (1950), PASSOUANT et al. (1960), VOLANSCHI (1960 a, b), VOLANSCHI et al. (1961), CAVENESS et al. (1962), PURPURA (1962), CRIGHEL et al. (1965), CRIGHEL (1971), PRINCE and GUTNICK (1971).

The correlation of such studies with data on the morphohistological, biochemical, and physiological characteristics of the brain structures at various ontogenetic evolutive stages supplies information which may provide for an understanding of both the peculiarities of the epileptic synchronization during different postnatal development stages and the complex mechanisms of epileptic synchronization in the mature brain.

The aim of this paper is to present some comparative data on the EEG synchronization of the epileptic type in the immature and mature brain and to attempt to establish some correlations with morphohistological data available in the literature and also with some histochemical and metabolic findings.

Our experimental research was carried out on kittens in the first two postnatal weeks and on adult cats. As is known, in the first two weeks after birth the kitten's brain presents a high degree of immaturity, specifically at the telencephalon and diencephalon level (MARINESCU and KREINDLER 1935, PURPURA et al. 1960, NOBACK and PURPURA 1961, PURPURA 1962, CRIGHEL et al. 1965).

The electroencephalographic data here reported concern the general characteristics of development of the neocortical penicillin focus at the above mentioned periods of life.

## Methods

Acute experiments were conducted on animals locally anesthetized with procaine for surgical procedure, immobilized with gallamine and artificially ventilated. Monopolar electrocorticograms were recorded from the sigmoid posterior, anterosuperior part of the sylvian and posterior part of the marginal gyri of both hemispheres by means of silver-silver chloride ball electrodes placed on the exposed surface of the cortex, the reference electrode being placed on the frontal or occipital bone. Also deep bipolar recordings were obtained by means of bipolar coaxial steel electrodes stereotaxically introduced into different structures. After recording the spontaneous activity, a 4 mm² piece of filter paper soaked with a total dose of 400 I.U. of potassium penicillin in aqueous solution was applied to a primary sensory neocortical area (somatic, auditory, or visual), and the local tracing, the tracings of the symmetric area and other neocortical areas and, in some experiments, also of some subcortical structures, were observed over the next two hours.

## Results

Since in our experiments the evolution during the two postnatal weeks of the neocortical penicillin focus in the three primary sensory areas explored was essentially the same, we shall only discuss here, as a representative example, the evolution of the sigmoid posterior focus.

In the first day after birth (Fig. 1), the cortical activity before the application of penicillin is expressed by low-voltage ECoG tracings and the appearance on this background of bursts of 3 to 6/second high-voltage waves of 4 to 6 seconds duration. These bursts appear simultaneously in different neocortical areas and recur 2 to 6 times per minute.

After the application of penicillin, the first graphoelements which could be considered as a focal discharge appear very late, after about half an hour, as a high-voltage biphasic slow wave, initially negative, superimposed upon the above-mentioned burst. Isolated penicillin discharges appear later (after about 1 hour) as monophasic negative or high-voltage (0.5 V) biphasic initially negative slow waves, recurring at variable intervals which were never less than 4 seconds. The ample discharges induce in the symmetric area slow and much smaller variations in potential. No additional changes were noted up to the end of the experiment (two hours after the application of penicillin).

The above data may be correlated with the observation of VOLANSCHI and SERVIT (1969) that the high degree of dependence of cortical

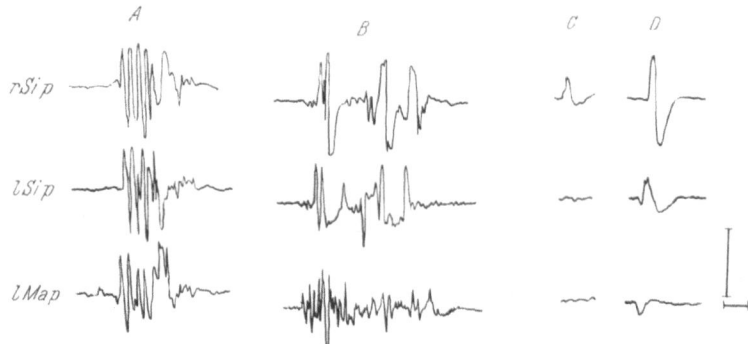

Fig. 1. Sigmoid posterior penicillin focus on the first day after birth. ECoG monopolar recordings. In this and the following figures, negativity is upwards. *rSip* = right sigmoid posterior gyrus; *lSip* = left sigmoid posterior gyrus; *lMap* = left marginal posterior gyrus. Application of penicillin invariably to *rSip* gyrus. A: Background activity with generalized bursts of 3 to 6 c/s high-voltage waves. B: 65 minutes after the application of penicillin; discharges of high-voltage, biphasic, initially negative slow waves superimposed on the background bursts. C: 70 minutes after penicillin; focal slow monophasic negative discharge in *rSip* gyrus. D: 75 minutes after penicillin; biphasic, initially negative, slow interictal discharge in the *rSip*; slow variations in potential induced in symmetrical area. Vertical bar indicates 200 microvolts in A and B and 400 microvolts in C and D. Horizontal bar: 1 second

interictal penicillin discharges upon non-specific facilitation of deep origin characterizes a cortical structure with poor epileptogenic reactivity, unable to pass from isolated interictal discharges to a self-sustained seizure.

On the following days of the first post-natal week (Fig. 2 *a*) the focal interictal discharges appear earlier, *i.e.*, within 4 to 10 minutes; sometimes they even appear from the start independent of the background bursts. Morphologically, the discharges resemble those of the first

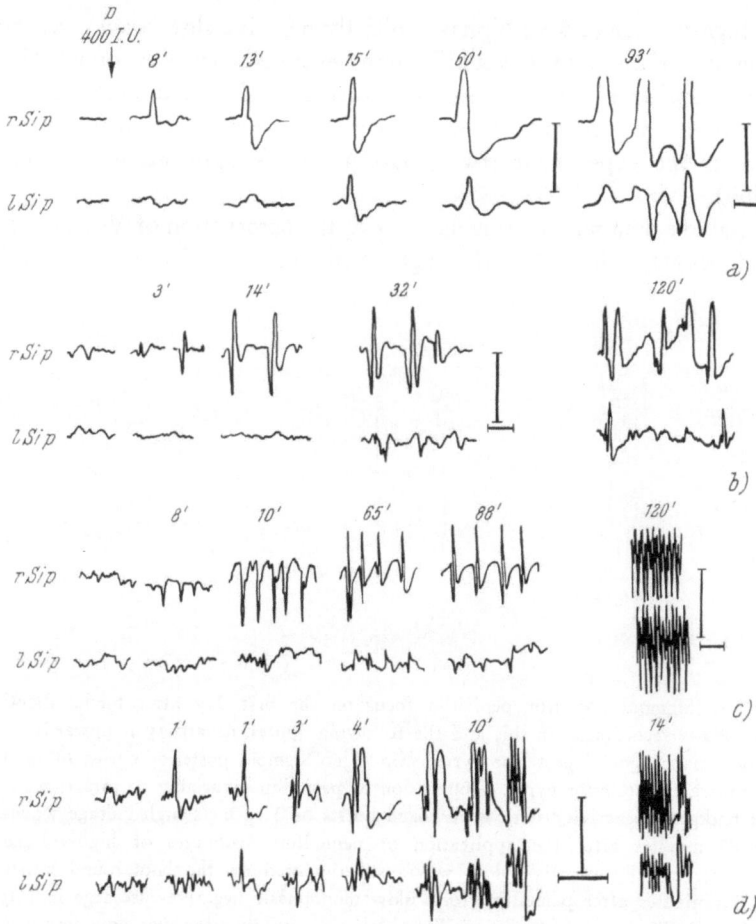

Fig. 2. Comparative development of the sigmoid posterior penicillin focus in kittens in the first (*a*), second (*b*), sixth (*c*) week after birth and in the adult cat (*d*). ECoG monopolar recordings. The first segment of each tracing corresponds to the background activity before penicillin application.

In *a* (4-day-old kitten): comparatively late (8 minutes after penicillin) appearance of the focal *rSip* interictal discharge as a monophasic negative wave; the fully developed discharge at 13 minutes and onwards consists of biphasic, initially negative slow waves, inducing slow variations in potential in the symmetric area; a group of discharges recurs at minimum intervals, 93 minutes after penicillin.

In *b* (14-day-old kitten): earlier appearance of the focal interictal biphasic discharges (3 minutes); initial positivity instead of the initial negativity observed in the first week; at 32 minutes fully developed discharges with spike-and-wave morphology recurring at minimum intervals, with afterdischarge-tendency during the negative wave; at 120 minutes, periods of slow discharges recurring at short,

day: slow, ample waves, biphasic with initially negative component. The median interval between two discharges was 6 seconds and the minimum was 1 to 2 seconds. Sometimes, a variable number of discharges occurred at minimum intervals but this phenomenon did not represent a true self-sustained seizure followed by electrical silence. As on the first day, the focal interictal discharges induce in the symmetric area slow variations in potential, of smaller amplitude and appearing with a latency which was easily perceptible even at low recording speed (7.5–15 mm/sec).

During the second postnatal week (Fig. 2 b), the first focal interictal discharges appear still earlier (3 minutes). At this age, the initial negativity characteristic of first-week discharges is, either immediate or soon afterwards, replaced by positivity. The fully developed discharge has a spike-and-wave morphology. Its comparatively slow negative spike temporarily interrupts the initial positive deflection. An afterdischarge tendency appears during the negative wave. The mean interval between the discharges decreases to 2 seconds. The length of the periods of rapid succession of discharges increases, but in this case either we cannot speak of a true self-sustained seizure followed by electrical silence. The amplitude of the variations in potential induced in the symmetric area approximates sometimes to that of the primary discharges.

Six weeks after birth (Fig. 2 c) the general morphology of the interictal discharge does not differ from that of the second week, but the duration of the spike-and-wave complex and of its different components tends to decrease, while the periods of rapidly succeeding discharges become longer. At a late stage of the experiment the

---

irregular intervals, with inconstant transmission to the symmetric area. In c (40-day-old kitten): first penicillin discharges (8 minutes) consist in monophasic positive waves (compare with a), the duration of the spike-and-wave complex is shorter (compare—65 minutes in c with 32 minutes in b); a focal selfsustained seizure with an initial tonic phase represented by 8 c/s high-voltage discharges transmitted to the symmetrical area appears at 120 minutes after penicillin. In d (adult cat): very early appearance of focal discharges as negative spikes symmetrically transmitted (1 minute), early reversal to positivity of initial polarity of discharges (3 minutes), later appearance of discharges with spike-and-wave morphology (4 minutes), of afterdischarges (11 minutes) and of the focal selfsustained seizure (14 minutes); the interictal and ictal discharges are faster than in kittens.
Calibration: in a and b 200 microvolts, except the 93-minutes portion of graph a, where it is 400 microvolts. In c and d 400 microvolts. Horizontal bar: 1 second

interictal discharges passed into a true focal self-sustained seizure with an initial tonic phase represented by comparatively fast (8 c/s) repetitive high-voltage discharges which were transmitted point-by-point to the symmetric area.

In the adult animal (Fig. 2 *d*) the first focal discharge and its trans-

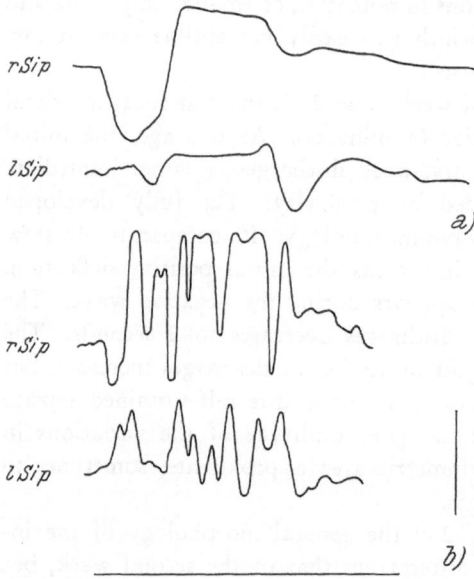

Fig. 3. Latency of appearance of variations in potential induced in the symmetrical area by the interictal discharge. ECoG monopolar recordings. *a*): 12-day-old kitten. *b*): adult cat. Calibration: 400 microvolts in *a*), 800 microvolts in *b*. Horizontal bar: 1 second

mission to the symmetric area may be recorded even as soon as 1 minute after the application of penicillin. The spike-and-wave morphology appears earlier and so does the after-discharge. The interictal discharges are evidently faster in cats than in kittens. Also the latency of the corresponding variations in potential induced in the symmetric area is significantly smaller in cats (Fig. 3).

The above described ECoG general characteristics of the evolution of the penicillin focus must be correlated with the morphohistologic information concerning the ontogenetic postnatal development of the cat's brain (Marinescu and Kreindler 1935, Crighel et al. 1965, and especially the detailed studies on the cat's neocortex performed by Purpura et al. 1960, Noback and Purpura 1961, Purpura 1962). From these studies it emerges that in kittens at birth the apical dendrites of pyramidal cells are well developed and densely packed in the molecular layer, but their tangential branches are only distal and short. The synaptic complexes in relation with superficial dendrites are very numerous. Most of the pyramidal neurons are devoid of basilar dendrites, axosomatic synapses and axon collaterals. Myelinated fibres cannot be found in the neopallium and diencephalon.

In the first postnatal week the superficial dendrites remain very dense despite the regression of the embryonic Cajal-Retzius cells, while the deep basilar dendrites and axosomatic synapses just begin to develop.

On account of the poor development of the internuncial neurons and of the horizontal connections, the neocortex appears as a palisade of vertical columns of apical dendrites and ascending and descending axons.

The histological findings suggest that in the first postnatal days the postsynaptic activity in the kitten's neocortical neurons is predominantly generated in the superficial neuropil in the apical dendrites, and that each of the above mentioned vertical columns is from the physiological point of view comparatively isolated from its neighbours.

In an attempt to correlate the ECoG characteristics of the neocortical penicillin focus in the first week after birth with morphohistologic data, the following is suggested:

The initially surface negative polarity of the interictal discharges indicates that the depolarization is generated by the superficial neuropil. The succeeding positive component probably corresponds especially with the propagation of depolarization in the depth of the cortex.

The late appearance of the first discharges, requiring at start a facilitation of subcortical origin, their low frequency and the slowness of the discharges probably are due to some functional peculiarities of the immature superficial neuropil. The high threshold of excitability and great fatigability were demonstrated by the studies of Purpura et al. (1960) and Crighel et al. (1965) on the superficial cortical responses to direct stimulation. Recent data by Prince and Gutnick (1971) show that the slow interictal discharges characteristic in kittens

13*

correspond in intracellular recordings to depolarization shifts of longer duration and greater variability than in the adult cat.

That no afterdischarge phenomena and no passage from interictal discharges to a self-sustained seizure were noted, may be to a certain extent explained by the poor development of axon collaterals, internuncial neurons, intracortical horizontal connections and some cortico-subcortical connections. Consequently, intracortical and corticosubcortical neuronal chains and reverberating circuits involved in the mechanism of initiation and maintenance of the epileptic seizure in adults are still undeveloped.

In the second week after birth the superficial dendritic system of the Cajal-Retzius cells continues to regress, an accelerated rate of growth of the basilar dendrites, axon collaterals and apical dendritic collaterals occurs, stellate cells with longer and more branched dendrites are more readily observed and dendritic spines become prominent and abundant. The development of the deep neuropil and consequently the deep seat of generation of the discharge can explain why from this stage on the polarity of the initial component of the interictal discharge becomes surface-positive.

As for the surface-negative spike of the spike-and-wave complex it is possible that its generation involves postsynaptic depolarization generated in the superficial layers and the propagation to the surface of the depolarization initiated in the deep layers.

The available information does not allow us to speculate more upon the site of origin and the significance of the later components of the interictal discharge.

The higher degree of functional maturation of the neurons, the lowering of the threshold of excitability and the lower degree of fatigability (CRIGHEL et al. 1965) may partly account for the earlier appearance of the discharges, the primal appearance of which is no longer dependent upon a nonspecific deep facilitation, and for their shorter duration and raised frequency.

The long latency of the variation in potential induced by the interictal focal discharge in the symmetric area (Fig. 3) reflects the low conduction velocity of non-myelinated or poorly myelinated interhemispheric pathways.

The still reduced degree of maturation of the intracortical and corticosubcortical connections (Fig. 4) may account for the reduced tendency to afterdischarges and for the impossibility to pass from the interictal discharges to true self-sustained seizures.

At the end of the third postnatal week the pyramidal cells of the neocortex appear almost completely developed, the number of basilar dendrites and apical dendritic collateral branches of these cells being essentially the same as in the mature cortex with numerous spines in all dendrites of the pyramidal and stellate cells (NOBACK and PURPURA 1961). The vertical columnization of the neurons is still

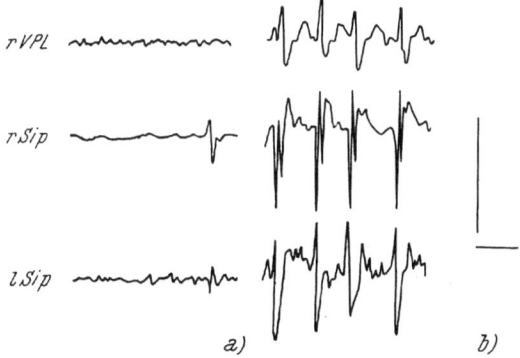

Fig. 4. Propagation of the *rSip* interictal discharges to the homolateral ventral posterolateral thalamic nucleus (VPL) and to the symmetrical cortical area in the 12-day-old kitten (*a*) and the adult cat (*b*). ECoG: monopolar recording; VPL: bipolar recording. Note in *a* the lack of propagation to VPL, 45 minutes after penicillin application in kitten, in contrast to the clear-cut propagation 10 minutes after penicillin application in the adult cat (*b*). Calibration: 200 microvolts. Horizontal bar: 1 second

clear-cut but the connections between the columns have multiplied (CRIGHEL *et al.* 1965). Obviously, then, such a degree of morpho-histologic maturation of the neocortex is a condition for the passage from the interictal focal discharges to true selfsustained focal tonic seizures, consisting of fast repetitive high-voltage oscillations in potential, which are transmitted point-by-point to the symmetric area. At this age, then, the mechanisms of epileptic synchronization approximates the complexity characteristic of the mature brain.

The above considerations on the correlations between the EEG synchronization of epileptic type and the morphologic data at different postnatal development stages may be criticized on the basis that they neglect certain important elements, as for instance the inhibitory phenomena etc., but our available material has not yet enabled us to approach all the aspects.

We wish to underline here that the epileptogenic functional immaturity of the nervous tissue has not only morphologic, but also histochemical correlates which have yet to be established. Such correlations regard, among other things, the development of the enzymatic equipment. As an example of the importance of the ontogenetic histoenzymologic approach we shall present some succinct data concerning the development of the alkaline and acid phosphatase in the kitten and cat neocortex.

As is known, during the brain ontogenetic development from the embryo to the adult stage, a gradual transition from growth to the function metabolism takes place (RICHTER 1955). The very important part played in this context by phosphatases in carbohydrate, phospholipoid and nucleotide metabolism is well known. Since in the adult brain cortex the acid phosphatase is found in larger amounts than the alkaline one, certain authors consider it more important for the metabolic function of the nerve cell.

We compared the alkaline and acid phosphatase reaction in the neocortex of the kitten (the first two weeks after birth) and cat. We did not find any significant differences in the alkaline phosphatase reaction between kitten and cat. These data are in agreement with those reported by BISHOP (1950) in the rabbit. The acid phosphatase reaction, on the contrary, showed wide differences between the immature and mature neocortex (Fig. 5). In adult cats there is a clearcut positive reaction in the cell soma and the neuropil and a very strong reaction in the white matter. When compared with cats, the reaction is, in kittens, somewhat weaker at the level of the cell soma, four times weaker at the neuropil level (histophotometric determinations) and exceedingly weak in the white matter.

The poor development of the acid phosphatase at the level of the neuropil and the white matter of the kitten's neocortex should be kept in mind when an interpretation of the differences in electrical reactivity of the epileptic type between the cat and kitten neocortex is attempted.

We shall conclude this paper by some succinct data concerning the relationship between the degree of ontogenetic development of the synchronization mechanisms and the intensity of metabolic activation during the seizure. By previous investigations (VOLANSCHI 1960) it was demonstrated that the electrical seizures induced by transcerebral

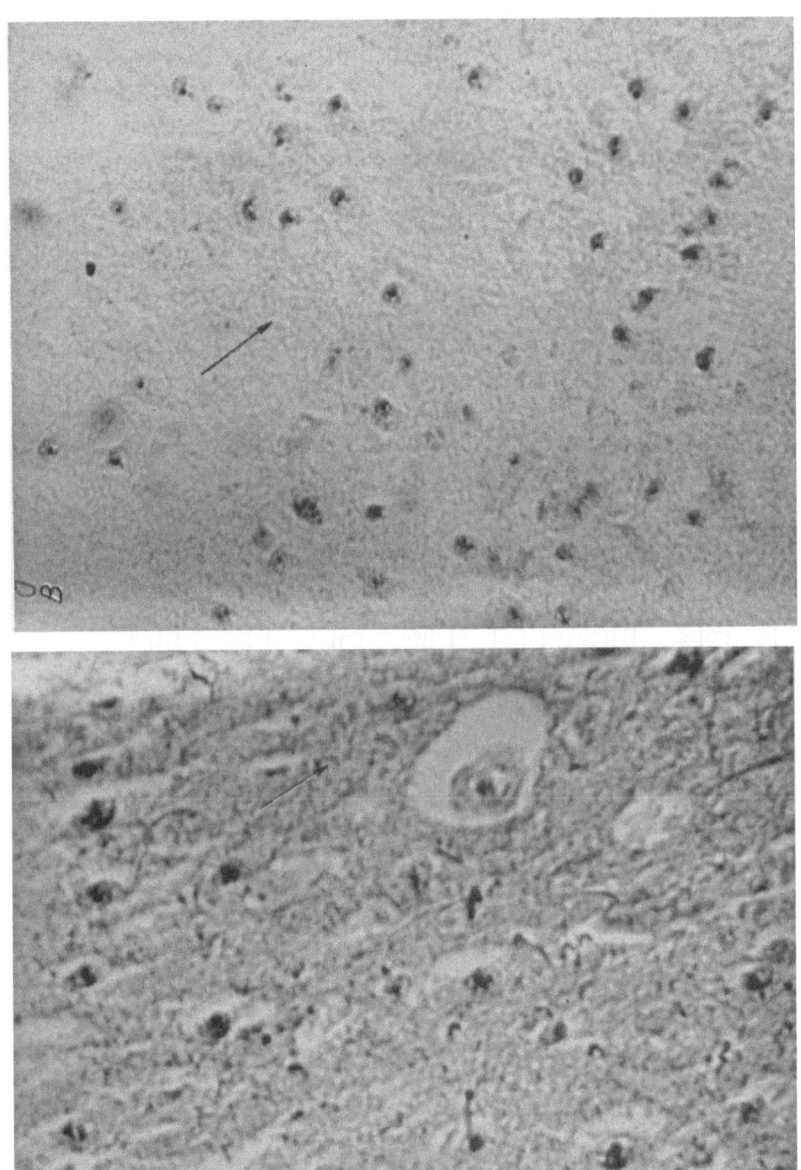

Fig. 5. Microphotograph of acid phosphatase reaction in pericruciate white matter in cat (*A*) and 8-day-old kitten (*B*). Note the visibly lower intensity of reaction at the neuropil level in kitten as indicated by arrows

electrical shock in kittens in the first week after birth differ from those induced in adult cats by a less manifest tendency of the waves in various cortical and subcortical structures to appear synchronized. Clinically, in such kittens, instead of the generalized tonico-clonic seizure, there appear irregular asynchronous motor responses especially in the head and forelimbs. The seizures in kittens may last as long as in cats.

As is known, the intensity of $^{32}$P-incorporation into the living tissue supplies information on the metabolic activity level. Experiments

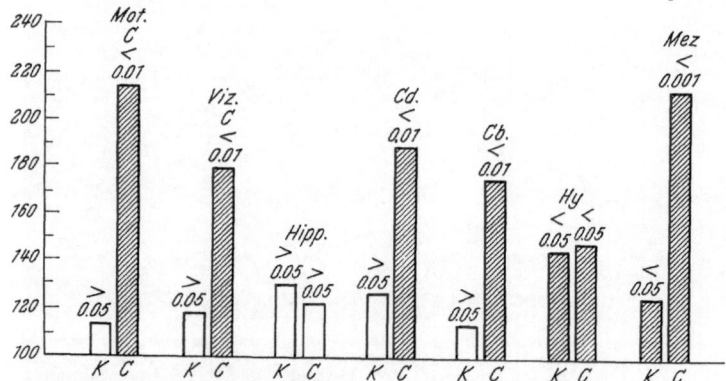

Fig. 6. Influence of transcerebral electrical shocks (7 stimulations at 5-minutes interval) on $^{32}$P-incorporation rate in different brain structures of the one-week-old kitten (K) and cat (C), versus controls (nonstimulated animals). Ordinate: $^{32}$P-incorporation rate in stimulated animals as percentage of control. Hatched and white columns: statistically significant and nonsignificant differences to control, respectively. Figures on top of columns = p value

were carried out on kittens in the first week after birth and on cats, exploring the $^{32}$P-incorporation rate in several cortical and subcortical brain structures in control and electroshock-stimulated animals. Essentially, the results showed that, in kittens, electrical shocks did not induce any significant increase in $^{32}$P-incorporation rate as compared with controls in any cerebral structure explored except the mesencephalon and hypothalamus. Contrarily, in adult cats the $^{32}$P-incorporation rate was significantly higher after electrical shock in all the structures studied (except the hippocampus), the peak values corresponding to the motor cortex and mesencephalon (Fig. 6).

According to these observations the degree of maturity of the mechanisms of critical epileptic EEG synchronization seems to run parallel to the degree of the cortical metabolic activation of the cerebral structures. In the adult cat's brain which is able to organize highly synchronized electrical and clinical seizures, a significant activation of metabolic processes takes place during the seizure in the majority of the cortical and subcortical structures. By contrast, in the kittens' brain which is not able to organize a major epileptic seizure but only an atypical, less synchronized, electrical and clinical one, the metabolic level during the seizure is significantly increased only in the phylogenetically older and ontogenetically earlier matured subcortical structures.

## Summary

The general electrocorticographic characteristics of the development of the neocortical penicillin focus were studied in kittens in the first postnatal weeks and in cats. In the first postnatal week only intercritical slow biphasic and initially negative discharges are recorded; they occur late and are frequently conditioned by an unspecific deep facilitation. In the second postnatal week the interictal discharges appear earlier, become initially positive and, when fully developed, display the spike-and-wave morphology. True self-sustained seizures do not yet occur. After the third week it is virtually possible to obtain, though with a longer latency than in cats, a true self-sustained focal seizure also. The latency of the propagation of paroxysmal excitation to symmetrical areas is much longer in kittens in the first weeks than in cats; in kittens no propagation to some subcortical structures was observed.

An attempt is made to correlate the EEG findings with the available morphohistologic information about ontogenetic postnatal development of the cat's brain and also with some histochemical data. The paper concludes with some considerations about the relationships between ontogenetic development of critical epileptic synchronization mechanisms and the degree of metabolic activation during the seizure in terms of $^{32}$P-incorporation rate in different brain structures.

## References

BISHOP, E. J.: The strychnine spike as a physiological indicator of cortical maturity in the postnatal rabbit. Electroenceph. clin. Neurophysiol. 2, 309—315 (1950).

CAVENESS, W. F., K. C. NIELSEN, P. I. YAKOVLEV, and R. D. ADAMS: Electroencephalographic and clinical studies of epilepsy during the maturation of the monkey. Epilepsia 3, 137—150 (1962).

CRIGHEL, E.: Experimental data concerning the epileptic activity in immature kittens. Rev. Roum. Neurol. *8*, 47—53 (1971).

— N. SOTIRESCU, and N. MARCOVICI: Reactivity of the cat immature neocortex to direct stimulation. Rev. Roum. Neurol. *2*, 45—52 (1965).

MARINESCU, G., and A. KREINDLER: Des réflexes conditionnels. Etudes de physiologie normale et pathologique, pp. 24—33. Paris: Felix Alcan. 1935.

NOBACK, C. R., and D. P. PURPURA: Postnatal ontogenesis in cat neocortex. J. Comp. Neurol. *117*, 291—308 (1961).

PASSOUANT, P., J. CADILHAC, M. RIBSTEIN, TH. PASSOUANT-FONTAINE, and L. MIHAILOVIC: Epilepsie et maturation cérébrale, pp. 62—73. Montpellier: 1960.

PRINCE, D. A., and M. GUTNICK: Cellular activities in epileptogenic foci of immature cortex. In: Proc. Internat. Union Physiol. Sci. Vol. IX, p. 460.

PURPURA, D. P.: Synaptic organization of immature cerebral cortex. World Neurology *3*, 275—298 (1962).

— M. W. CHARMICHAEL, and E. M. HOUSEPIAN: Physiological and anatomical studies of development of superficial axodendritic synaptic pathways in neocortex. Exptl. Neurol. *2*, 324—347 (1960).

RICHTER, D.: The metabolism of the developing brain. In: WAELSCH, H. (ed.), Biochemistry of the developing nervous system, pp. 225—250. New York: Academic Press. 1955.

VOLANSCHI, D.: Experimental research on the convulsive activity of the immature brain. Electroenceph. clin. Neurophysiol. *12*, 753 (1960 a).

— Cercetări experimentale asupra reactivității convulsivante a creierului imatur. St. Cerc. Neurol *5*, 505—515 (1960 b).

— and ZD. SERVIT: Epileptic focus in the forebrain of the turtle. Exp. Neurol. *24*, 137—146 (1969).

— N. STERESCU, N. VOICULET, and MIOARA LECCA: Studiul comparativ al capacității convulsivante a creierului imatur și matur (cercetări cu $P^{32}$). St. Cerc. Neurol. *6*, 291—302 (1961).

# Discussion

SCHERRER: In the newborn rabbit's or cat's brain spontaneous activity or transmission of messages at high rate do not work in a satisfying way in the first month of life. In the first weeks the oxydative mechanisms are rather primitive. At the same period much energy is needed for the growth of brain tissue and for the maintenance of the membrane potentials of neurons, the number of which already equals almost that of the adult animal.

VOLANSCHI: I agree with you. Obviously, the particularities of the epileptic type of electrical activity during the various stages of ontogenetic brain development depend not only on the differences in the morphological maturity of the neurons and synaptic links within the neuronal network, but also on intimate biochemical and metabolic characteristics. My intention was to underline the importance of all these factors. As far as metabolism is concerned, it is well known that in the adult animal, the counterpart of critical epileptic electrical activity is a strong metabolic activation with increased oxygen consumption and glucose utilization etc. Besides it is also well known that during the process of growth from embryo to the adult stage of the brain, growth metabolism is gradually replaced by function metabolism, which presupposes changes in enzymatic activity. The fact that during the first two postnatal weeks metabolic activity during paroxysmal epileptic activity was less strong than in adults may partly be due to the above mentioned metabolic differences. That this may be so, also emerges from the data on acid phosphatase activity which I reported on. I wish to emphasize that determination of the $^{32}$P-incorporation rate in kittens showed differences in critical metabolic activation among the various brain structures. The significantly increased rate at only mesencephalic and hypothalamic levels corresponds to an earlier maturation of these structures.

FLEISCHHAUER: May I just add a few observations made in the cortex of newborn and very young cats. Not only are the neuronal and glial elements immature and little differentiated, but they are also less numerous than in the older animals. During the early postnatal period there are still groups of cells, most likely neuroblasts, moving up from the periventricular matrix towards the cerebral cortex. And in brains fixed by perfusion you also find numerous mitoses in the various layers of the cerebral cortex. Although it is most likely that the majority of these mitoses gives rise to glial cells, they may greatly influence the physiological events taking place in the developing cerebral cortex.

VOLANSCHI: In my paper, I referred only to those aspects of the histomorphologic development of the kitten postnatal neocortex which have been better defined in the literature. Obviously, there are also other brain tissue constituents among which glia as you pointed out, that can answer for the differences in the epileptic type of electrical activity among the various stages of ontogenetic development. To say more about the morphofunctional correlations requires further study.

LEHMANN: Did you conduct any experiments with enriched environment?

VOLANSCHI: No.

# The Measurement of Synchronization

J. C. Shaw and C. Ongley

MRC Clinical Psychiatry Unit, Graylingwell Hospital, Chichester, Sussex, England

The high degree of spatial organization in the cortex makes it reasonable to look for a relationship between cortical function and the EEG in terms of the spatial distribution or topography of electrical activity. Although this is implicit in almost all EEG research, it is seldom stated explicitly. Synchronous activity is so named because of its specific topographic characteristics and its measurement may be useful to help understand the relationships between the obvious EEG phenomenon and the underlying physiological process.

There are four main approaches to the measurement of EEG topography. In the simplest of these, observations or measurements are made of the activities present in the individual channels of the conventional EEG record. The way in which activity is distributed over the head and its variability with time or recording conditions can then be described. An example using spectral analysis is the work of Walter, Kado, Rhodes, and Adey (1967). The second method is to display the distribution of potential derived from a matrix of electrodes as a corresponding pattern of light modulated by the potential amplitude. Examples are the Toposcope of Grey Walter, and Shipton (1957) and the 400 channel system of DeMott (1966).

In the third method, primary EEG signals are converted into contour maps. These may depict potential or potential gradient over a two dimensional electrode array, in which case separate maps are required for each time instant considered [e.g., Petsche, Rappelsberger, and Trappl (1970)]. Alternatively they may represent the time course of potential gradient distribution along a line of electrodes as in the familiar chrono-topograms of Rémond (1964). The fourth is concerned with the degree of dependence or independence of the activities from different electrode sites. It makes use of cross-correlation analysis and can be used to derive information about the degree of spread of selected activities and the way in which their topography relates to physiological and psychological variables.

The visual recognition of synchronous activity in the EEG requires at least two channels recording from different areas, and a high degree of linear correlation between the derived signals. Cross-correlation analysis is therefore the appropriate method for the study of this phenomenon and it is the purpose of this paper to review developments of this technique. Although these methods have been in use in the EEG field for many years (Brazier 1965), it is our opinion that a review of their basic mechanisms would be useful for the following reasons. Firstly, there is a developing interest in them because of the increasing use of small laboratory digital computers in the EEG field. Secondly, papers which use these methods rely on a mathematical description or assume considerable insight into their mode of operation. It is our aim to describe these techniques on a purely physical basis.

Two variables are exactly correlated if the fluctuations in one are exactly matched by the fluctuations in the other. Their correlation coefficient is then unity. This remains true if the fluctuations are time dependent as in the case of EEG signals and the standard regression theory of discrete random variable statistics is still applicable. We must emphasize that at this point we are talking about a single correlation coefficient and not the so-called cross-correlation function. The value of the correlation is independent of the relative magnitude of the two signals because a true correlation coefficient is always normalized with respect to the r.m.s. value or standard deviation of the two variables.

This is in fact a common source of misuse and misunderstanding of the application of correlation to signals like EEG's. Many electronic devices described as correlators do not carry out this normalization process. The magnitude of their output will then depend on the magnitude of the individual signals. This is appropriate for some purposes but not for others. The correlation coefficient relating two variables is their covariance normalized by dividing by the product of their standard deviations. If this normalization is not carried out, the result is just the covariance of the variables, that is, the average cross product of two signals.

Although the two concepts of covariance and correlation as applied to time series were originally defined as separate functions by statisticians, unfortunately they were accidentally made synonymous by Norbert Wiener (1948) when he developed ideas based on the covariance function. Since electronic devices for measuring the covariance function arose from his work, they were often erroneously named correlation function instruments (Shaw 1967, Ziskin 1967).

These points are emphasized in Figs. 1 and 2. The first labours the fact that the covariance or average cross-product is an important measure when two signals are being compared [1]. Similar signals have

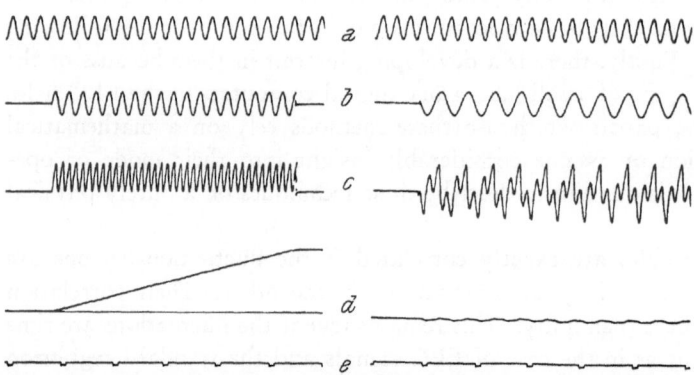

Fig. 1. This illustrates the computation of the cross-product and covariance of two signals. The case of two identical signals is on the left and two dissimilar signals on the right. (a) signal $x(t)$, (b) signal $y(t)$, (c) their cross-product $x(t) \cdot y(t)$, (d) the integral of the cross-product $\int x(t) \cdot y(t) dt$, and (e) a time marker. The final value of the integral in (d) is proportional to the covariance over the interval of integration

a finite covariance, dissimilar signals have a covariance of zero because their cross-product has both positive and negative deflections and has a mean value of zero. The second shows two signals of identical pattern together with their cross-product and running estimates of their covariance and correlation coefficient. This demonstrates that the covariance is dependent on amplitude, but the correlation coefficient is constant because of the normalization with respect to the r.m.s. values of the signals.

It seems likely then that a correlation measure could be used to

---

1 In this paper, signals are assumed to have zero mean as when they are recorded by conventional capacity-coupled EEG amplifiers. If this is not the case, the covariance is the average cross-product of the deviations of the signal amplitudes from their means.

measure synchronization. However, the same problems arise in the interpretation of correlation coefficients applied to time series as in the case of the discrete random variables of conventional statistics. This is illustrated in Fig. 3 which shows two quite independent signals —one is narrow band noise, the other is a sine wave—which have similar dominant frequencies. There are times when the signals have a high value of covariance and correlation. This can be compared

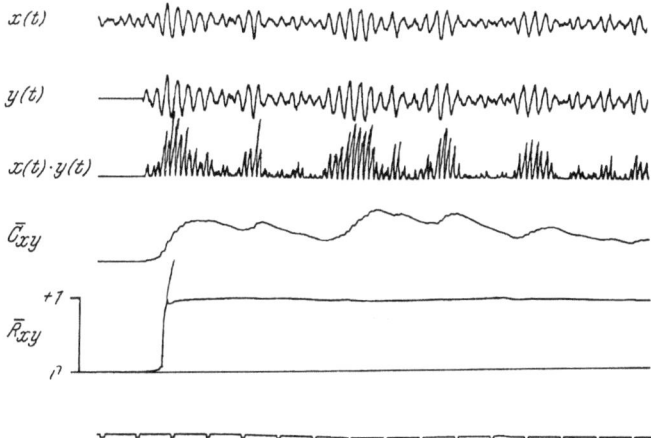

Fig. 2. This shows, from above down, two identical but fluctuating signals, their cross-product, a moving average estimate of their covariance $(\overline{C}_{xy})$, a moving average estimate of their correlation coefficient $(\overline{R}_{xy})$

with the opposite case of two signals which appear to be quite independent but which in fact do have some underlying correlation. This correlation can be detected by computation, but is not visable to the eye, as shown in the upper part of Fig. 4.

There is another characteristic of EEG signals which must be taken into account when considering their degree of correlation and synchronicity. Two signals recorded from different electrode sites may have similar patterns of fluctuation, but they are usually displaced in time. They are therefore not truly synchronous according to the dictionary definition of the term. This time delay or apparent propagation seems to be a fundamental feature of the genesis of electrophysiological activity. Now two signals which have an identical pattern of fluctuation will have a reduced correlation coefficient as one is displaced in time with respect to the other. Thus there are at least two parameters which are required to define the degree of synchrony between pairs of EEG signals. One is the similarity of their patterns of fluctuations, the other is the time difference between them. This time difference may be determined by reversing the time displacement procedure just described. The signals are stored in such a way that one can be displaced in time with respect to the other.

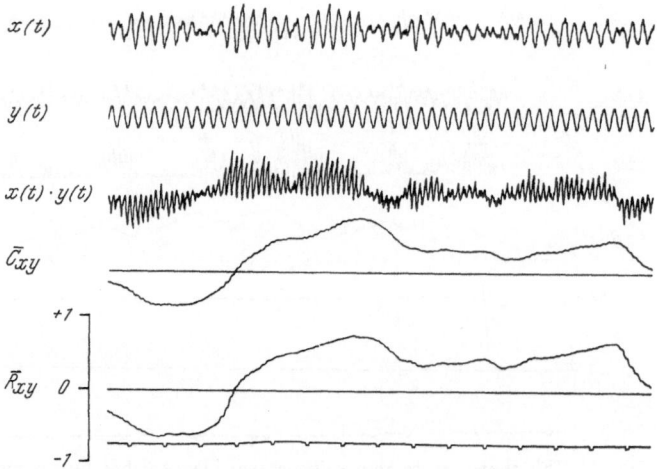

Fig. 3. The two signals $x(t)$ and $y(t)$ have similar dominant frequencies, but are otherwise quite independent. The moving average correlation estimate shows quite high values of correlation

If the correlation coefficient is now determined for a number of different values of time displacement between the signals, the resultant graph of correlation as a function of time difference will indicate the time displacement which corresponds to the maximum correlation coefficient. In this way both the correlation and time difference may be determined. This graph is of course the familiar cross-correlation function which is often the first introduction we have to the application of correlation to signal analysis. It is the correlation coefficient as a function of the time displacement between the two signals. We have already indicated that this function is often derived by an electronic device which does not normalize the result with respect to signal amplitude. In this case it is a covariance function.

The covariance function is still a useful function because it indicates time differences between signals and also its shape gives some information about the pattern of signals being compared. This is illustrated in Fig. 4 which shows two identical signals, $x(t)$ and $y(t)$, with

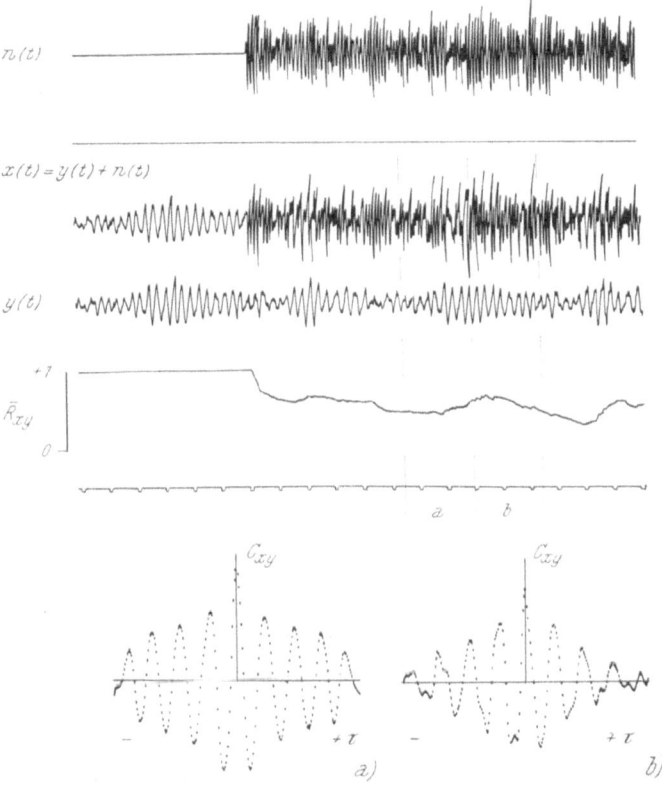

Fig. 4. The upper part shows the effect on the running correlation $\overline{R}_{xy}$ of adding the "noise" $n(t)$ to one of two channels recording identical signals. The lower part shows covariance functions from the epochs indicated. The time displacement scale $\tau$ is 10 ms per point

noise $n(t)$ added to $x(t)$. The moving average correlation coefficient shows a corresponding drop in correlation; but its value is still high. Visually one cannot see that $y(t)$ is contained in $x(t)$. However the covariance functions below show that the dominant frequency in $y(t)$ is still present in $x(t)$, and with no time shift. In Fig. 5 part of an

EEG recording shows a synchronous discharge preceded by low voltage activity with eye movements in the two frontal channels. A moving average correlation coefficient record for the first two channels shows a correlation close to unity throughout with no indication of the changing pattern of activity. The covariance function for epochs of 2.5 seconds duration from these two channels illustrates that the changing pattern is also detected. If these were correlation functions and not covariance functions, their maximum values would be close to unity as in the moving average correlation record. One difficulty with the interpretation of these functions is that they contain all the frequency components that are present in both of the original signals and so can themselves have quite complex patterns.

Another way of looking at signal correlation is in terms of variance. Consider the standard regression equation applied to signals. If we have a reference signal $y(t)$ and a test signal $x(t)$, $x(t)$ is the dependent variable and $y(t)$ is the independent variable. The linear association between $x(t)$ and $y(t)$ can be represented by the regression equation:

$$x(t) = b\, y(t) + n(t)$$

where $b$ is the regression coefficient, sometimes called the transfer ratio (Walter 1963), $n(t)$ is a component of $x(t)$ not correlated with $y(t)$. The variance of $x(t)$ is then compounded of a proportion accounted for by linear correlation with $y(t)$ and a residual. Now the square of the correlation coefficient is equal to the following ratio. The proportion of the variance of $x(t)$ accounted for by linear correlation with $y(t)$ to the total variance of $x(t)$. These notions are of course familiar in the context of conventional discrete random variable regression theory, but are less familiar in relation to signal analysis. They suggest that the correlation between pairs of signals may be analyzed in terms of covariance components just as the power spectrum analyzes a single channel into components of variance as a function of frequency. To see how this may be done consider the following case.

The spike makes very little contribution to the variance of a spike and wave discharge. However, removal of the spike does seem to have an effect on the covariance or correlation function which becomes less pointed on the positive peaks. Also, if the spikes have a different time relationship across different channels from that of the waves, correlation or covariance functions computed separately for these differing components would detect it. In this way the

distribution or synchrony of a complex pattern can be analyzed by
separating it into visually obvious components and examining the
correlation and time relations of each. Now of course distinct com-

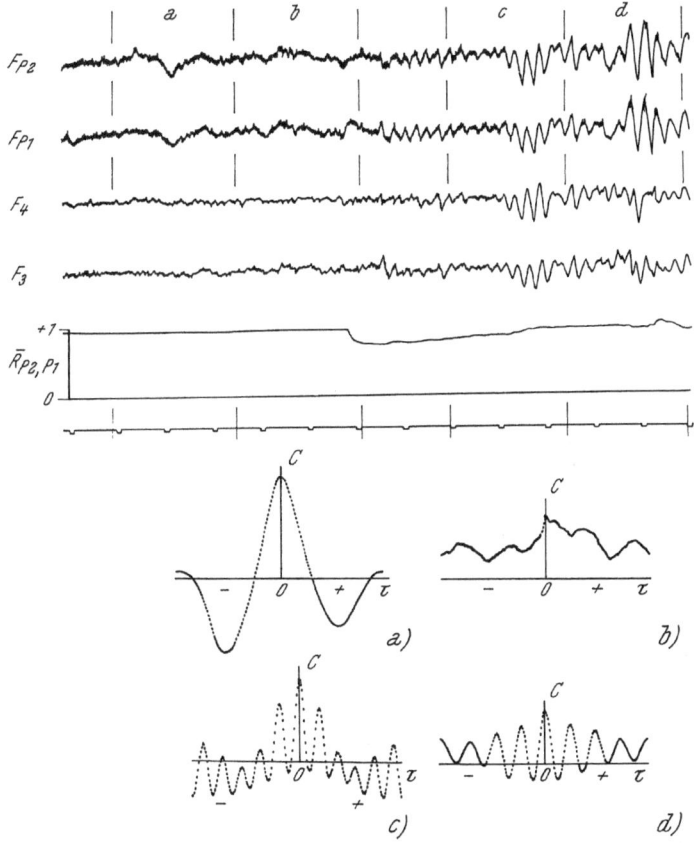

Fig. 5. The upper part of the figure shows a four channel average
reference EEG recording, together with a moving average estimate of
the correlation coefficient for channels 1 and 2. Below are four covariance functions for the epochs indicated. The time displacement
scale $\tau$ is 10 ms per point

ponents, as in the case of spike and wave are not always detectable
or separable. However, one can extend this process so that the correlation and time relationships are determined for chosen frequency
ranges—or in the limit, for each frequency component in the

14*

activity being examined. A rigorous analysis of this concept of measuring the correlation function as a function of frequency leads to two new functions which are currently of interest. These are the coherence function and the phase function. Their derivation is illustrated in Fig. 6. This shows correlation functions that could be

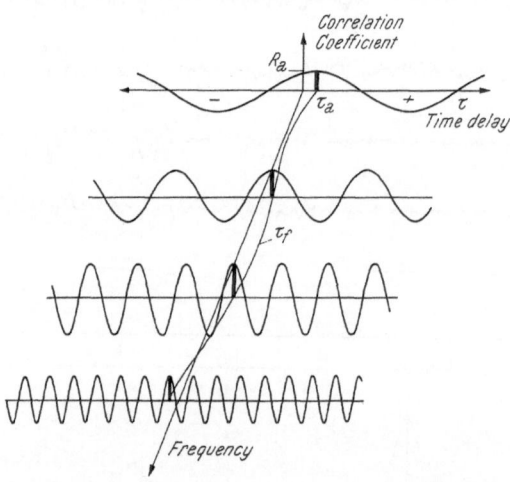

Fig. 6. A diagram to illustrate the derivation of the coherence function and phase spectrum. Correlation functions derived from hypothetical signals at four different frequencies in the shared spectrum are shown. The amplitudes of all possible such functions for the given signals represent correlation as a function of frequency—the coherence function. Their phases represent phase as a function of frequency— the phase spectrum. The phase spectrum for the case shown could be derived from the delay spectrum $\tau_f$ shown on the diagram

derived for some of the frequency components in the spectrum shared by two signals. They are pure cosine waves because they relate to the power shared between the two signals at discrete frequencies. If a series of such functions were obtained for all frequencies in the shared spectrum, a graph of their amplitude values would indicate the magnitude of correlation as a function of frequency. This graph is called a coherence function. Because the correlation functions are pure cosine waves, time can be expressed as phase. A graph drawn through the corresponding phase displacements is then a phase function or phase spectrum. If the correlation functions are replaced by covariance functions the coherence function becomes the cross power spectrum. Thus the coherence function and phase spectrum can, in theory, give all the information about the degree of synchrony between signals, their time differences, and indicate changing patterns as well.

The coherence function and the phase function are not normally computed via a series of correlation functions. This is why a more rigorous mathematical exposition gives little insight into the physical basis of these methods. Programming techniques for computing these functions have changed radically as a result of the introduction of the Fast Fourier Transform (FFT) algorithm which enables the Fourier transform of a function to be computed very rapidly (BERGLAND 1969, BINGHAM, GODFREY, and TUKEY 1967, GOLDBERG, SAMSON-DOLLFUS, and GREMY 1969). The preferred method is now to first obtain the sine and cosine terms of the spectrum of each signal using the FFT. The covariance function can then be obtained by a further transformation. One way to obtain the coherence function, is to use the sine and cosine terms to calculate the average inphase product and the average quadrature product, each normalized for amplitude, at each frequency. These are vectors from which a resultant vector can be computed as their vector sum. Its magnitude is the correlation, and its angle the phase, at the corresponding frequency.

One important problem with these functions is the determination of the significance of the derived values. This is made more difficult than in the case of correlating discrete random variables, because each signal has a certain amount of autocorrelation. This means that successive amplitude values of the signal are not independent, and it is difficult to assess the appropriate degrees of freedom. In the case of the cross-spectrum and the coherence function, some assistance with this problem may be derived from the accompanying phase spectrum. If, for example, a broad peak in the coherence function is accompanied by a random distribution of phase values at the corresponding frequencies, little significance can be put on the coherence. If, however, it is accompanied by an orderly phase spectrum at the relevant frequencies, particularly a linear phase frequency relationship, then some significance may be placed on the coherence. An example of this approach is given in the paper by PERONNET, LAVIRON, and SINDOU (1972) in this volume.

The computational aspects of the methods that have been discussed are described more fully by Dumermuth and Flühler (1967) and by Kleiner, Flühler, Huber, and Dumermuth (1970). They have already been exploited by Adey and colleagues in studies on both humans and animals (1969); by Brazier (1968, 1972) and by Peronnet, Laviron, and Sindou (1972) in studies from intracerebral recordings in man; and by Caille and coworkers in psychophysiological studies in man (Caille, Lelievre, and Monteil 1970), to name but a few workers using these methods. Our own research programme aims to use the coherence function to study EEG topography as a function of directed attention in differing sensory modalities. It would appear to be an appropriate method of analysis for the study of synchronization.

## Summary

Synchronization is a characteristic of the topography of the EEG. Its visual detection requires the observation of similar patterns of activity in several recording channels. Methods for analysing EEG topography are enumerated and it is suggested that correlation methods are the most appropriate for the measurement of synchronization. The various developments of correlation methods, including the coherence function, are described in a non-mathematical way. The difference between correlation functions and covariance functions is emphasized.

## References

Adey, W. R.: Computer analysis in neurophysiology. In: Stacy, R. W., and B. D. Waxman (eds.), Computers in biomedical research, Vol. 1, pp. 223—263. London: Academic Press. 1965.

Bergland, G. D.: A guided tour of the Fast Fourier Transform. IEEE Spectrum, pp. 41—52 (1969).

Bingham, C., D. Godfrey, and J. W. Tukey: Modern techniques of power spectrum estimation. IEEE Trans. $AU—15$, 56—66 (1967).

Brazier, M. A. B.: The application of computers to electroencephalography. In: Stacy, R. W., and B. Waxman (eds.), Computers in biomedical research, Vol. 1, pp. 295—315. New York: Academic Press. 1965.

— Electrical activity recorded simultaneously from the scalp and deep structures in the human brain. J. nerv. ment. Dis. *147*, 31—39 (1968).

Caille, E. J., D. M. Lelievre et A. C. Monteil: Information apportée par l'intégration en temps réel des spectres électroencéphalographiques au cour de la vigilance. L'Onde Electrique *50*, 423—428 (1970).

DUMERMUTH, G., and H. FLÜHLER: Some modern aspects in numerical spectrum analysis of multichannel electroencephalographic data. Med. and Biol. Engng. *5*, 319—331 (1967).

GOLDBERG, P., D. SAMSON-DOLLFUS, and F. GREMY: Analyse spectrale, Fast Fourier Transform. Agressologie *10*, 541—552 (1969).

KLEINER, B., H. FLÜHLER, P. J. HUBER, and G. DUMERMUTH: Spectrum analysis of the electroencephalogram. Computer Programs in Biomedicine *1*, 183—197 (1970).

DeMOTT, D. W.: Cortical micro-toposcopy. Med. Res. Engng. *5*, 23—29 (1966).

PERONNET, F., M. SINDOU, A. LAVIRON, F. QUOEX, and P. GERIN: Human cortical electrogenesis: stratigraphy and spectral analysis. See this book, pp. 234—262.

RÉMOND, A.: Level of organisation of evoked responses in man. Ann. N. Y. Acad. Sci. *112*, 143—159 (1964).

SHAW, J. C.: On the application of correlation theory to signal analysis. Med. and Biol. Engng. *5*, 407—409 (1967).

SHIPTON, H. W.: A new frequency-selective toposcope for electroencephalography. Med. and Biol. Engng. *1*, 403—495 (1963).

WALTER, D. O., R. T. KADO, J. M. RHODES, and W. R. ADEY: Electroencephalographic baselines in astronaut candidates estimated by computation and pattern recognition techniques. Aerospace Medicine *38*, 371—379 (1967).

— Spectral analysis for electroencephalograms: Mathematical determination of neurophysiological relationships from records of limited duration. Exptl. Neurol. *8*, 155—181 (1963).

WIENER, N.: Extrapolation, interpolation and smoothing of stationary time series. London: Chapman and Hall. 1949.

ZISKIN, M. C.: On the application of correlation theory to signal analysis. Med. and Biol. Engng. *6*, 453—455 (1968).

# Discussion

ELUL: Could you say a few words about the estimation of statistical significance of the results obtained with your method?

SHAW: The problem of estimating statistical significance is raised as soon as one considers correlation methods. What significance can one place on the results of correlation methods? Just to generalize this a bit for the people less familiar with the subject, the difficulty is to determine the degrees of freedom because you have considerable autocorrelation within each signal so that the values of signal samples are not independent of neighbouring ones. My own experience is not sufficient to be able to answer this one except to say that I try to overcome the difficulty by appropriate experimental design. There is some published material about this problem however.

FIRNEIS: Is there a connection between the concept of coherence functions and the usual power- and cross-powerspectrum? Which insight does the coherence function give, that the power-spectrum does not? Can the computation of the coherence function be facilitated when the power-spectrum is known?

Shaw: Yes, there is indeed a connection. The power spectrum relates to one signal; the cross-power spectrum relates to a pair of signals and expresses the covariance as a function of frequency, whilst the coherence function gives the correlation between two signals as a function of frequency. The power spectrum itself gives no information about the degree of dependence or independence between two signals. There are several ways of computing the coherence function. If the power spectrum is computed from the sine and cosine terms (for two signals), then these can also be used to compute the coherence function.

# Invited Discussion

## The Fast Walsh Transform:
## A New Method for Analyzing EEG Data

R. Trappl

Institute of General and Comparative Physiology
of the University Vienna, Austria

I want to draw your attention to a new method for analyzing continous recordings which has not—as far as I know—been applied to EEG-data. The only medical application published up to now which came to my knowledge is the recognition of types of waveforms in the electrocardiogram by Morgan (1971).

In the commonly used Fourier analysis amplitudes and frequencies of sine- and cosine-functions are calculated the summation of which approximates a given curve. In 1923 Walsh found other orthogonal two-valued functions ($+1$ or $-1$) which can also be used to approximate (nearly) all functions in a given interval, the quality of the approximation in a least squares sense being as good as by trigonometric functions. These functions which nowadays are called "Walsh-functions" can be ordered in a rather similar way like trigonometric functions, the difference being the use of "sequency" ($=$ average number of zero crossings per second divided by 2, abbr. "zps"; Harmuth 1968) instead of "frequency". The value of the Walsh-function wal $(k, j)$ of order $k$ ($=$ number of sign changes in the interval with length $N = 2^p$) in the $j$-th place of the interval may be defined by

$$\text{wal}\,(k, j) = \prod_{r=0}^{p-1} (-1)^{(k_{p-r} + k_{p-r-1})j_r}$$

$$j = 0, 1, 2, \ldots, N-1 \qquad k = 0, 1, 2, \ldots, N-1$$

where $j_r$, $k_r$ are the binary bits of $j$, $k$; i.e., $j = \sum_{r=0}^{p-1} j_r 2^r$, $j_p = k_p \equiv 0$, (Pratt et al. 1969). It is also possible to use particular names for even and odd Walsh-functions ("cal- and sal-functions", Pichler

1967); a comparison of both Walsh- and trigonometric functions is shown in Fig. 1 taken from a paper of BÖSSWETTER (1970).

Only few mathematicians took care of Walsh-functions until 1965 when COOLEY and TUKEY demonstrated an algorithm which reduced the average number of calculations necessary for performing a Fourier

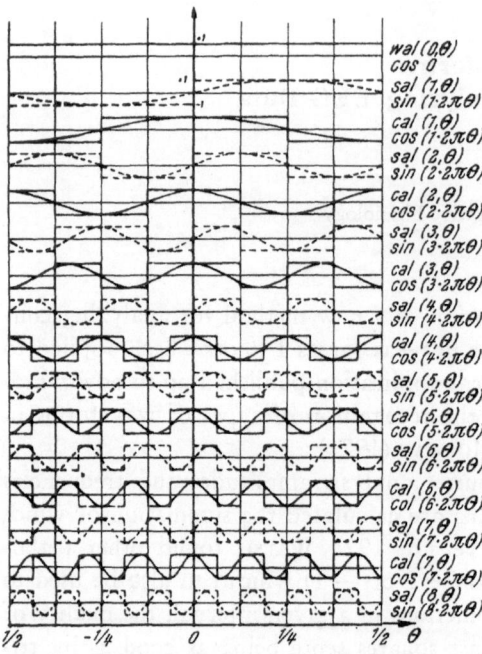

Fig. 1. Orthonormal systems of Walsh functions sal $(i, \theta)$, cal $(i, \theta)$ and trigonometric functions sin $(i \cdot 2\pi\theta)$, cos $(i \cdot 2\pi\theta)$ (BÖSSWETTER, 1970)

transform from $N^2$ to approximately $N \cdot \log_2 N$ ($N$ = number of sample points = number of coefficients). Though this "Fast Fourier Transform" (FFT) established an important progress (computation time reduced up to one fiftieth of the former one), the reduction in computation time was by far increased by basing the transform not on trigonometric but on Walsh-functions. The algorithm for this "Fast Walsh-Fourier Transform" (now "Fast Walsh Transform", FWT) finally given by SHANKS (1969) simplified all necessary computations to additions and subtractions which allow the use of integer variables in the computer programs, thus reducing computation time even more. The FWT therefore very often permits real-time and on-line processing, and this even on computers with smaller core memories, since the FWT occupies less data storage than the FFT (only $N + 1$ instead of $2 N + 1$). When comparing these important advantages with the only drawback, the loss of information about phase, the "success" of this sequency analysis performed by the FWT becomes clearly understandable.

Time does not permit to inform you about the possibility of calculating Walsh power spectra, dyadic autocorrelation functions, etc., for further details, besides computer programs for the FWT (*e.g.*, KREMER 1970), I only can recommend the book of HARMUTH (1970), the report of PICHLER (1970) and the proceedings of the Walsh Symposia (one in 1970, two in 1971).

## Acknowledgement

I want to thank Doz. Dr. PICHLER, Kepler-University of Linz, for his permanent most valuable advice on Walsh-functions.

## References

BÖSSWETTER, C.: Analog sequence analysis and synthesis of voice signals. In: Walsh Symp. *1*, 220—229 (1970).

COOLEY, J. W., and J. W. TUKEY: An algorithm for the machine calculation of complex Fourier series. Math. of Comp. *19*, 297—301 (1965).

HARMUTH, H. F.: A generalized concept of frequency and some applications. IEEE Trans. Inf. Theory *IT-14*, 375—382 (1968).

— Transmission of Information by Orthogonal Functions, 2nd printing. Berlin-Heidelberg-New York: Springer. 1970.

KREMER, H.: Algorithmen für schnelle Walsh-Transformationen. Forsch. Ber. Nr. 12 d. Inst. f. allg. Nachrichtentechnik d. Techn. Hochschule Darmstadt; Dez. 1970.

MORGAN, D. G.: The use of Walsh-functions in the analysis of physiological signals. In: Walsh Symp 3 (1971).

PICHLER, F.: Das System der sal- und cal-Funktionen als Erweiterung des Systems der Walsh-Funktionen und die Theorie der sal- und cal-Fouriertransformation. Phil. Diss. Univ. Innsbruck, 1967.

— Walsh-functions and linear system theory. Dept. of Electr. Eng., Univ. of Maryland, Rep. T-70-05 (1970).

PRATT, W. K., J. KANE, and H. C. ANDREWS: Hadamard transform image coding. Proc. IEEE *57*, 58—68 (1969).

SHANKS, J. L.: Computation of the Fast Walsh-Fourier Transform. IEEE Trans. Computers *IC-18*, 457—459 (1969).

WALSH, J. L.: A closed set of normal orthogonal functions. Amer. J. Math. *45*, 5—24 (1923).

*Walsh Symposia*

1. Applications of Walsh Functions.  Symposium and Workshop.  Held at Naval Res. Laboratory, Washington, D. C. March, 31–April 3, 1970.
2. Applications of Walsh Functions.  Symposium.  Held at Departmental Auditorium, Constitution Ave, Washington, D. C. April 13–14, 1971.
3. Theory and Applications of Walsh Functions.  Symposium.  Held at The Hatfield Polytechnic. June 29–30, 1971.


# Discussion


Shaw: Walsh functions are also being used in England, but I understand that they are even more difficult to interpret in physiological terms.

Firneis: If the continuous sine and cosine functions are dispensed with in favor of Walsh-functions, do not the discontinuity-jumps introduce numerical ill-conditioning of the problem?

Trappl: No, they don't. (To Shaw): I do not think that Walsh-functions may raise additional difficulties of interpretation in physiological terms since both Walsh and sine-cosine functions are uninterpretable.

Firneis: But if you consider error-estimates, then the existence of higher derivatives is presumed in order to obtain them.  And exactly that fails at the discontinuity-jumps.

Trappl: First of all, not being able to calculate error-estimates when using Walsh-functions would not seem to be a great drawback since nobody seems to have done that for Fourier coefficients when analyzing EEG. However, it is possible to calculate error-estimates as has been done by Polyak and Shreider even in 1962; besides this Pichler has introduced a continuous representation of Walsh-functions.

# Temporal Distribution of Interictal and Ictal Discharges from Penicillin Foci in Cats

F. ANGELERI, S. GIAQUINTO, and G. F. MARCHESI

Clinica delle Malattie Nervose e Mentali dell'Università di Perugia, Perugia, Italia, and Laboratorio di Cibernetica del C.N.R. Arco Felice, Napoli, Italia

Aside from AJMONE MARSAN's recent findings on the phenomenon of the transition at a cellular level (1965), there has, until now, been no definitive study of the temporal relationship between random interictal spikes and rhythmic paroxysms.

In 1958 RALSTON stated that a favourable condition for rhythmic ictal organization is established when spikes are followed by an after-discharge. However, it is well documented that in a penicillin focus the spike and after discharge complexes may go on for quite a while without provocation of a seizure. In certain experimental foci an intensification of interictal spikes has been observed when a seizure is about to begin (WALKER 1950). In other cases, however, this pattern does not appear. For example, strychnine spikes maintain a constant interictal recurrence even though they may appear in bursts of two or more elements.

Our earlier work (ANGELERI et al. 1970) was aimed at the quantification of interictal and ictal activity as well as the evolution in time of EEG foci in forty patients with documented temporal lobe epilepsy. In these cases we counted interictal spikes in twenty second increments during the entire period of nocturnal sleep. In every case a clinical seizure occurred during the period of sleep.

Although the interictal discharge counted at a distance in time from the onset of the electroclinical seizure consistently presented a cyclic trend, the spikes close to the seizures presented the following three patterns in the 5 to 10 minutes just preceding the onset of the seizure: an increase in frequency observed in 20% of the total crises examined; a decrease in frequency in 10%; a cyclic trend similar to the one observed two hours before and after the seizure in 70%.

The present work is aimed at greater precision in defining the evolution and the time distribution of epileptic focal discharges. We have, therefore, decided to make use of experimental penicillin foci since their anatomical locations can be exactly defined; we have also introduced a spike detecting and counting system in order to obtain a greater number of data and to make possible statistical elaboration through a digital computer.

## Methods

Our monopolar recordings were carried out in the anterior portion of the lateral gyrus in both hemispheres of six adult cats under slight nembutal narcosis. An epileptic focus was produced in one hemisphere through the topical application of penicillin. In each case regular interictal spikes were well established within thirty minutes. The electrical activity of the primary and secondary foci was recorded without interruption for the entire duration of the epileptic activity of the focus.

The evolution of a penicillin focus is shown in Fig. 1. Statistical calculations on interictal and ictal spike trends were subsequently carried out and the following classes were studied: 1. the interictal spikes in the period prior to the first seizure, the pre-ictal period; 2. the interictal spikes between one seizure and another; 3. the interictal spikes following the seizure period; 4. the rhythmic spikes of the seizure periods. As we shall mention later, the counting was carried out in some cases on the interictal spikes of the pre-ictal period and on the remaining spikes.

Fig. 2 shows a schematic representation of our apparatus for detecting and counting spikes. The electrical activity was simultaneously

---

Fig. 1. Four fragments from the continuous recording of an experimental penicillin focus in cat. 1. Pre-ictal tracing forty minutes after the application of penicillin. 2. Appearance of the first seizure approximately 120 minutes from penicillin application. 3. Period of maximum ictal activity approximately 140 minutes after the appearance of the first seizure. 4. Post-ictal period showing that about 300 minutes from the beginning of the experiment the ictal activity disappears while interictal spikes continue for as long as 1 or 2 hours before vanishing. L.G.r.: recording from the anterior portion of the lateral gyrus. Second channel: the same tracing filtered for high frequency and a time costant before input to the detecting and counting system. Third channel: monitor of total pulses detected. L.G.l.: recording from the anterior portion of the lateral gyrus as a control mirror focus

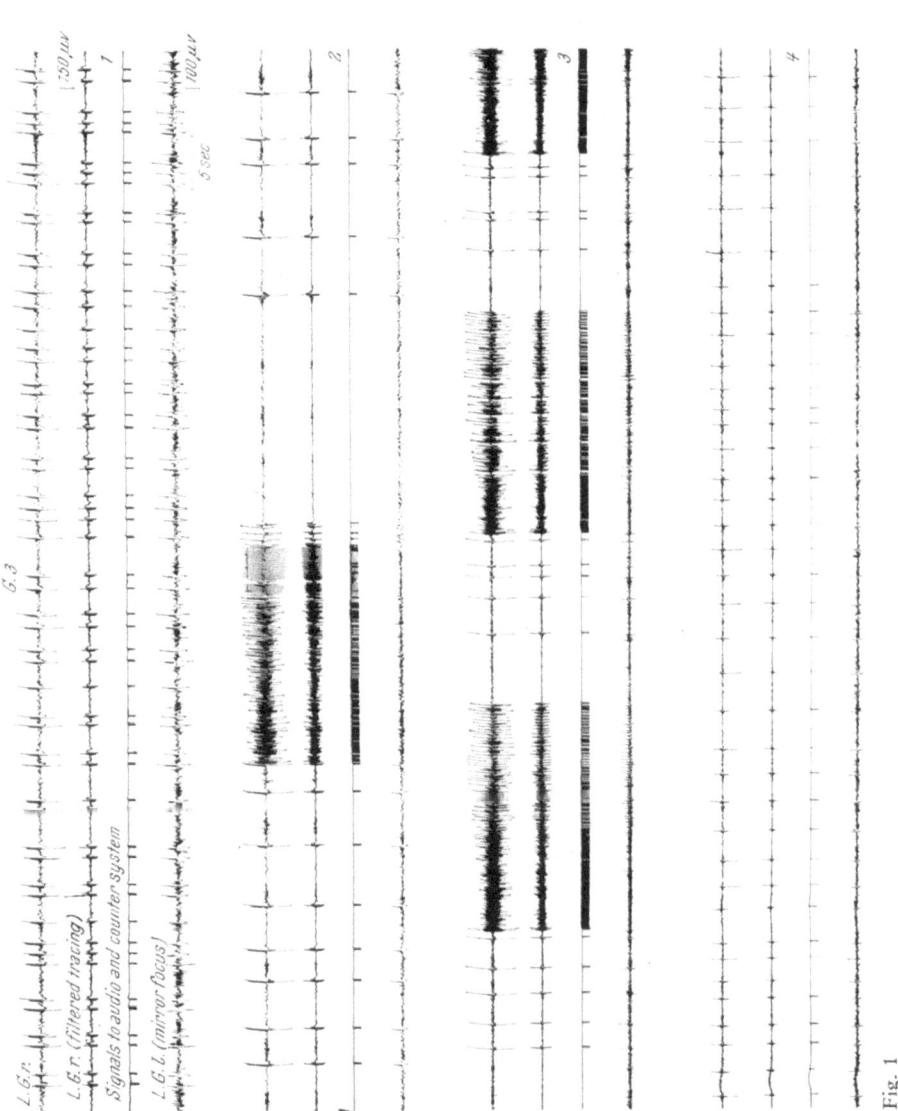

Fig. 1

recorded on paper and tape. The analyses were carried out after having set up the number of windows and the millisecond duration for each window.

At the bottom of the figure is a diagram of the function performed by our spike detector. This instrument recognizes the maximum and minimum troughs and crests of each wave and determines the rise

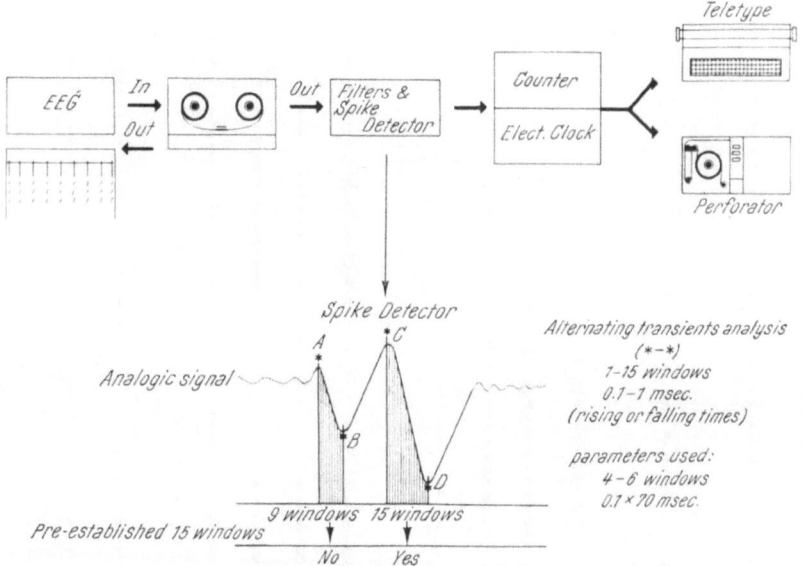

Fig. 2. A schematic representation of the instrument used in these experiments and an example of the electronic analysis system used on the tracings for spike recognition

or fall times of every other transient by means of a series of windows.

The windows are pre-selected so that we can discriminate fast potentials through their frequency and amplitude. A hypothetical example of such an electronic selection is also illustrated.

Although segments $\overline{AB}$ and $\overline{CD}$ have the same fall time, the circuit can discriminate them when the fifteen windows are pre-estabilished. Thus the first segment passes through only nine windows while the second passes through all fifteen and so is transformed in a single pulse.

Counts were continuously collected from ten or sixty second periods in all recordings. However, in the ictal period, the sixty seconds prior to the crisis were counted at five second intervals. The precise durations of each seizure, of each post-ictal spike silence, and of each interictal period were measured. Analyses of these data were performed on a digital computer.

## Results

*a) Statistical Analysis of Spike Frequency during the Period of Focus Evolution*

More than two hundred thousand interictal and ictal spikes were counted in the six recordings of our experiments. The figures obtained by ten second counting in each experiment were divided according to the evolution of the focus in the three classes of numbers we have described.

The following analyses were made for each class: 1. frequency distribution; 2. serial array (for the first class only); 3. run test; 4. joint frequency distribution; 5. ninetieth order correlogram.

Interictal Spikes Preceding the First Seizure

The histogram of the frequency distribution is generally unimodal with a peak corresponding to four to six spikes per ten second period. We observed a direct relationship between the frequency distribution and the intensity of the successive ictal periods. Peaks at six spikes per ten seconds, for example, occur in experiments in which numerous and long seizures are observed. The overall evolution, as seen in the serial array, consists of quite a steady activity, which begins to oscillate five or ten minutes before the onset of the seizure, with a bottom as low as zero. The run test consistently gave negative results for the hypothesis of stationarity. The lack of the randomness is confirmed by the joint frequency distribution and by the correlogram in which high correlation coefficients slowly decay to zero.

Ictal Spikes

Although we performed the same statistical tests on the class of numbers obtained in counting ictal spikes, because the samples fell into mixed periods of ictal and interictal discharges, the elaboration of these data cannot be considered significant. However, for the sake of completeness, we have included the following results. The frequency distribution histograms appeared to be either symmetrically bimodal or frankly gaussian. The run test gave weak negativity for the hypothesis of stationarity or even positive results. The correlation among data, as seen from the joint frequency distribution and the correlogram, was much lower than in the first class.

Remaining Interictal Spikes

The frequency distribution histograms have always shown the highest peaks corresponding to the lowest classes of values. Generally the predominant peaks are two, one collects post-ictal silence, that is, the zero values, and the other represents the characteristic norm of firing, which was found to be around two spikes every ten seconds. The run-test was negative, but, as seen in the previous class, the joint frequency distribution and the correlogram yielded correlation values which are much lower than those found for the interictal spikes preceding the first seizure.

We compared the results of the above mentioned statistical analyses using information we obtained from the first class of numbers (pre-ictal spikes) and from the third class (remaining interictal spikes) since the data were homogeneous both with respect to the samples and physiopathology.

In all experiments we found a high correlation in the first class of numbers. These correlations were far less pronounced in the third class. This indicates that when the first seizure appears and its activity develops, the interictal spike trend undergoes a change in its temporal distribution (Fig. 3).

*b) The Occurrence Distribution and Duration of Seizures in Penicillin Foci*

The second point we examined concerned the interictal spike trend in the moments just prior to each seizure. Our findings in this respect differ from those of Ralston. In fact, in our tracings spike and afterdischarge complexes may be observed long before the appearance

---

Fig. 3. Joint distribution, 90th correlogram and run test results on the ten second counts of the preictal period (top). Results of the same statistical tests on the ten second counts of the second class of numbers, ictal spikes, (middle). Results of the same statistical tests on the ten second counts on the interictal spikes after the onset of the first seizure (bottom). A comparison may be made between the first and the third classes of numbers, indicating that after the onset of the first seizure the correlation of the figures of the third class is decreased with respect to the first

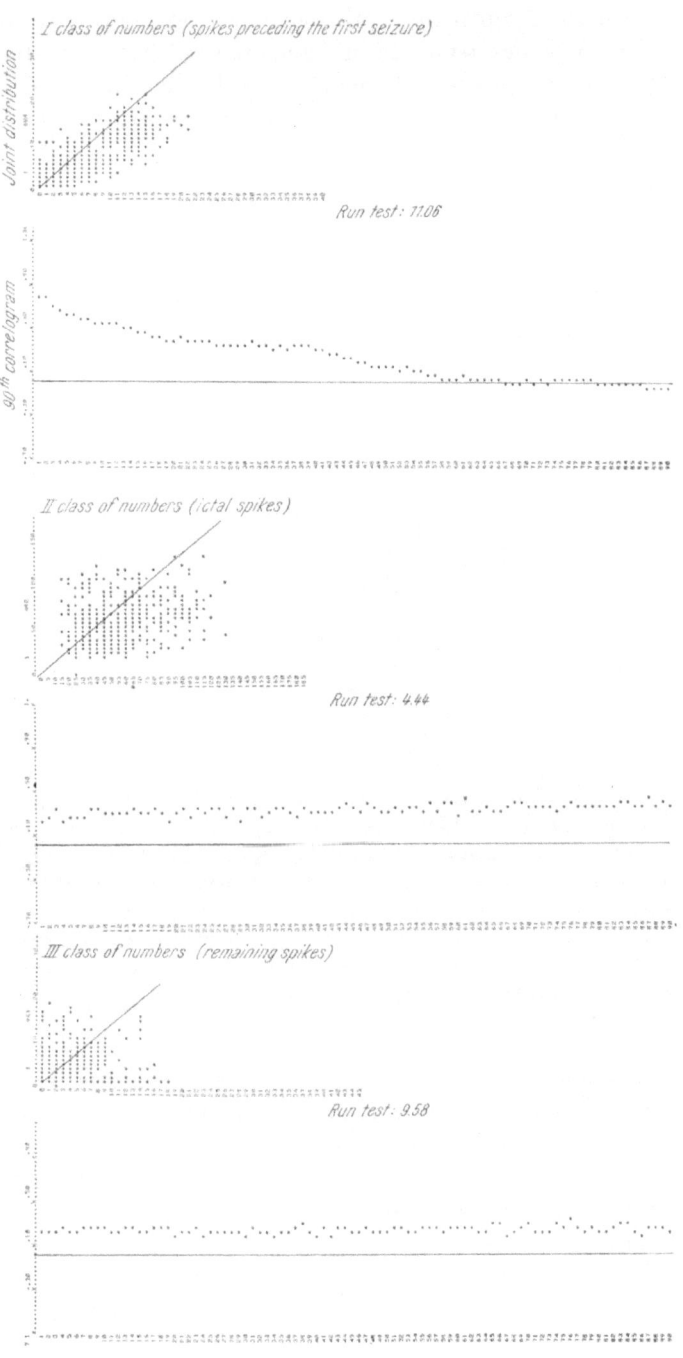

Fig. 3

15*

of a seizure. Furthermore, they may also occur in mirror foci having no autonomous ictal activity and may be absent in the intervals between one seizure and another. On the other hand, we found that two to three spikes constantly appear before the onset of each seizure.

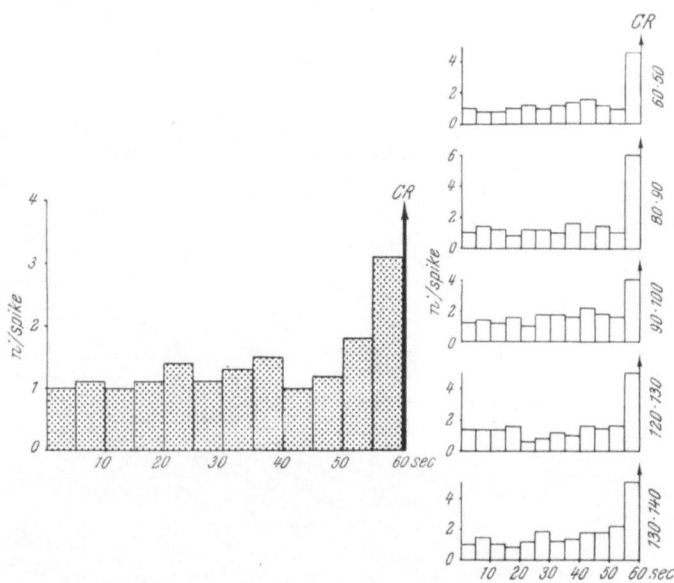

Fig. 4. The dotted histogram represents the average interictal spike frequency per 5 second increment in a total of three hundred seizures from four penicillin foci. The blank histograms represent the average of interictal spike frequencies from the 5 second increments of ten-seizure groups as indicated on the right. The increase of spike frequency in the 5 seconds preceding the onset of the seizure is evident in the dotted histogram and such an increase is a constant phenomenon as we can see from the averages of the ten-seizure groups

This finding led us to a closer evaluation of the frequency of spikes occurring in the sixty seconds immediately preceding the seizure. In Fig. 4 the dotted histogram on the left summarizes the averages of the five second spike increments in the sixty second pre-crisis period. We can see that the spike averages of the first fifty-five seconds of the interval are relatively constant, but that in the five seconds before the moment of crisis there is a significant elevation of spike frequency. To confirm this characteristic numerical property we did not find it necessary to perform more rigorous analyses since, as we can see on the right, these averages formed a constant pattern.

In order to come to a better understanding of overall trends in our tracings we next compared the duration of interictal periods, the

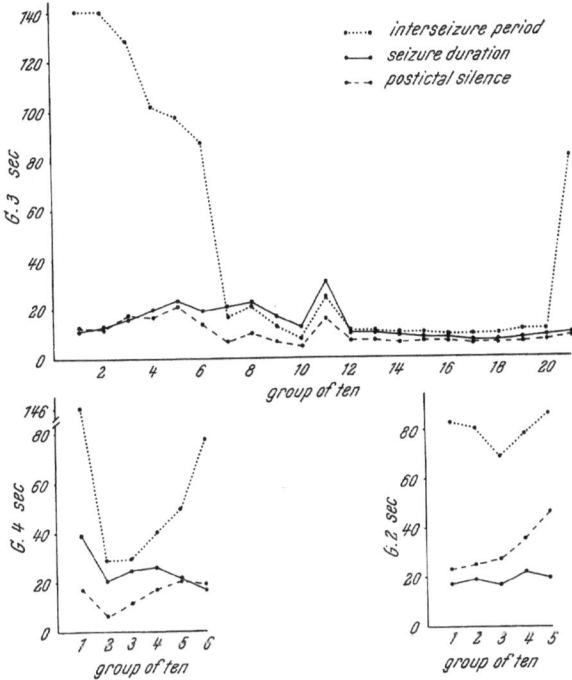

Fig. 5. The curves were traced from the average values taken from ten-seizure groups measured sequentially from the first to the last seizure in three experiments according to the three parameters indicated in the graph

duration of the seizures and the duration of the post-ictal spike silences. These comparisons are summarized in Fig. 5.

In all experiments, as in the three shown here, there was a moment of maximum ictal activity represented by a drop in the dotted line which corresponds to the shortest time interval between two seizures. The units of the abcissa are tens of seizures in the order of their occurrence. The ordinate's units are taken as the duration of each interval in seconds. The period of maximum ictal activity is both preceded and followed by an attenuation which is expressed by longer interictal intervals. The lengths of the seizures represented by the solid line and the duration of the post-ictal spike silence, dashed line, have proportional values if we consider their averages.

## c) Interval Analysis

Another approach to defining the ictal-interictal relationship lies in the analysis of the time intervals between spikes. Fig. 6 illustrates

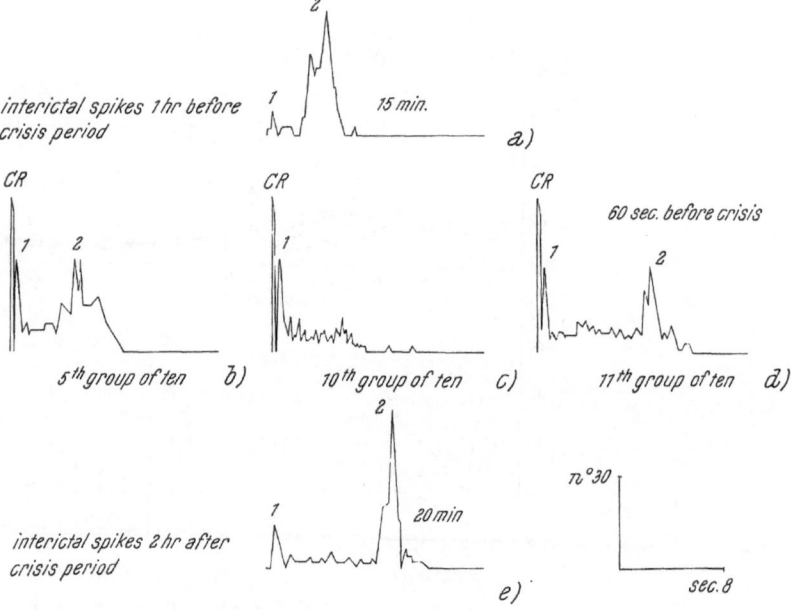

Fig. 6. *a*) and *e*) show the interval histogram of spikes taken from periods before and after those characterized by the presence of the seizures [*b*), *c*), *d*)]. These last three histograms were obtained by summing ten 60 second periods before the seizures. In *a*), *b*), *d*) and *e*) it is possible to recognize a characteristic trend showing two peaks in frequency (1 and 2), absent in *c*)

three patterns of spike intervals: the immediate pre-ictal period, the relative pre-ictal period and the post-ictal period. Random fifteen or twenty minute samples of tracings were examined for the relative pre-ictal (curve *A*) and post-ictal (curve *E*) periods, while in the immediate pre-ictal period (curves *B*, *C*, *D*) tens of sixty second groups were used. The significance of these data may be simplified by comparing curves *A* and *E* which are similar in their having two peaks. The first one represents the same interspike interval density in the two periods which were examined. It is a short time interval (less that 0.4 seconds) that totals the temporal distance between double spikes. The second peak may be considered the norm of spiking in that period. Before the first seizure (curve $A$) the distance in time between interictal spikes is about 1.5 to 2.5 seconds; after the seizure period (curve $E$) such a norm decreases to 4 to 5 seconds. This confirms the results obtained by the frequency distribution of spike counts in 10 second increments before and after the seizure period.

In the middle of the figure we can see three curves which represent the interspike time distribution preceding several crises. The two curves $B$ and $D$ also have two peaks. The first one corresponds to the intervals between coupled spikes immediately preceding the seizure. The second peak demonstrates that in the interictal period, too, there is a norm of interictal spiking, a norm which is lower than in the pre-ictal period. Yet when the interval analysis (curve $C$) is carried out on very short interictal periods the second peak is absent; this may imply that the spiking norm cannot be reached in such conditions.

## Discussion

The use of penicillin in experimental epilepsy insures a precise anatomical focus. Such a precise localization is lacking in the EEG alterations in human epilepsy. However, even under experimental conditions we find limitations which, for the most part, involve recording technique.

For example, spikes recorded from the depths of a cortical focus may be found absent when the electrode is placed on the surface. Another difference in the tracings lies in the fact that although surface and deep recordings may both register simultaneous spikes, a discrepancy in their amplitude and shape is frequently found (PETSCHE and ŠTERC 1968, PETSCHE and RAPPELSBERGER 1971).

Although it is very likely that other synchronization phenomena occur in the focus, we are concerned here only with the spikes recorded from surface macroelectrodes. Spike recognition and counts discriminated only their occurrence rather than their form and amplitude. We have taken into consideration the distinct possibility that the mechanisms generating surface recognizable spikes can be independent of other mechanisms of synchronization which are not producing surface recordable discharges.

The mechanism sustaining surface spike activity may at times recruit varying numbers of neurons, resulting in a greater or lesser spike amplitude even though the generation of activity may remain unchanged. The form in which the spike fires in the tracing suggests that this is a unique phenomenon.

We have, therefore, examined the temporal trends more closely in order to determine whether their activity is presenting a characteristic rhythm and time distribution, and whether a relationship exists between interictal spikes and rhythmic ictal spikes. If we were to find a norm within this system we could propose an autonomy in its function which would constitute indirect support of our hypothesis.

Although our present data have been obtained from only six experimental foci, the consistency of our results and their analogy in all the experiments permits the following interim physiopathological interpretation.

1. The temporal distribution of interictal spikes is not statistically stationary. Furthermore there is a characteristic frequency distribution of the discharges counted in ten second periods. When the seizure first appears in the focus two characteristic changes are seen in the temporal distribution of interictal activity. The correlation within the numbers of spikes counted in the ten second intervals decreases with respect to the pattern of firing prior to the seizure. The frequency distribution and the interval analysis show that after the first seizure the norm with which the interictal spikes occurs decreases. Furthermore, in the serial arrays, we found oscillations in the spike frequency occurring five to ten minutes before the first seizure.

These results indicate that the tests we employed were sensitive in showing changes in our experimental epileptic activity. Such results indicate relationships between interictal spikes and ictal activity. Within certain bounds, when the seizure mechanism is activated, a disorder in the occurrence of interictal spikes is realized. It is significant that, under certain circumstances, this firing disorder of interictal activity can anticipate, for some minutes, the appearance of the first seizure.

2. The interictal spikes of the sixty pre-ictal seconds maintain a fairly constant average temporal distribution for the first fifty-five seconds. However, their frequency increases, doubling or even tripling itself, in the five seconds preceding the seizure; a burst of two to three spikes very often precedes the seizure's rhythmic discharge.

It is possible that the increase in spike frequency acts as a trigger for the seizures, or that this intensification depends on the rhythmic mechanism which both initiates and maintains the seizure. In this second case the mechanism which starts the seizure may actually precede its onset by a few seconds, acting also on the generation of interictal spikes.

3. The postictal silence of interictal discharges shows another relationship between their occurrence and the appearance of seizure activity. Such silence is constant after all the seizures and its duration is proportional to the length of the seizure. This silence does not affect the background activity. It is likely that the brake mechanism is acting not only on the rhythmical discharge of the seizure but also on the interictal activity for several seconds after the crisis has ceased.

4. A comparison of interictal and ictal discharges shows that when there is a predominance of rhythmic ictal activity the frequency of interictal spikes decreases with respect to the spiking prior to the seizure period. This decrease in the frequency of interictal activity is linked to the progressive diminution in the length of inter-seizure periods. These intervals become shortest when the maximum focal activity is recorded. If the inter-seizure period is longer than one or two minutes the spike pattern returns to normal firing.

From our present findings we can foresee the possibility of designing a model for the behaviour of epileptic discharge from EEG recordings. Our aim is directed toward a standard for comparing foci of differing origins and for evaluating the properties of a particular focus under variations in pharmacological and physiological conditions.

## Summary

The present study is based on continuous recordings of epileptic activity from penicillin foci in the anterior portion of the lateral gyrus of six adult cats. These recordings followed the entire temporal course of evolution and disappearance of discharges. Tape recordings permitted subsequent analysis and short term counts of interictal and ictal spikes as well as an interval analysis. The data obtained were transcribed by a teletype machine and perforated on tape for a series of statistical tests subsequently performed on a digital computer. Our results are as follows:

1. The interictal activity preceding the first seizure has a high degree

of correlation which diminishes in the subsequent interictal periods. There is a parallel drop in the predominant firing norm.

2. The first seizure is heralded by oscillations in the frequency of interictal spikes which begin to occur five to ten minutes prior to seizure activity. In the same way, each seizure is coupled to bursts of interictal spikes.

3. The post-ictal spike silence, which does not reflect an electrical extinction, is constant for every seizure and appears to be a function of seizure duration.

4. The appearance of the ictal period induces a decrease in the interictal spiking norm as indicated by the analysis of the frequency distribution of spikes and of interspike time intervals before and after the onset of the seizure.

## References

AJMONE MARSAN, C.: Micro-structural mechanisms of seizure susceptibility. In: SERVÍT, Z. (ed.), Comparative and Cellular Pathophysiology of Epilepsy, pp. 47—59. Prague: The publishing house of the Czechoslovak Academy of Sciences; Excerpta Medica Foundation. 1966.

ANGELERI, F., G. F. MARCHESI, A. FERRONI, and P. BERGONZI: Registrazione elettroclinica delle crisi temporali e scariche intercritiche nel sonno notturno. Riv. di Neurol. 40, 321—327 (1970).

PETSCHE, H., and P. RAPPELSBERGER: Spatio-temporal and laminar analysis of self-sustained cortical activity. (In press.)

— and J. ŠTERC: The significance of the cortex for the travelling phenomenon of brain waves. Electroenceph. clin. Neurophysiol. 25, 11—22 (1968).

RALSTON, B. L.: The mechanism of transition of interictal spiking foci into ictal seizure discharges. Electroenceph. clin. Neurophysiol. 10, 217—232 (1958).

WALKER, A. E.: Convulsive activity. Quart. Phi. Beta Pi. 47, 108—115 (1950).

# Human Cortical Electrogenesis:
# Stratigraphy and Spectral Analysis

F. Peronnet *, M. Sindou, A. Laviron *, F. Quoex *, and P. Gerin *

Unité de Recherches de l'INSERM sur la physiopathologie du système nerveux, Bron, France, et Service de Neurochirurgie, Hôpital Neurologique, Lyon, France

## Introduction

The most important results concerning the mechanisms of cortical electrogenesis have been achieved with animals thanks to transcortical bipolar investigations (Bishop and Clare 1952, Calvet and Scherrer 1961 a, b, Calvet et al. 1964, Jami et al. 1965, Fourment et al. 1965). These researches revealed that only the technique of transcortical derivation could give an accurate account of cortical activity. This is in accordance with the histological radiate structure of the cortex. The best proof of this is the parallelism between the phenomena of EEG type that this technique records and the unitary activity of neuronal volume explored. Moreover stratigraphic explorations enable these phenomena to be linked with the existence of generators. Experiments on human beings have been made with simple transcortical derivation by Hirsch et al. (1961, 1965, 1966) in particular. To our knowledge there have not been stratigraphic experiments with humans.

This work offers a strati-corticographic study of the human cortex, using electrodes with poles in tiers, limited to the study of slow EEG type phenomena and to their relation with cortical histological structures; the results are evaluated on a small computer, with a technique of spectral analysis.

## 1. Technique

### 1.1. Neuro-Surgical Operations

These strati-corticographic explorations were made during *neurosurgical operations* for cerebral tumours, intracerebral hematomas,

---

* Chercheurs INSERM.

traumatic contusions, vascular ligatures and biopsies. We made every effort not to be prejudiced in the choice of patient. We never modified the operative indication, the anaesthetic technique or the type of operation. The only drawback was the increasing duration of the surgical act from fifteen to twenty minutes. The patients underwent the operation under general anaesthesia with barbiturates and trifluoro-chloro-ethane (Fluothane[R]).

## 1.2. The Electrodes

The *electrodes* are made with enamelled silver wires the diameter of which is 150 micron. We made four- and six-pole electrodes. The extremity of each wire is bevel-edged so as to increase the area of contact and thus reduce the resistance of the electrode. We use a dissecting microscope for the cutting, the verification of the isolation and the disposition of the different wires at interpolar distances of 1.3 mm for the 4-pole electrode and of 1 mm for the 6-pole electrode (the two extreme poles are 4 mm apart for the 4-poles electrode and 5 mm apart for the 6-pole electrode).

The electrode is inserted into a support, a metallic tube 3 cm long so that the upper pole of the electrode is level with the end of the tube (see inset Fig. 1). The wiring is stiffened with a varnish; the electrode is completed by the addition of a Plexiglass plate, 1.5 mm thick, 10 mm long and 7 mm wide. A hole is made in its center in which the tube is inserted so that the lower end is level with the plate. The tube and the plate are held in position only by friction so that the Plexiglass plate, which maintains the stability of the electrode on the cortical area when the approach is adequate (cases of craniotomies with osteoplastic volet) may be removed when the zone to be recorded is too narrow (craniotomy with trephine hole only, for example).

The electrode-plug connection is made with enamelled copper-conducting wires the diameter of which is 150 micron and the length is from 1.5 to 2 m; this distance is sufficient for the connecting box to be placed far enough from the operative field. The connection is very flexible, so that the setting in motion of the connection zone is not transmitted to the electrode.

The whole electrode is protected by a cap fixed on the socle with rubber bands, put in a test-tube, obturated with a wad of gauze and placed in an envelope in polythene. Sterilization is carried out with oxide of ethylene, its efficiently being controlled by a colored indicator. This must be done some days before, so that the disinfectant product, which is toxic, is eliminated. This method seems satisfactory to us since we have never had post-operative infection. The electrode is checked with an ohm-meter and its chlorination is regularly renewed.

### 1.3. The Recording

It is made on the one hand on a portable ECEM T 3 type electro-encephalograph, on the other hand on FM-magnetic-tape with a four-channel thermionic recorder T 3000. The recording parameters are usually those of clinical electroencephalography (high frequency filter 35 or 50 Hz; time constant 0.3 sec; speed: 15 mm/sec; calibration: 50–100 $\mu$V/cm).

The recordings are made in the middle of a gyrus and at a distance from the sulci that limit the convolution in question. For each im-planted electrode, the activity of the cortex is recorded only after one or two minutes at least, so as to obtain a stabilization of the electric phenomena. Each recording includes three fractions of one minute, separated by a "word" (e.g., a square wave) so that they may be perfectly located on the tape.

Whenever it was possible, we associated with the strati-corticographic exploration a pre-operative EEG study, a peroperative EEG record-ing being obtained with pin-electrodes inserted into the scalp.

This record enables us to judge electroencephalographic modifications induced by anesthesia. The degree of anesthesia usually corresponds to the second stage, sometimes to the third stage, of Courtin's classifi-cation (in BRECHNER et al. 1962, SADOVE et al. 1967). The second stage is distinguished by rhythms that are distributed over the whole scalp, of an amplitude from about 100 to 150 $\mu$V and of a frequency ranging from 6 to 12 c/s. In some cases, and for comparison, electro-corticographic recordings using the technique of mono- or bipolar derivation on the surface were made.

### 1.4. Histological Study

Whenever the region in question could not be retained, a cerebral biopsy of the recorded zone was made.

Electrical data were in all cases compared with histological data of the region studied. They result either from the biopsy of the region under observation whenever possible, or, in other cases, from in-formation given by VON ECONOMO's (1925) atlas. In this connection it is to be noticed that the cerebral parenchyma undergoes a retraction due to a fixation. We measured this coefficient of retraction, it is about 6%.

## 1.5. Evaluation of the Data

This is made both on the record through direct observation and on the tape with a computer.

### 1.5.1. Direct Observation

In the first stage the study of cortical activities includes: a descriptive analysis of the principal activities that have been encountered and the determination of their frequencies and their respective amplitudes concerning the whole transcorticographic activity as well as each of the stratigraphic activities.

In a second stage, this information is compared with anatomo-histological parameters, which are the precise cortical area, the thickness of the grey matter, the structural type and the functional state of the recorded zone and its situation in relation to the lesion.

### 1.5.2. Spectral Analysis on Computer

With the object of specifying and concentrating the results and in the hope of getting further information about the rhythms and their mutual relations, we analyzed parts of the recordings with a technique of spectral analysis on the basis of correlation functions. We made use of a programme written by one of us and using a PDP 8/I. This programme is subdivided into programmes of auto- and of cross-correlation (Laviron 1971 b).

The programme of auto-correlation computes on-line the function of auto-correlation of a signal for a given period of integration and the power spectrum (Fig. 1, spectra) by means of Cooley-Tukey's algorithm. Before computing the spectrum, the auto-correlation function is weighted by hanning in order to avoid the "window" effect (Laviron 1971 a).

The programme of cross-correlation is similar to the programme of auto-correlation. It gives the cross-correlation function and the cross-spectrum of two signals (Fig. 1). It is thus possible to study the spectral components common to two channels as also their differences in phase.

It is adivsable to limit the band of frequencies analyzed to the informative frequencies so as to get a better definition to each of them.

The number of channels simultaneously analyzed is limited to 4 by

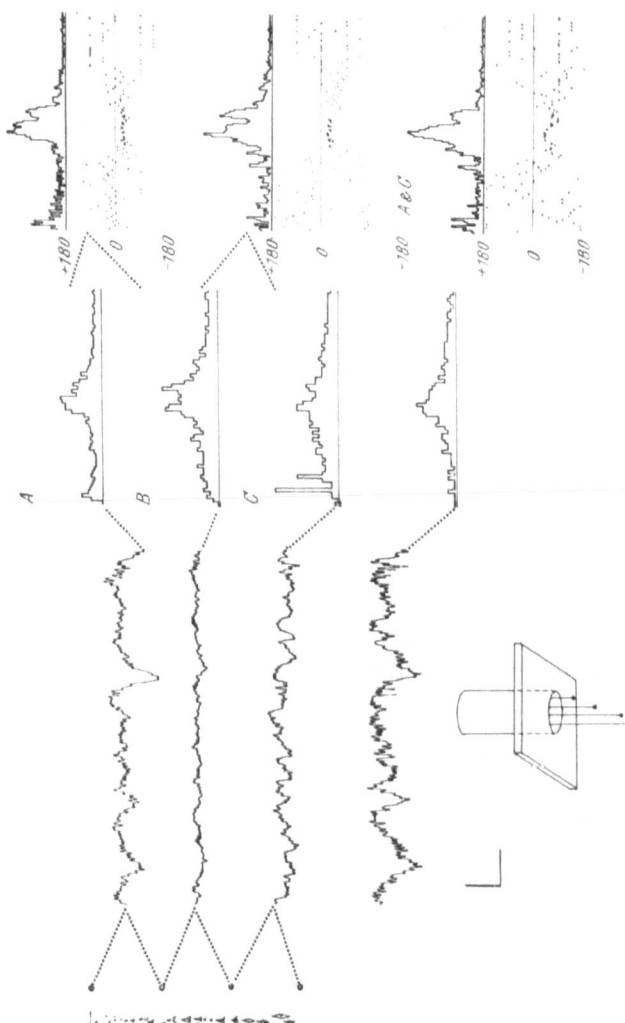

Fig. 1. Method of data exploitation. Propagated activities, Type P-behaviour. From left to right: histo-logical structure (biopsy or VON ECONOMO's atlas). Localization of electrode (4 pole-type). EEG bipolar recording (lowest run is transcortical). Calibration: 1 second, — 100 μv. Frequency power spectrum (from zero to 25 c/sec). Cross spectra and phase relationships (down to — 180° and up + 180°). The peaks of frequency appear in all spectrums round 10 c/sec, with high coherence. This "electrophysiological behaviour" is typical (1) of normal anesthetized cortex, with same frequencies as on the simultaneous scalp EEG, and (2), of propagated type activity according to the — 30° value of phase. (Maximum speed of radial propagation: 19 cm/sec). Inset: schematic drawing of the 4-pole electrode

the possibilities of the magnetic tape-recorder. Three channels most often correspond to three adjacent couples of electrodes, the fourth channel records the total transcortical potentials.

Coherence function: on the basis of the different spectra which were obtained and the ordinates of which appear on punched tape, it is possible to calculate the coherence function given by the formula:

$$\text{Coherence} = \frac{\text{cross-spectrum } 1, 2}{\sqrt{\text{spectrum } 1 \times \text{spectrum } 2}}$$

This function gives an estimate of the degree of connection between two channels for each frequency: coherence approaches closer to 1 as the signals compared are better correlated, and closer to 0 as they are less correlated.

## 2. Results

The recordings were made on 17 patients. Most often several recordings were made on the same person, which gave a total of about 60 recordings, only 46 of which were of any use. Automatic analysis with a computer was used for 10 patients with a total of 12 recordings, 5 of which included a study of 3 periods totalling one minute, so as to check the constancy of rhythms as time passed. Concerning the topography of the graftings, the different cerebral lobes, at least on the level of their convexity, were explored, but this was only by the merest chance since the choice of the recording site was limited by the operative spots.

### 2.1. Electro-Physiological Results

Preliminary remarks: most intracortical recordings are conducted on the basis of a "de visu" observation of the record on paper. It is interesting to notice that spectral analysis with a computer always confirmed visual observations. Another essential remark (Elul 1972) is the repetitive character of the results obtained with computer whenever it has been possible to make an analysis of three one-minute periods of the same recording.

### 2.1.1. Descriptive Analysis

2.1.1.1. *Polarity.* Sometimes we found very typical aspects (the polarity of which can be defined in relation to the base-line) as for example a series of positive waves on the surface (Fig. 2). But most of the time the records offer a more complex aspect for which it would be difficult to specify polarities because of the indetermination of the zero level. In this case we can only speak of differences of phases among channels.

2.1.1.2. *Spectral Analysis.* The impossibility of analyzing records in terms of specified grapho-elements led us to undertake this study in terms of spectral analysis, all the more so as on most records there

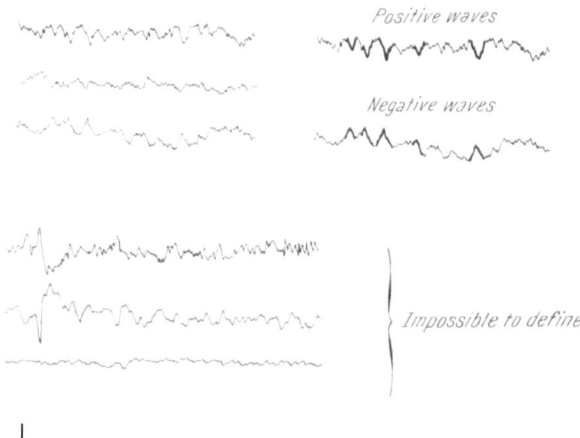

Fig. 2. Straticorticographical records, polarity. *Upper part:* example of slow waves with definite polarity. Same record on the left and the right side, but some waves are traced with Indian ink on the right side. *Lower part:* example of a recording without definite polarity. Positivity of the more superficial electrode of each run is recorded downwards. Calibration: 1 sec, — 100 μv

were rhythms with different frequencies. The complementary relation between direct observation and computer analysis, as also the high degree of homogeneity of each record when taken separately, allow different cortical rhythms to be distinguished and a certain number of conclusions to be drawn.

a) Spectral analysis of a channel. Although in certain cases the frequency spectrum is distributed at random over the whole scale of frequencies, in most cases a certain number of frequencies stands out either in a relatively widespread way on each side of the central value (Fig. 1 *a*) or in a very narrow band (Fig. 1 *c*). These different results may be found jointly in the same spectrum but they are always concordant with the direct observation of the records.

16  Synchronization

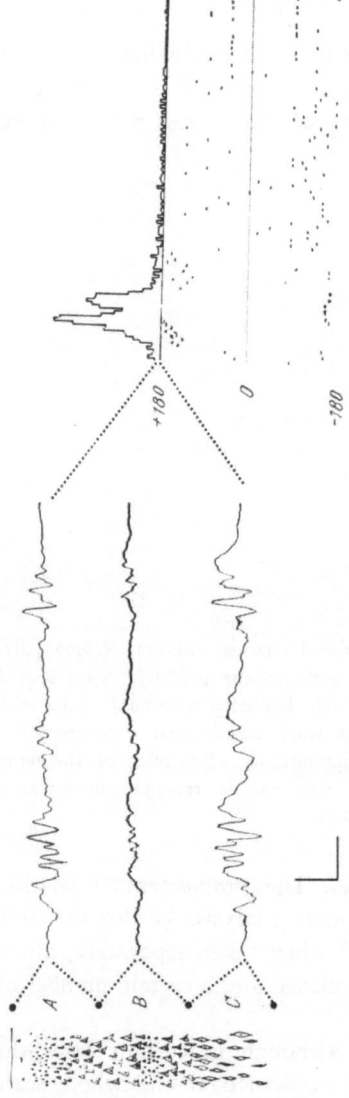

Fig. 3. Straticorticography of the grey matter. The 4-pole electrode is entirely in the grey matter. The same frequencies are present in the three channels, but with lower amplitudes in the middle one and with phase reversal between the two extreme ones, confirmed by the 180° phase-relationship of the cross-spectrum. Calibration: 1 sec, — 100 μv (Type 2-cortex)

b) Cross-spectra. It is exceptional for the peaks of frequency to be limited to one channel. They are usually found in two or more channels and thus also in the cross-spectrum (Figs. 1–5).

c) Phase relations. The readings of the phases corresponding to the peaks of the cross-spectra are usually grouped round a certain value. This reading can be 0° (Fig. 5), 180° (Figs. 3 and 7) or any other intermediate reading (Fig. 1). Sometimes these readings are arranged along a certain slope dependent on frequency (Fig. 6). As far as the non-significant frequencies of the cross-spectra are concerned (see below), the phase varies widely and cannot be taken into consideration.

d) Coherence. The maxima of coherence confirm the peaks of the cross-spectra in most cases. But their practical value is limited by the lack of precision at low readings of the spectrum.

### 2.1.2. Notion of "Type of Electro-Physiological Behaviour" for a Band of a Given Frequency

It is difficult to make a decision which results are to be considered as significant. For although the frequency peaks in the spectra and cross-spectra have precise phase relationships in a great number of records, there are others in which this is not the case. Then, with the object of clarifying the reading of the results, we attempted to define the notion of "type of electro-physiological behaviour". Estimated on the basis of a simultaneous study of the record and the spectral analysis, the sufficient and necessary conditions of the type of electro-physiological behaviour being uniform in different recording epochs are as follows (Fig. 1):

1. the frequency band must be common to several channels and thus appear in the corresponding cross-spectrum,
2. the reading of coherence must be close to 1,
3. the phase connections must be grouped.

The behaviour thus retained may—in terms of phase connections—be divided into 4 groups:

1. phases grouped round 0°,
2. phases grouped round 180°,
3. phases of an intermediate reading (30° for example),
4. phases shared out on a certain slope.

In one and the same record there can be two different types of behaviour corresponding to two distinct frequency spectra.

16*

## 2.2. Research on Anatomical, Histological, and Lesional Factors

### 2.2.1.

The confrontation of electrical data with cortical thickness enables electro-anatomical correlations.

The utilization of the data of the stratigraphic study must necessarily begin with the differentiation of the activity which may be attributed to grey matter on one side, to underlying white matter on the other side; the latter, by definition does not contain neuronal somata. In a great number of cases one or several of the lower poles of the electrode are in the white matter. The level of the electrode tip is either estimated on the basis of the information of Von Economo's atlas, or determined by biopsy. When simultaneously available, both approaches are concordant.

2.2.1.1. *Reversal of Phases.* In the present context, we consider only the behaviour of electrical activities that exhibit one or several reversals and these are the commonest activities. Indeed we found them isolated or associated with other bands of frequency, the behaviour of which is identical or different, in four fifths of the records.

2.2.1.1.1. On the level of grey matter. In this case the same rhythms are found on the level of the superficial layer and of the deep layer (Fig. 3). They have the same morphology and the same spectrum of frequency. But between the superficial three-fifths and the deep two-fifths of grey matter, there is usually a reversal of phases, creating a mirror image of superficial rhythms in relation to deep rhythms. In short, it seems possible to distinguish 3 layers in grey matter: a superficial layer and a deep layer in which rhythms have the same frequencies and are synchronous, but phase-reversed, and an intermediate layer in which the reversal of polarity takes place.

2.2.1.1.2. On the level of white matter. White matter was explored in almost all of our recordings (44 times out of 46, the exceptions corresponding to a cortex measuring 4 mm). Most often (32 instances) there were two poles in it and thus one bipolar derivation. Sometimes (12 instances) there was only one pole, the derivation being located on each side of the junction between grey and white matter. The different aspects can be schematically grouped according to two possibilities:

On the one hand, when the deepest 2 poles of the electrode are in white matter (the upper one at the junction between grey and white matter, the lower one from 1 to 2 mm below) the rhythm recorded

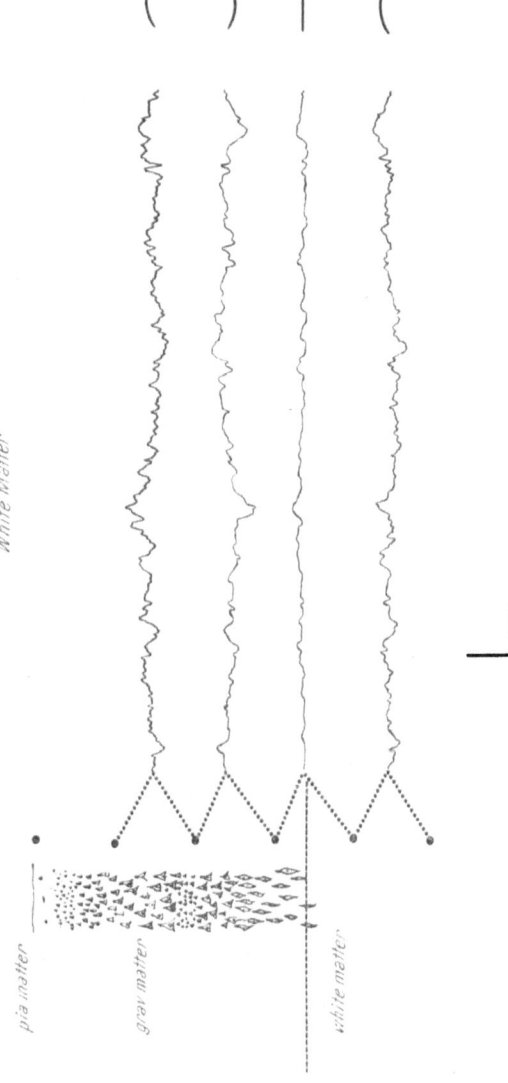

Fig. 4. Straticorticography including white matter. Two of the six poles are in the white matter. As schematically illustrated at the right, the main rhythms recorded have the same morphology in the white and the grey matter, but with (1) two phase reversals, one of them in the grey matter, similar to the one in Fig. 3, the other between the white matter and the deep grey matter; and with (2) a low amplitude at the level of the junction between the grey and the white matter. Calibration: 1 sec, — 100 μv

has the same spectrum of frequency and the same morphology as the one recorded in grey matter. Its polarity is inverse to the deep grey matter rhythm. The reversal of polarity seems to occur at the junction between grey and white matter. Most of the time the amplitudes are lower here than in grey matter.

On the other hand, when the lowest 2 poles of the electrode are on each side of the junction between grey and white matter, the rhythm recorded has a low amplitude. It is usually difficult to analyze its morphology with the naked eye and to recognize its polarity.

Moreover, one of the recordings has the advantage of having been made so as to join both the preceding data (Fig. 4). The last derivation in the white matter records a rhythm identical to the grey matter rhythm, but in contrast of phase with the deep grey matter rhythm. The penultimate derivation, on each side of the junction gives a record with an iso-electric tendency.

It is to be noted, finally, that on the level of white matter we never recorded activities different, on a morphological and spectral point of view, from those recorded in the level of cortex.

2.2.1.2. *Other Phase Relations.* Beside activities with phase relationship of 180° as previously described and in which stratigraphic study shows pure reversals in relation to cortical depth, the electro-anatomical parallelism of activities with different and more complex types of behaviour must be dealt with. As far as activities without a phase reversal are concerned, several types of behaviour in relation to strati-corticographic studies can be distinguished:

2.2.1.2.1. On the level of grey matter. The first type deals with the bands of frequency for which the phase relations among different channels are all of 0°. This aspect was particularly found with records of an activity called "de volet" (Fig: 5).

The second type deals with the relations of phase which are—from one channel to another—different from 0° and 180°. Two types may be distinguished: 1. in the first one, the phase relations are all equal among the different channels. We found this aspect particularly in two cases: in one of them with phase relations of 30° (Fig. 1), in the other of 90°; 2. the second type shows relations of phase different from one channel to the other, for instance 60° and 110° (in one of these cases, the relations of phases between the channels changed during the three successive recordings).

The third type corresponds to a mixed aspect, in so far as there was an in-phase activity between two of the three channels and an activity

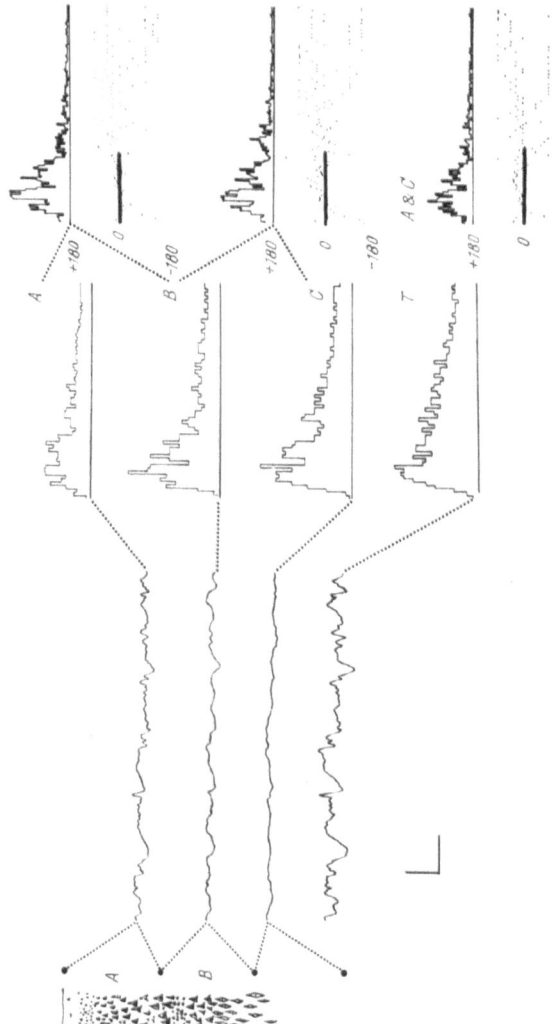

Fig. 5. Non-propagated activities—Type Z-behaviour. As appears in the cross spectra at the right, there is no difference in phase between the different channels. This behaviour is encountered with non-propagated activities, and does not make it possible to locate a generator. The lower recording is transcortical. Calibration: 1 sec, — 100 $\mu$v

30° out of phase between the other two channels. There were 3 instances of this aspect; for 2 of them the phase relation of 0° was superficial, while for the third one it was deep.

2.2.1.2.2. On the level of white matter. We never found electrical activity in phase with the deep grey matter. Besides, in several cases electrical activities of the upper part of white matter were connected

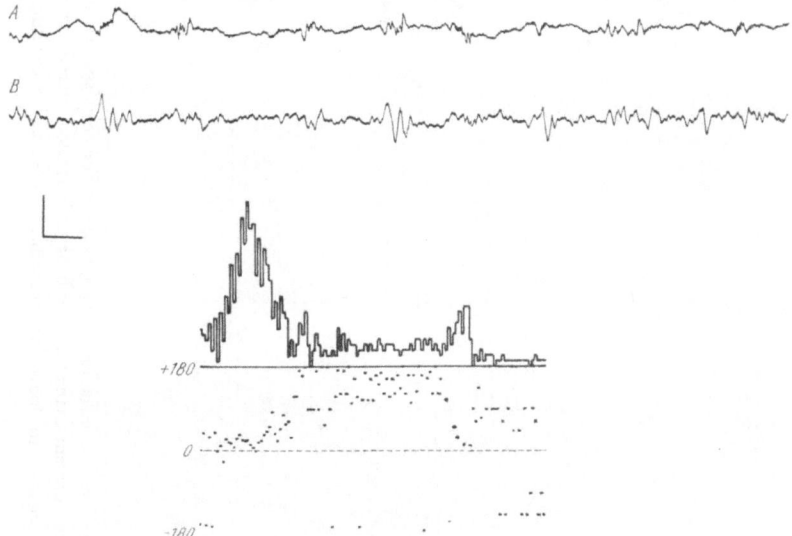

Fig. 6. Propagated activities—Type DS-behaviour. *a* and *b* correspond to the two upper bipolar recordings of the 4-pole electrode. Calibration: 1 sec, — 100 μv. Below is their cross-spectrum. The higher frequency activity (around 18 c/sec) shows a "slope aspect" if phases are considered. This is found with discontinuous activities which are phase-shifted between the channels. The maximum speed of radial propagation is 0.8 cm/sec in this case

with those of the deep grey layers with phase relations between 0 and 180°.

2.2.1.3. *Aspects Called "Slope Aspects"*. Finally there is a very particular case which is worth special mentioning in so far as bands of frequency sometimes show phase relations between channels which are grouped in a linear way; not along a horizontal direction as in all other cases but on a slope either positive or negative. The slope is one to be taken into consideration if it spreads from one end of the base of the given band of frequencies to the other (Fig. 6).

This slope may be found in one as well as in both cross-spectra, and in the latter instance the slope may be in parallel obliqueness. The significance of these slopes is difficult to explain. We tried to solve this problem by examining theoretical electrophysiological patterns which might exhibit the same phenomenon (see below).

These different types of behaviour may be found in one and the same recording, and this seems to be a fact of fundamental importance.

### 2.2.2. Electro-Histological Correlations

Comparison of electro-physiological data with the structure of the cortex: *electro-histological correlations*.

For this study we used VON ECONOMO's (1927) classification with its five structural types:

type 1 or granular pyramidal which educes itself to one pyramidal layer,

type 2 or pyramido-granular in which the 6 layers exist but in which the 2 pyramidal layers are more important than the 2 granular layers,

type 3 or homotypic type with 6 layers equally arranged,

type 4 or granulo-pyramidal with a granular predominance,

type 5 or granular in which pyramidal cells are almost absent.

At the first sight the cortical structural type does not seem to determine the stratigraphic activities recorded. However, a few observations seem to be called for, especially as far as the activities with phase reversals within the white matter are concerned: in the first type, stratigraphic activities are similar from one derivation to the other and most of them have only one phase reversal. In the second and third types, which include 2 pyramidal layers alternating with two granular layers, in most cases the same aspects were encountered. In type 4 and 5 with granular predominance (3 recordings), the rhythms observed do not usually show the same features as in the other structural types, at least when observed directly (they were not submitted to spectral analysis). These records are difficult to classify.

### 2.2.3. Functional Correlations

Comparison of electro-physiological data with lesions: *electro-functional correlations*.

2.2.3.1. *Healthy Cortex* (3 Recording Sites): spectral analysis, made on these recordings, results in a frequency peak most often between

8 and 12 cycles per second. This peak corresponds to the prevailing frequency in the per-operative EEG, related with anaesthesia (Fig. 1). 2.2.3.2. *Pathological Cortex:* spectral analysis shows that there is still a peak of lower frequency, between 1 and 6 cycles per second; this frequency is different from that of the pre-operative EEG (Fig. 3). It often is accompanied by one or several peaks of higher frequencies some of which may correspond to the main frequency of the per-operative EEG in a healthy zone, but others may be different. As far as the phase relation is concerned, the different frequency bands usually do not behave in the same way.

## 3. Discussion

### 3.1. Evaluation of the Results

3.1.1. Technical Conditions for Recording

In most cases the subjects recorded are patients and the recordings take place close to the cerebral lesion; this, of course, prevents us from drawing conclusions concerning the electrogenesis of a normal cortex although some of Hirsch' results support our conclusions (Hirsch et al. 1965). Moreover, all the subjects were explored under general, though light, anaesthesia and this prevented us from considering the recordings from a healthy cortex (8 instances) as made under physiological conditions. On the other hand, however, these rhythms (slower, higher, more synchronous than under normal conditions) can be more easily analyzed also.

As far as the electrode is concerned the main drawback is in the fact that its poles are at a fixed distance, which does not permit a progressive descent into the cortex; but the latter procedure would have taken too much time.

Besides, if the resolving power is, as a whole, satisfactory, it is not always sufficient with the 4-pole electrode. Therefore, electrodes with 6 poles were used too.

3.1.1.3. *Artifacts.* During per-operative recordings artefacts are common: operating rooms are not meant for electrophysiology, the portable recording instruments are comparatively rudimentary, so the recording must be short.

The movements of the brain by respiration and pulsation may simulate graphoelements which are, however, easily recognizable by their regularity and synchronism with thoracic movements or the ECG. If the socle of the electrode rests softly upon the cortex they usually can be be avoided by putting on the socle wads of cotton-wool imbued with sterile saline. The electrode then follows the movements of the brain instead of moving in relation to it as would be the case if one resorted to an anchorage of the skull.

Once the electrode has been implanted, a certain time elapses before the electrical activity stabilizes. The question should be considered as to whether lesions are made by the implantation. In general, this does not seem to modify the electrical activity. Whenever trans- and straticorticographic monopolar and bipolar recordings are compared, the electrical activities recorded turn out to be superimposable. "Epileptic" activities were not observed.

3.1.1.4. When recording the EEG the corresponding unit phenomena which could have been obtained with microelectrodes were not considered. But this problem was studied by several authors and in particular by HIRSCH et al. (1965, 1966) in human beings.

3.1.1.5. If the location of the different poles of the electrode in relation to histological structures turns out to be relatively correct in cases of biopsies, this does not hold good when simply the sections of VON ECONOMO's atlas were used for orientation. It turned out to be difficult to ascertain the exact situation of the lowest pole and thus also of the others. This may be partially due to peri-lesional alterations such as cerebral edema which may cause a variation of cortical thickness.

### 3.1.2. Technical Conditions of Evaluation

3.1.2.1. *Choice of the Parameters of Correlation Functions* (Bibliography in LAVIRON 1971). Band of frequency analyzed: it is advisable to limit the frequency band in question, for the maximum frequency analyzed depends on the sampling frequency. Thus it is necessary to use a low-pass filter to cut off all frequencies higher than the permissable maximum frequency (in our case 22 c/s).

Sampling frequency: SHANNON's theorem requires the sampling frequency to be more than twice the maximum frequency of the signal. To facilitate calculation, we chose a frequency of 51.2 c/s.

Frequency resolution: within the limits of this work, auto-correlation was calculated for period of 2.5 seconds, which gives a frequency resolution of 0.4 c/s and of 0.2 c/s for cross-spectrum. This rather high resolution was deemed necessary to get as much information as possible. This, however, has the drawback of increasing the number of data, particularly for phases corresponding to the different peaks of the cross-spectrum. But if frequency resolution is too low, the presence of two independent frequencies in one band may simulate phase-shifts and thus falsify the data. Optimization of frequency resolution is a problem to be tackled in another work.

Integration period: the error in correlation functions is inversely proportional to the square root of the product of the integration period multiplied by the band of frequency analyzed. It is, therefore, advisable to choose the largest possible integration period with respect to the stationarity of the signal. Our upper limit was 40 seconds, a time compatible with the practical requirements of recording in an operating room. The condition of stationarity is not contradicted by the results, as indicated above.

3.1.2.2. *Remarks on Spectra.* Fourier's transformed correlation functions result in the power spectrum of the signal analyzed, which can be read more easily from correlation functions. The fact, however, that spectra display the different frequencies of a signal as the squares of their voltages, leads to a comparatively minor representation of the signals of low voltage which may even be hidden by noise. This was the case in a great number of signals faster than about 5 c/s.

We were thus led to accentuate higher frequencies by using a capacitor at the computer input. The result is indeed a distortion in amplitude which was controlled by the following procedures: 1. we calculated the distortion curve which enabled us to restore, if needed, the true amplitude readings of the spectrum; this was also experimentally checked; 2. we thus made the frequencies appear most difficult to estimate from the written record, without losing the lowest ones; 3. we did not take into account the relative ratio of amplitudes between the different frequency bands.

The capacitor mentioned above not shift phase, since it dephases the two signals to be analyzed by the same quantity. This was ascertained by calculating the auto-spectrum of a sine wave. The readings of the phases corresponding to the peaks were close to zero and never exceeded $4°$ which, moreover, also gives an idea of the reliability of phase readings in this context.

3.1.2.3. *Remarks on a Certain Particular EEG Aspect.* Some records have a particular aspect, the interpretation of which by means of spectral analysis comes up against difficulties. To overcome these difficulties we made an experimental approach with square waves that were analyzed by the same technique. These are distorted by the capacitor at the computer input, and all the more so as the frequency is lower. Analysis results in a main peak at the basic frequency of the wave, with low amplitude harmonics. We have been bearing this in mind when interpreting results. When studying brain activities, however, we never encountered such harmonics; they may be covered by noise.

3.1.2.4. *Coherence.* A priori, coherence seemed most interesting to us as it answers, in a quantitative way, the question as to which frequencies are common in two channels. In practice, however, its application turned out to be limited by two principal causes of errors: on the one hand, the locating of the beginning of each period of analysis may, if imprecise, lead to false readings. On the other hand, calculation becomes inaccurate for very low frequencies. This was estimated for each reading of coherence. Therefore, coherence was only used to verify the interpretation of cross-spectra.

### 3.2. Attempt of a Hypothesis on Cortical Generators

The possibility of distinguishing between different types of electro-physiological behaviour led to the question of the possible relations between strati-corticographic activities and cortical structures, and so to the problem of cortical generators.

This question is approached here in the new classical view of the radiate organization of cortical electrogenesis, supported by histological and electrophysiological arguments (bibliography in BREMER 1960, ANDERSEN 1960, GLOOR *et al.* 1963, CALVET and SCHERRER 1961 a, b, CALVET *et al.* 1960, 1962, 1965, CALVET 1962, SCHERRER 1965, HOLUBAR 1964, OCHS 1965, RALL and SHEPHERD 1968, NICHOLSON and LLINAS 1971).

The generator concept is used here in a topographical way and without considering the electrographic aspects.

### 3.2.1. Phase Reversal and the Hypothesis of the "G" Type of Generators (Figs. 3 and 7)

The level of phase reversal probably corresponds to the site of the generator. In the great majority of cases it is located at three-fifths of the depth of the cerebral cortex.

Each physical model of this kind of generators has to take into account the existence of phase reversal both in the grey matter and at the junction between the grey and white matters. The only physical pattern which seems compatible with these facts is the generator to be located somewhat below the middle of the grey matter. If depolarized, in the upper part of the cortex a potential difference will take place due to the "source" and "sink" phenomenon. The same will happen in its lower part, but in the inverse direction. Besides, the junction between the grey and white matter will be positive when the underlying white matter has a zero potential; so a new gradient of potential will be created in the superficial part of the white matter, which is in the opposite direction and of lower amplitude than that located in the deeper part of the grey matter. This is a passive phenomenon. At the time of a polarization of the generator the phenomena are inverse. On the whole, therefore, there would be 2 "active" dipoles on each side of the generator in the grey matter, and one "passive" dipole in the underlying white matter.

The potential difference as found in the white matter pleads in favour of the active nature of the deep grey dipole, otherwise no potential difference could be found there.

This pattern is concordant with Calvet and Scherrer's (1964) "B" and "C" generators; the phase reversal in the white matter permits us to ascribe an active role to the deep dipole, a question that was raised by the authors quoted above.

3.2.1.1. *Histo-Electrical Parallelism: the Role of Pyramidal Cells* (Fig. 7). A certain number of arguments leads us to ascribe these aspects to the activity of pyramidal cells. Phase reversals are particularly often found in cortex of type 1, 2, and 3 in which there are plenty of cells in a radiate orientation (this is even true of "improperly" oriented pyramidal cells; Van der Loos 1965). The level of the generator corresponds to regions where pyramidal somata are found.

This confirms the data of the papers mentioned above and concerning neocortex (Li et al. 1956 a, b, Calvet et al. 1964, Holubar 1964, Holubar and Fischer 1964), immature neocortex (Garma and Verley 1965, Verley 1965), olfactory bulb (Rall and Shepherd 1968), hippocampus (Andersen 1960) and cerebellar cortex (Eccles et al. 1967, Nicholson and Llinas 1971).

The two intracortical active dipoles opposite one another are located on each side of the region which corresponds to pyramidal cell bodies. They probably underlie the "sink and source" phenomenon between somatic and dendritic structures; so the phase-reversal type of activity seems to correspond to a "G" generating system of a somato-dendritic nature.

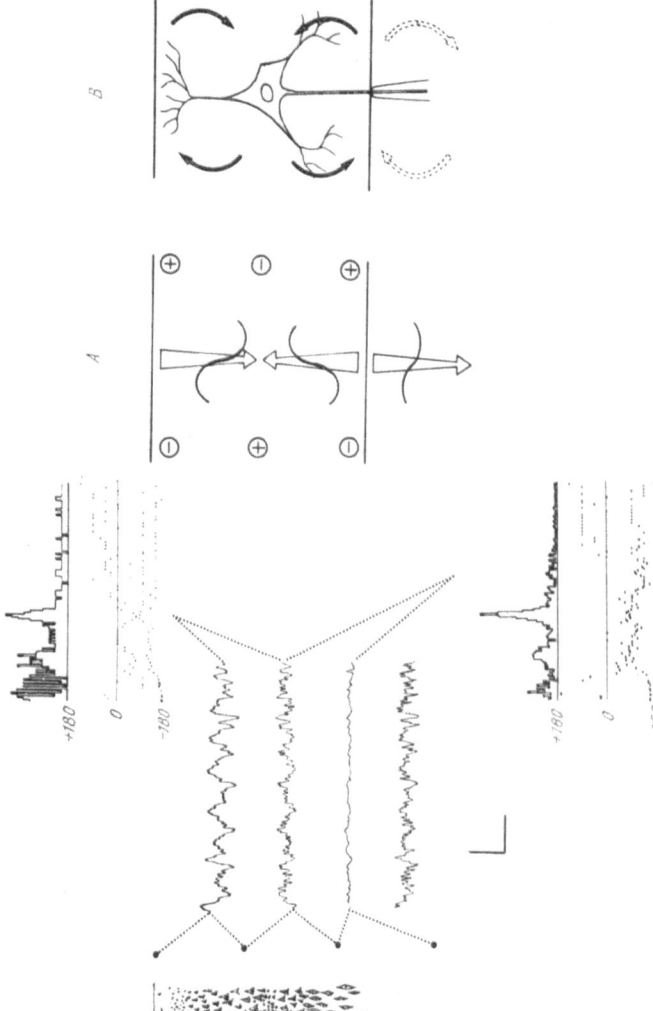

Fig. 7. Physical and histological models for Type G-behaviour. For frequencies simultaneously recorded in the white and the grey matter and characterized by two phase reversals (cf. Fig. 4), as here for the low frequencies (2 c/sec), the physical model (A) consists of three successively opposite and synchronous dipoles. The hypothetical histological model (B) would suggest the passive nature of the one in the white matter. Calibration: 1 sec, — 100 μv

This hypothesis has still to be checked by unit recordings, since one cannot conclude from mere extracellular recordings the amount of current passing the membrane, as shown by several authors (BUSER and ALBE-FESSARD 1955, CREUTZFELDT 1960, RAABE and LUX 1972, SPECKMANN et al. 1972, KUHNT 1972) and put on a theoretical basis by RALL (1962) and RALL and SHEPHERD (1968). Unless such recordings are available, we can propose such a hypothesis only by referring to the comparable data by CALVET and SCHERRER (1964).

### 3.2.2. The "0°" Type of Behaviour and the Hypothesis of Type "Z" Generator

Activities without any phase difference in two channels (Fig. 5) could correspond either to passive transmission from a generator at a distance or to a local generator. The former hypothesis would, however, require too powerful a source to be thinkable. The question of the localization of the generator is difficult to answer since there are three different types of in-phase behaviour:

1. different peaks of frequency in-phase throughout the entire cortex,
2. bands of frequency in-phase which are only found in half or in the upper three-thirds of the cortex,
3. in-phase bands of frequency which are only found in half or in the lower two-thirds of the cortex.

The pattern may be theoretically represented by a single dipole which, in (1) occupies the whole cortical thickness, in (2) occupies only the superficial part of the cortex, in (3), occupies, on the contrary, its deeper part.

The study of histo-electrical relations did not permit any satisfactory conclusion to be drawn from these findings. However, case (2) could be compared with CALVET and SCHERRER's "A" generator (CALVET et al. 1964).

### 3.2.3. Propagation

The existence of phases different from 0° or 180°, i.e., asynchronous activities, can only be explained by propagation.

3.2.3.1. *Properties of Propagated Activities.* Let us first consider the case in which the sum of the differences of phases between diverse derivations differs from the phase difference between the two extreme derivations. This situation indicates either the presence of two independent activities of the same frequency but generated at different depths of the cortex, or the presence of two independent activities with almost identical frequencies which spectral analysis does not discriminate, or even only experimental errors. These latter cases cannot be explained and are therefore eliminated; they were rarely encountered.

If the relations of phase are consistent two possibilities may be distinguished:

1. phases remain almost constant over the entire interval of the

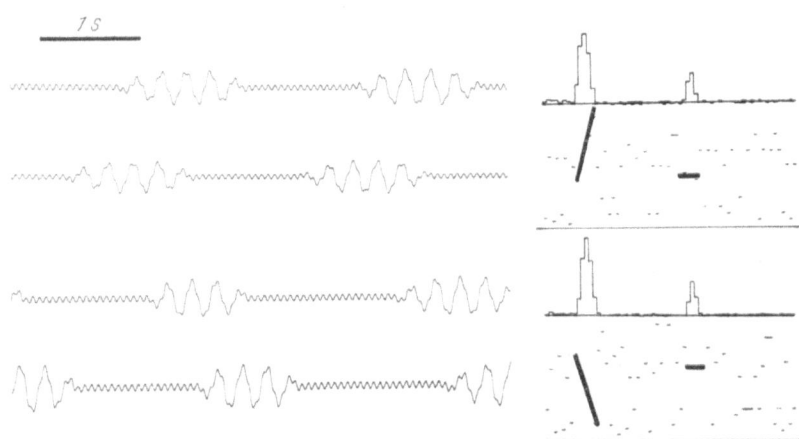

Fig. 8. Experimental model of DS-type-behaviour. Cross-spectra are obtained from two channels with the same signal consisting of two sine frequencies only, the slower being discontinuous; one channel is shifted positively in the example above, negatively in the one below. In contrast to the continuous higher frequency, the discontinuous one presents phase relations arranged along an oblique line, the direction of slope of which being dependent on which channel displays the leading pattern

frequency considered (Fig. 1). The time difference between the two channels can be calculated from the formula for phase-speed

$$t = \frac{1}{2\pi} \frac{\varphi}{\nu}$$

($\nu$: peak frequency; $\varphi$: phase). If the distance $x$ between the two pairs of electrodes is known, the phase speed may be calculated (LAVIRON 1971) according to the formula:

$$v = \frac{x}{t}$$

17 Synchronization

It should be noted that the phase speeds in this work are about from 0.5 to 8 cm/sec (calculated in 7 cases);
2. the degree of the phase shift depends, in linear way, on frequency.

This aspect is difficult to interpret. It may be understood, however, by studying some experimentel pattern. For this purpose we produced a 15 c/s low voltage sine wave and superimposed it on a larger sine wave of 4 c/s. The autospectrum of this pattern gives two peaks at 4 and at 15 c/s respectively. The cross-spectrum between this activity and the same activity when time-shifted leads to the same result, but the phase readings at the 4 c/s peak are shared out on a slope positive or negative according to which channel is leading (Fig. 8). This result accords with the studies on phase and group speeds in connection with the propagation of EEG waves (Laviron 1971 b).

The following conclusion may be drawn: in the case of discontinuous and time-shifted activities, the cross-spectrum between two channels reveals not only time-shifting but also its direction. Calculation was made according to the formula:

$$t = \frac{1}{2\pi} \frac{d\varphi}{d\nu}$$

This "phase-slope", therefore, demonstrates both the discontinuous nature of the activity and the shifting between the two channels which we summed up by the phrase "discontinuous activity-shifting time" or briefly "DS".

3.2.3.2. *Interpretation.* These time differences mean propagation, the word being understood in its broader meaning, without prejudice to the neurophysiological mechanism concerned.

This interpretation of our findings is however subject to various unknown factors as far as the orientation, the direction and the speed of propagation are concerned: the orientation, since only the speed component along the axis of the electrode can be measured; the direction, because the above theoretical considerations hold true only for ohmic resistors. This latter condition is met with in the brain: the rate of change of current flow underlying EEG activity is rather slow compared with the time constant of cellular membranes. As far as the speed of propagation is concerned, phase speed and group speed are only maximum limits as there are unknown factors caused by phase-relation and by the modulations of amplitude; it is most likely however that these maximum readings are very close to the actual readings, since those are rather low (about 1 cm/sec), and most similar to those reported on propagation in deep evoked responses, for instance non-reversing field potential of locally evoked potentials in the alligator cerebellum (0.2–0.07 m/sec; NICHOLSON and LLINAS 1971), hippocampal responses evoked by perforant path excitation (0.01–0.1 m/sec; GLOOR et al. 1963), or "epileptic" spikes (0.06– 0.17 m/sec; PETSCHE et al. 1972).

It seems premature to consider the mechanisms underlying this kind of slow propagation; the hypothesis suggested by NICHOLSON and LLINAS (1971) and by GLOOR et al. (1963) however seems worth considering. According to this hypothesis this slow type of propagation corresponds to synaptic mechanisms rather than to active propagation along dendrites. It may be noticed that this propagation is at least ten times slower than that noticed by different authors in a tangential direction (PURPURA et al. 1960; NICHOLSON and LLINAS 1971).

The existence of this phenomenon adds to the prevailing concept of cortical electrogenesis of spontaneous and evoked potentials; moreover it demonstrates the practical significance of straticorticographic spectral analysis and the concept of group speed.


## Summary

Human electrocorticograms were recorded with multipolar transcortical electrodes during surgery, and spontaneous activities analyzed by spectrum analysis on a small computer, with special emphasis upon phase relationships correlated with histological structure.

Besides an influence of pathological conditions on these data, the main result is the finding of two different types of electrophysiological behaviour: 1. synchronous activities with $0°$ or $180°$ phase relationships which may be produced by a stationary source; the eventual reversal zone is roughly related to the pyramidal layer; 2. a radially asynchronous activity with intermediate phase relationships requiring a propagated source the speed of which does not exceed several cm/sec.

Both these types of electrophysiological behavious can be compared with data on phasic spontaneous activity and evoked potentials.

17*

# References

Andersen, P.: Interhippocampal impulses. Acta Physiol. Scand. *48*, 178—208 (1960).

Bishop, G. H., and M. H. Clare: Sites of origin of electric potentials in striate cortex. J. Neurophysiol. *15*, 201—220 (1952).

Brechner, V. L., Walter, and Dillon: Practical electroencephalography for the anesthesiologist, American lecture series. Springfield, Ill.: Charles C. Thomas. 1962.

Bremer, F.: L'interprétation des potentiels électriques de l'écorce cérébrale. Structure and function of the cerebral cortex. Elsevier Publishing Company. 1960.

Buser, P., and D. Albe Fessard: Explorations intracellulaires au niveau du cortex sensorimoteur du chat. In: Microphysiologie Comparée des Éléments Excitables. Paris: C.N.R.S. 1955.

Calvet, J.: Comparaison de l'activité électroencéphalographique dérivée par électrodes de surface et par électrodes transcorticales. J. Physiol. *54*, 308—309 (1962).

— M. C. Calvet, and J. M. Langlois: Diffuse cortical activation waves during so called desynchronized EEG patterns. J. Neurophysiol. *28*, 893—907 (1965).

— — et J. Scherrer: Etude stratigraphique corticale de l'activité EEG spontanée. Electroenceph. clin. Neurophysiol *17*, 109—125 (1964).

— M. C. Holingue, A. Guillard et J. Scherrer: Etude par électrodes transcorticales de certaines réactions d'arrêt du lapin. Revue Neurol. *102*, 316—318 (1960).

— et J. Scherrer: a) Activité bio-électrique de l'écorce cérébrale à ses différents niveaux. C. R. Soc. Biol. *155*, 275—278 (1961).

— — b) Relation des décharges unitaires avec les ondes cérébrales spontanées et la polarisation corticale. C. R. Acad. Sci. *252*, 2297—2299 (1961).

Creutzfeldt, O. D., and S. Watanabe: Relations between EEG phenomena and potentials of single cortical cells. Electroenceph. clin. Neurophysiol. *20*, 1—8, (1960).

Eccles, J. C., K. Sasaki, and P. Strata: Interpretation of the potentials fields generated in the cerebellar cortex by mossy fibre volley. Exp. Brain Res. *3*, 58—80, (1967).

Elul, R.: Randomness and synchrony in the generation of the electroencephalogram. See this book, pp. 59—77.

Fourment, A., L. Jami, J. Calvet et J. Scherrer: Comparaison de l'EEG recueilli sur le scalp avec l'activité élementaire des dipoles corticaux. Electroenceph. clin. Neurophysiol. *19*, 217—229 (1965).

Garma, L. et R. Verley: Générateurs corticaux étudiés par électrodes implantées chez le lapin nouveau-né. J. Physiol. *57*, 811—818 (1965).

Gloor, P., C. L. Vera, and L. Sperti: Electrophysiological studies of hippocampal neurons. Electroenceph. clin. Neurophysiol. *15*, 353—378 (1963).

Hirsch, J. F., J. Buisson Ferey, M. Sachs, J. C. Hirsch et J. Scherrer: Electrocorticogramme et activités unitaires lors de processus expansifs chez l'homme. Electroenceph. clin. Neurophysiol. *21*, 417—428 (1966).

HIRSCH, J. F., M. SACHS et G. ARFEL: Enregistrements transcorticographiques chez l'homme. Revue Neurol. (Paris) *105*, 230—235 (1961).

— — J. BUISSON FEREY, J. C. HIRSCH et J. SCHERRER: Etude par microélectrodes du rapport des ondes électrocorticales et des décharges unitaires chez l'homme dans cetains processus pathologiques. Revue Neurol. (Paris) *113*, 229—235 (1965).

HOLUBAR, J.: Mechanisms of the primary cortical responses (PCR) of the somatosensory area in rats. Physiol. Bohemoslov. *13*, 385—396 (1964).

— and J. FISCHER: Histological localisation of the components of primary cortical responses (PCR). In the somatosensory area in rats. Physiol. Bohemoslov. *13*, 484—494 (1964).

JAMI, L., A. FOURMENT, J. CLAVET et J. SCHERRER: Activité EEG étudiée en transcorticographie et en dérivation monopolaire classique. Revue. Neurol. (Paris) *113*, 251—252 (1965).

KUHNT, U.: Neuronal correlates of the visual evoked response and disinhibition in the visual cortex during flicker response. See this book, pp. 78—92.

LAVIRON, A.: Filtrage numérique des rythmes d'origine biologique. Med. and Biol. Engng. *9*, 97—120 (1971 a).

— Interprétation automatique, en ligne, de l'électroencephalogramme. Critique méthodologique. Réalisation pratique sur petit ordinateur. Thèse de sciences N⁰ 48. Université de Lyon (1971 b).

LI, C. L., C. CULLEN, and H. H. JASPER: Laminar microelectrodes studies of specific somatosensory cortical potentials. J. Neurophysiol. *19*, 111—130 (1956 a).

— — — Laminar microelectrodes analysis of cortical unspecific recruiting responses and spontaneous rhythms. J. Neurophysiol. *19*, 131—143 (1956 b).

NICHOLSON, C., and R. LLINAS: Field potentials in the alligator cerebellum and theory of their relationship to Purkinje cell dendritic spikes. J. Neurophysiol. *34*, 509—531 (1971).

OCHS, S.: Cortical potentials and pyramidal cells. A theoretical discussion. Perspect. Biol. Med. *9*, 126—136 (1965).

PETSCHE, H., P. RAPPELSBERGER, and Zs. FREY: Intracortical aspects of the synchronization of self-sustained bioelectrical activities. See this book, pp. 263—284.

PURPURA, D. P., M. N. CARMICHAEL, and E. M. HOUSEPIAN: Physiological and anatomical studies of development of superficial axodendritic synaptic pathways in neocortex. Exp. Neurol. *2*, 324—347 (1960).

RAABE, W., and H. D. LUX: Studies on extracellular potentials generated by synaptic activity on single cat motor cortex neurons. See this book, pp. 46—58.

RALL, W.: Electrophysiology of a dendritic neuron model. Biophysics *2*, 145—167 (1962).

— and G. M. SHEPHERD: Theoretical reconstruction of field potentials and dendrodendritic synaptic interactions in olfactory bulb. J. Neurophysiol. *31*, 884—915 (1968).

SADOVE, M. S., A. D. BECK, and F. A. GIBBS: Electroencephalographic for anesthesiologists and surgeons. London: Pitman Med. Publ. and Philadelphia: J. B. Lippincott, Cie. 1967.

SCHERRER, J.: Analyse de l'activité électrocorticale spontanée. Actualités Neurophysiologiques, 6ème série, pp. 201—221. Paris: Masson. 1965.

SPECKMANN, E. J., H. CASPERS, and R. W. JANZEN: Relations between cortical DC shifts and membrane potential changes of cortical neurons associated with seizure activity. See this book, pp. 93—111.

VERLEY, R.: Recherches sur le developpement des activités électrocorticales avec des électrodes corticales radiaires. J. Physiologie *57*, 407—436 (1965).

VAN ECONOMO, C.: L'architecture cellulaire normale de l'écorce cérébrale. Paris: Manson. 1927.

VAN DER LOOS, H.: The "improperly" oriented pyramidal cell in the cerebral cortex and its possible bearing on problems of neuronal growth and cell orientation. Bull. J. Hopkins Hosp. *117*, 228—250 (1965).

# Discussion

SCHERRER: As I already mentioned we assume that most of the waves are usually formed by two inversely oriented dipoles. One has to be very cautious as to wave polarity when using the classical surface derivations because of the curvature of the cortex.

PERONNET: We had really many difficulties: once the electrode passed through the white matter into the grey matter of the opposite part of the implanted circonvolution. The result were two simultaneous phase reversals caused by the two parts of the circonvolution recorded at the same time.

BRAZIER: I believe Dr. SCHERRER is suggesting an explanation based on volume conduction. The principle of coherence with determination of phase-lag between similar electrical activities can be used as a test for volume conduction. In the latter there is, of course, no phase difference as electrodes at both sites are sampling the same field.

GERIN: As a matter of fact, slow activities frequently do not show any phase difference which supports the hypothesis of a volume conductor model; but the opposite is more frequent for faster activities.

SCHERRER: What is the 18/sec activity you were describing?

PERONNET: As was demonstrated in the slide with the DS-type behaviour we found in many cases that, in one electrode and at one instant, two distinct frequencies may present two different types of electrophysiological behaviour.

PETSCHE: These two types of activity propagated and non-propagated in the direction of the Z-axis of the cortex, were observed by us too. With locally applied penicillin, the spikes propagate from the pyramidal layer to the cortex whereas the "waves" in penicillin seizures don't show any vertical propagation but occur in phase-reversal between deep layers and surface.

# Intracortical Aspects of the Synchronization of Self-Sustained Bioelectrical Activities[1]

H. Petsche, P. Rappelsberger, and Zs. Frey [2]

Brain Research Institute, Austrian Academy of Sciences,
and Neurological Institute of the University of Vienna, Austria

For studying cortical electrogenesis and its synchronization to what is commonly called "EEG", the rabbit has turned out to be the most suitable animal. Owing to his unfolded, evenly stretched-out cortex this animal offers optimum conditions for studying the geometry of potential distribution in an electrically active, structurally fairly homogeneous, multi-layered accumulation of generators. In all but lissencephalic animals, the laws described by Woodbury's angle theorem (1960) make the phenomena still more complex.

The important role of the cortex for the origin and spreading of EEG activity was pointed out in several previous papers (Petsche and Šterc 1968, Petsche and Rappelsberger 1970a, Petsche et al. 1970). Moreover, seizure waves and—more generally—all kinds of so-called synchronized EEG activities were shown to be not stationary events but to move in different directions and at different speeds. This property, however, does not seem to depend on subcortical pace-makers since it was even found in chronically isolated hemispheres. Moreover it was shown that for the spreading of brain waves, cortical morphology seems to be of primary importance: at the borderlines between two areas of different structure, the waves of activity either stop or are delayed or their travelling properties are different. Finally, the observations resulted in the conclusion that the travelling phenomenon seems to be a cause rather than a result of the

---

1 This work was carried out with support from the Fonds zur Förderung der wissenschaftlichen Forschung, Österreich.
2 Visiting scientist of the Institute for graduate medical education, Budapest, Hungary.

fact that EEG waves of uniform shape may be seen in large areas, *i.e.*, that there exists a phenomenon which was chosen as the main topic of this symposium.

One of the most important fibre systems underlying the task of maintaining synchronization is to be looked for in the deepest layers of the cortex, since only when these layers are severed by a cut do the two resulting halves of the cortex display incoherent seizure patterns.

The main purpose of this paper was to enter, by means of microelectrodes, into the third dimension of the cortical generator layer in order to establish whether, in the depth of the cortex, the phenomena of synchronization prove to be the same as on the surface. This step into the third dimension resulted in findings which were unexpected from the point of view of the experience obtained with surface recordings alone and may offer some solution to problems concerning synchronization.

## Methods

The technique of intracortical recording with multiple microelectrodes was described in detail by PETSCHE and RAPPELSBERGER (1972). Here, only a brief account may be given concerning the arrangement and the kind of electrodes used, as well as the ways the seizures were produced.

The experiments were performed on unanesthetized, curarized, artificially respirated rabbits. After exposure of the sensorimotor cortex, an electrode carrier with 4 moveable steel tubes of 0.5 mm diameter at 2 mm distances is fastened on the skull. These four steel tubes (isolated except of the tip) are lowered under microscopic control until just touching the exposed arachnoidea. The tubes serve as both surface electrodes and guides for the steel semi-microelectrodes which are inserted through them and lowered in the same way until just touching the cortex. The four semi-microelectrodes, the impedance of which is kept at round 5 meg at 10 c/sec, are fastened then in such a way that they may be lowered altogether by a microdrive. By this device it is possible to study cortical layers of different depths by means of an equidistant rank of electrodes in line and to compare these recordings with simultaneous epicortical activity.

Recordings were made by FET-source-followers connected to a conventional Schwarzer-EEG-machine. 7 of the outputs of the EEG machine were simultaneously recorded on tape. Unipolar, bipolar, and transcortical recordings were made.

The microelectrodes have only been used to record the gross EEG activity; unit activity has not been taken into account.

Seizures were elicited by the following means: 1. by direct local application of ouabaine (PETSCHE and RAPPELSBERGER 1970 b), 2. by intravenous Metrazol, 3. by contralateral electric stimulation, 4. by locally applied Penicillin. Since this paper is not intended to deal with the differences of seizures produced by different epileptogenics but rather with the general features of synchronization, most of the results dealt with here were obtained with Penicillin. The reason is that these spikes offer, by their relatively constant shape an adequate subject for studying the question of the mutual interrelationships of small generators located at different levels of the cortex.

## Results

The results obtained clearly demonstrate that the problem of cortical EEG generation and synchronization is far more complicated if considered under its spatial aspect than if approached in the conventional way by mere epicortical recording. At first sight, one is surprised by the great variety of phenomena observable in the different layers of the cortex which are not seen with epicortical recording. The most characteristic intracortical features of seizure activity are the following:

1. Intracortical activity has, as a rule, much more high-frequency components than activities recorded from the surface of the cortex. It looks as if the epicortical activity has passed through a low-pass filter. This property, however, is not caused by the microelectrodes having smaller recording surfaces than the epicortical steel tubes, because when recording from upper layers by microelectrodes the same low-frequency activity is found. It seems as though the properties of the superficial cortical tissue itself were the reason for this filter effect. We shall come back to this point in the discussion.

2. A general observation is the presence of a phase-reversal between surface and depth. This fact, well known since the studies of LI et al. (1956), however, does not concern all elements of intracortical activity, as will be shown in extenso with the example of Penicillin spikes. The concept of "phase-reversal" has to be taken cum grano

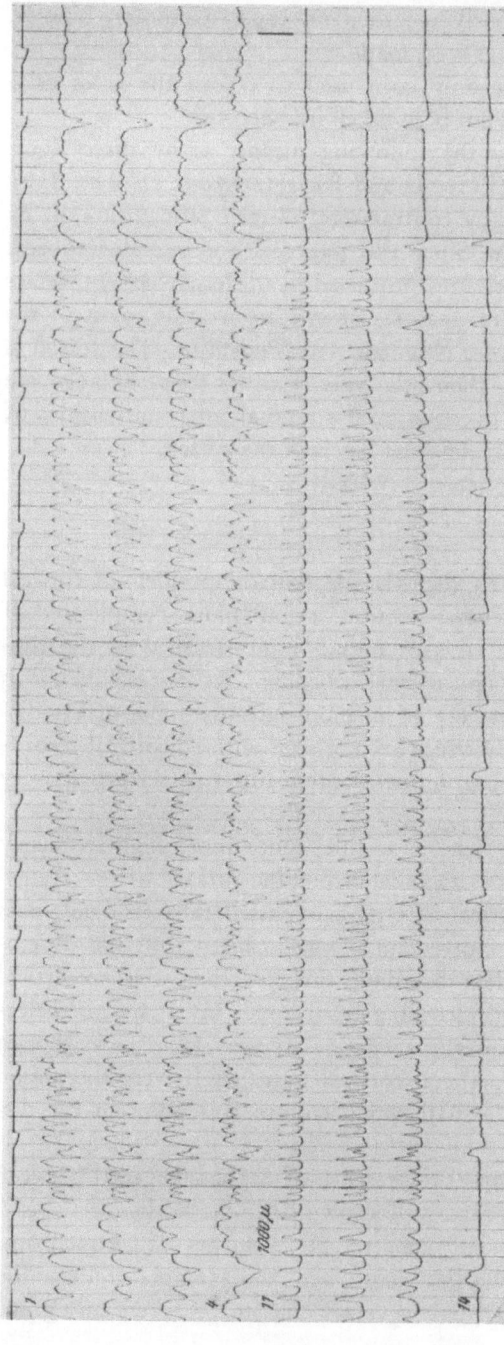

Fig. 1. End of a seizure induced by contralateral electrical stimulation. Surface electrodes 1–4 are in an equidistant longitudinal row (2 mm interelectrode distances). The microelectrodes 11–14 are at a depth of 1000 $\mu$, respectively. Common reference: nose bone. Calibration 1 mV.

The main differences between surface and depth activity are clearly seen: the intracortical pattern has higher amplitudes, is less homogeneous if the entire recording area is considered, and is "phase-reversed" with respect to the cortex salis, however. If one looks at the trace, there is certainly some correspondence between the wave shapes in the depth and on the surface (Figs. 1 and 2): deviations towards negative in deep cortical

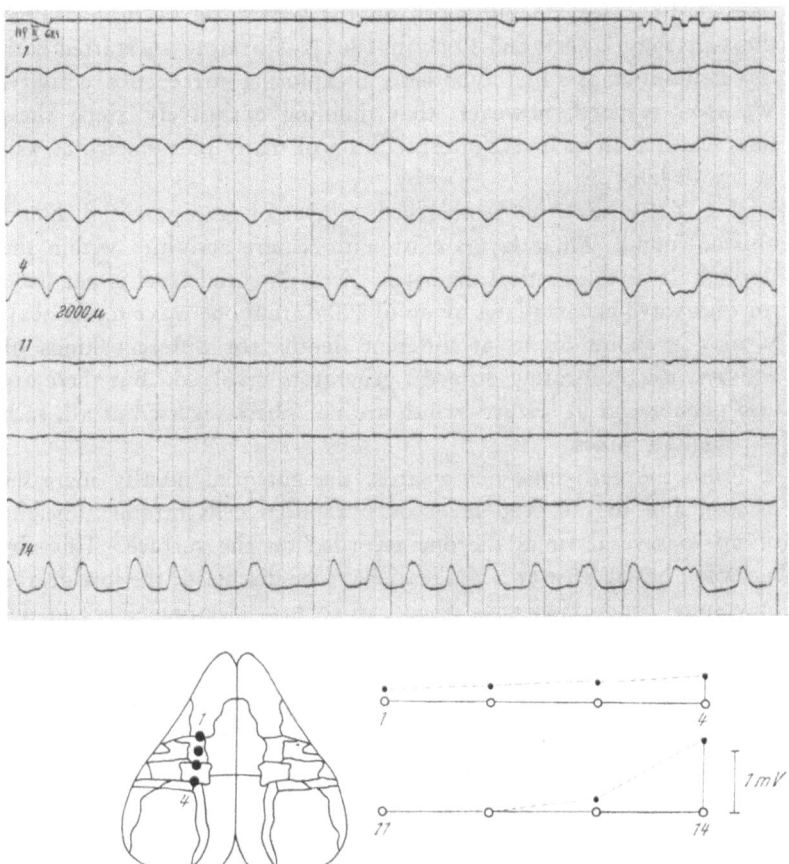

Fig. 2. Seizure activity produced by locally applied Ouabaine ($10^{-4}$ M). The potential gradients are steeper in deep cortical levels than on the surface, the amplitudes are larger and a phase reversal is present

layers are roughly coordinated to positive deflections on the surface and vice versa. The deeper the recording sites, the more frequently deep and surface records constitute true mirror images so that the concept of "phase-reversal" seems justified to describe this feature. Nevertheless, this is not quite the right term if records are compared other than those between surface and the deepest layers of the grey matter, namely records from within the layer of generators.

The concept of "phase-reversal" implies the existence of dipoles placed vertically within the cortex. Such an idea of dipoles arranged palisade-like arose in our work on the generation of hippocampal theta rhythm (Green and Petsche 1961). Therefore we started with this assumption as a first hypothesis in exploring intracortical activity. We soon realized, however, that thinking exclusively along these lines leads to an incorrect interpretation of many phenomena, as will be shown later on.

3. If a "phase-reversal" is observed, a layer of zero potential should also be found. These layers exist with seizure activities within the upper half of the cortical diameter. Zero potential level is not fixed for each wave but oscillates by up to 200 $\mu$ from one wave to the next. It may even be found at different depths for different kinds of activity, thus indicating different generators involved. But there are also phenomena in seizures which are not phase-reversed at all, such as penicillin spikes.

4. From the zero-zone downwards, the potential usually increases in size, and in the deep layer of pyramidal cells attains a height of up to several times the one recorded on the surface. This observation has at least two implications to be discussed: the low degree of volume conduction with respect to cortical electrogenesis and the primary importance of the large pyramidal cells of layer V for the generation of seizure potentials.

5. As far as synchronising force is concerned, there are also clear-cut differences between epi- and intracortical recordings: the similarity between two adjacent epicortical traces is much greater than between the two corresponding intracortical records. This difference is already noticeable in layers closely above the zero potential layer, but it is most pronounced at the level of maximum amplitude below this zone. This finding of epicortical records with much higher degrees of coherence than intracortical ones is common to all the bioelectrical activities we have studied thus far. For the reasons just mentioned, the size of the electrodes used cannot account for this finding.

This feature is clearly demonstrated in Figs. 1 and 2. Both the surface and the depth electrodes are 2 mm apart, and one electrode is 1 mm below the other. The declining seizure pattern of Fig. 1 was recorded from a longitudinal row of electrodes from the parietal cortex of the rabbit, electrode 4 being on the border-line between parietal area and striata. Surface activity is characterized by a pattern of almost equal shape in all four recordings whereas the activity below the surface proves to consist of at least three different phenomena: 1. spikes with a maximum at electrode 12 and a smooth gradient towards 11 and 13; 2. spikes, somewhat smaller than under 1, with a maximum at 12 too and a smooth gradient towards 11 but a steeper one towards 13; and 3. broader and more complex transients at 14 with a steep gradient towards 13. These three populations of EEG spikes are probably produced by different sets of neurons because they die out at different times. A careful study of this record also shows that the degree of synchronization between the different recording sites is not constant.

In contrast to this great variety of phenomena in the depth of the cortex, the epicortical activity is fairly uniform.

With the homogeneous seizure activities as usually seen after ouabaine, the above-mentioned behaviour is even more pronounced. In Fig. 2, the deep recordings were made from the layer of the deep pyramidal cells. Obviously both the amplitudes and the gradients are larger intra- than epicortically. This observation, however, does not justify concluding that there is a higher degree of volume conductivity at levels close to the surface.

Apart from these consistent observations on the differences between epi- and intracortically recorded seizure potentials, there are a few more general findings as to synchronization. There is a general relationship between the height of a certain epicortical potential and the probability that it will be detected in recordings from adjacent sites. The higher the voltage of a transient, the more likely is it also detectable at more distant electrodes. This rule, however, does not thoroughly hold good on the boundaries of cytoarchitectonic structures where extremely steep potential gradients may be found.

Another common finding is that slow transients usually reveal a greater tendency than faster transients to occur with equal shape over different areas.

Finally, differences in shape are most conspicuous along the Z-axis of the cortex. Consistent with this observation is that—with Penicillin—the steepest voltage gradients are also found along this axis.

When the phenomenon of phase-reversal was studied in more details, and particularly at levels close to the potential zero level, it turned out that there are events which are not reversed in phase. This is shown in Fig. 3 where, at each depth of the cortex recorded from,

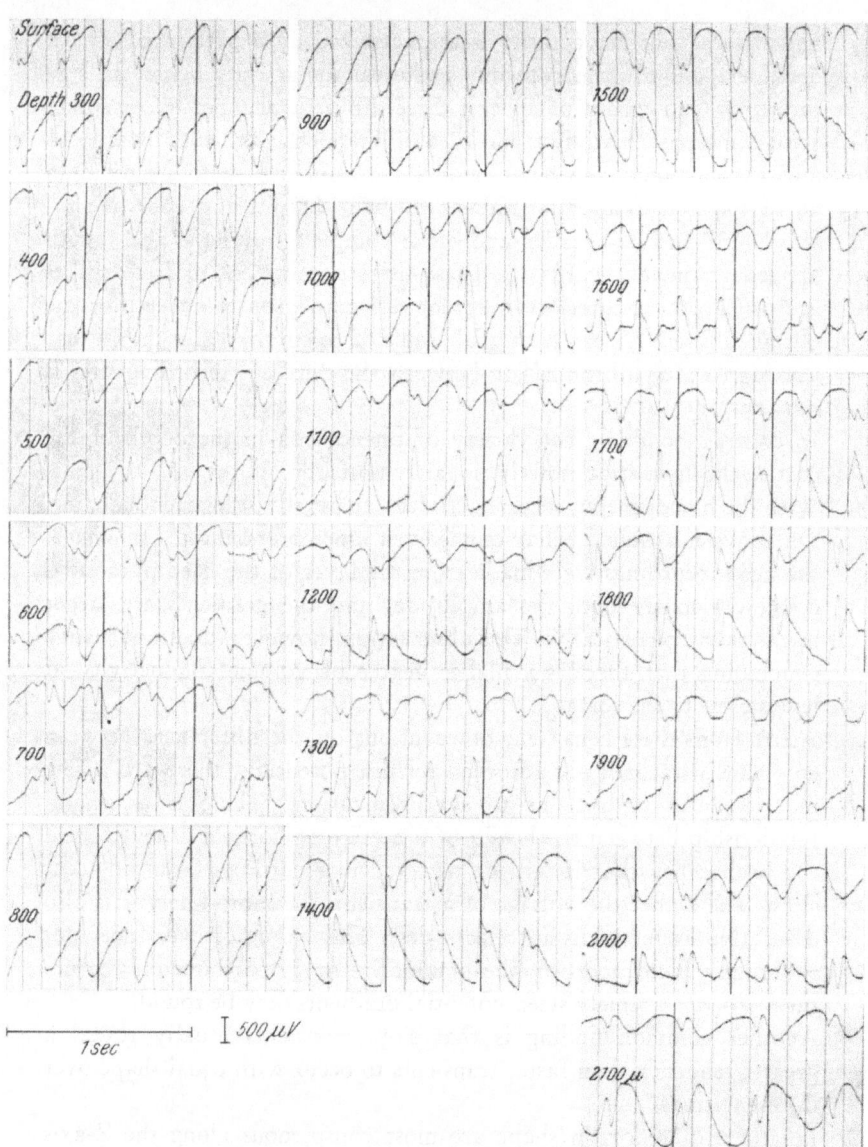

Fig. 3. Spike-and-wave patterns, elicited electrically (contralateral stimulation) and recorded at steps of 100 μ depth. Simultaneous recordings from surface and depth. Note the characteristic differences between the "wave" and the "spike": "Wave": remains in-phase with respect to surface down to 1000 μ. At about 1200 μ it is obviously phase-reversed. Potential zero-zone is at about 1100 μ. "Spike": has no potential zero-layer, but rather shifts in phase with respect to surface spike (as far down as about 1000 μ seems to climb up the wave)

regular spike-and-wave discharges were elicited by contralateral electrical stimulation. The electrode track was in the striata. Spike-and-wave activity is shown here in steps of 100 $\mu$ down to a depth

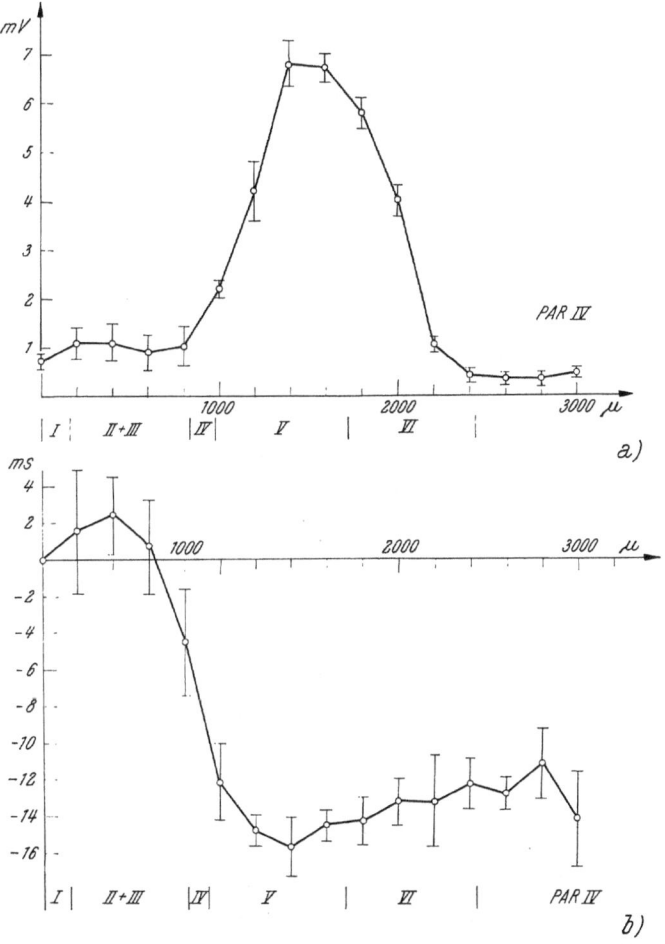

Fig. 4 *a*. Penicillin spikes, parietal area. Depth profile of the voltage of 20 sequential spikes at each depth. Mean and standard deviations. Maximum amplitude in layer V

Fig. 4 *b*. Average latencies of the same population of spikes with respect to surface. Spikes appear at the earliest in layer V

of 2000 $\mu$. At first sight, there is an obvious phase-reversal between superficial and deep recordings. This seems to be rather distinct for both waves and spikes. But if one follows the position of the spike with respect to the wave from the top of the cortex downwards one will notice that the wave seems to remain roughly in phase with respect to surface down to a depth of about 1100 $\mu$, and from then on it reverses its phase. But the spike behaves in a quite different way: starting from the surface and proceeding towards deeper layers, the spike seems to climb further and further up to the crest of the wave and thus to be more and more phase-shifted with respect to the surface spike. Moreover, at a depth of 1100 $\mu$, the wave seems to be close to zero level whereas the spike reaches a maximum.

This observation suggests a propagation of the "spike" from intra-cortical regions up to the surface, whereas the "wave" seems to better correspond with the concept of a vertical dipole. This finding of two kinds of cortical activities, of propagated and non-propagated type, is in agreement with the results of Peronnet et al. (1972).

In order to study the phenomenon of the intracortical vertical spreading of spikes in more detail, penicillin foci were produced. The penicillin spikes offer by their fairly constant shape the best and most constant conditions for investigating time relationships between intracortically and epicortically recorded spikes. In the present context only the time relationships along the Z-axis are to be considered. Fig. 4 a presents, from an electrode track in the parietal area, the average potentials of 20 successive penicillin spikes in steps of 200 $\mu$. Down to a depth of about 800 $\mu$, the potentials remain roughly constant; from this depth on, however, potentials rise steeply up to a summit of almost 7 mV at a level corresponding to the layer of deep pyramidal cells.

In Fig. 4 b, the average latencies of the spikes, recorded in steps of 200 $\mu$, with respect to the spikes recorded epicortically, are drawn. Again, the averages of 20 successive transients at each depth were chosen. This graph shows that, at the place where the spikes have their highest potentials, namely within the deep pyramidal layer, they originate and propagate towards both the surface and the lower parts of the cortex. The speed of propagation towards the surface is much slower (about 3 cm/sec) than the speed of propagation towards the deeper layers of the cortex (about 30 cm/sec).

It is not yet clear whether the propagation towards deeper layers is only apparent and simulated by the fact that different spikes may originate at different levels of the pyramidal layer. It is also uncertain whether or not the apparent surface-fugal propagation in the upper 400 $\mu$ is due to another set of generators linked with the deep ones or merely caused by their great dispersal near the surface. The shape of the spikes when observed with fast sweep-speed on CRO is not at all uniform but seems to be composed of several potential

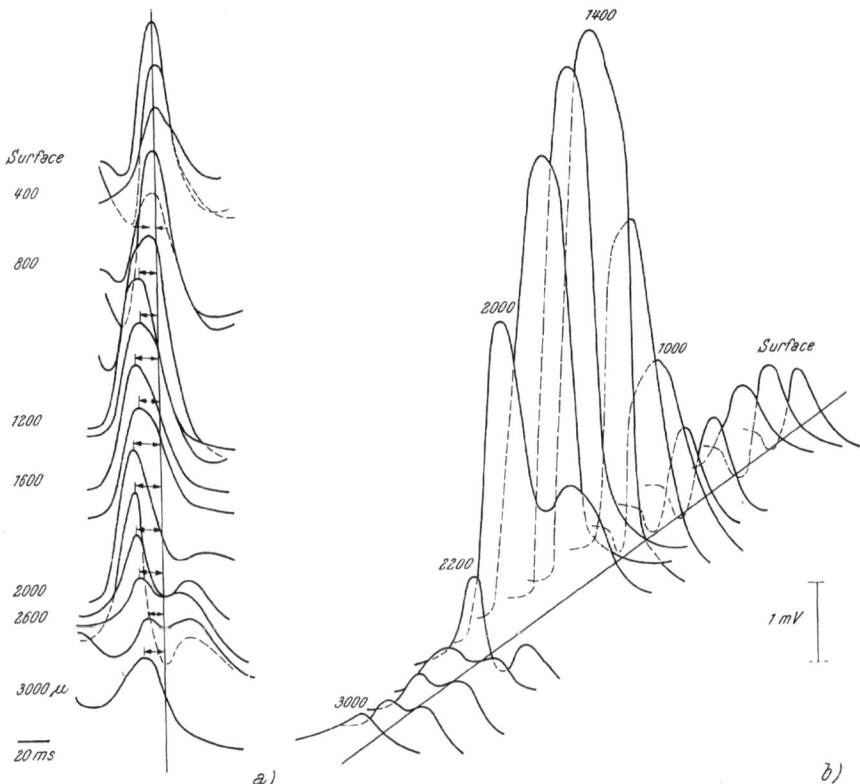

Fig. 5 a. Individual spikes of the same experiment as in Fig. 4 recorded at steps of 200 $\mu$ and arranged in correct time with the simultaneous surface spike. Spikes in a depth of about 1200 $\mu$ precede the other spikes. The amplitudes are not reproduced at the same scale. This was done in Fig. 5 b

components (PETSCHE et al. 1972). This is shown in Fig. 5 a, where single transients of the experiment of Fig. 4 are drawn at different depths, with their exact time relations to the epicortical spikes. Since the spikes change somewhat in shape from one instant to the other, the vertical propagation is not so clearly seen in this representation as in the one derived from an average of 20 transients at each depth. Fig. 5 a does not take into account the amplitudes in order not to make the drawing too difficult to survey. The exact potential relationships are sustained in Fig. 5 b.

Another observation stresses the probability of a propagation of spikes along the Z-axis: in the deep pyramidal layer, the spikes rise immediately from the potential zero line whereas, towards the surface, they are preceded by a positive deflection. The reason for this difference in potential shape between depth and surface recordings may be that the propagation of the potential takes place in an actively conducting tissue (nerve cells and its processes) embedded in a volume conductor. The potential, being initially negative at the zone of primary excitation (the "sink"), becomes initially positive as it travels.

These observations demonstrate that, at least for some fast transients, there is no zero potential zone within the cortex; this means that the origin of these potentials is quite different from that of the most prominent lower-frequency seizure waves.

The material presented thus far gives the impression that, in seizures, potentials with a certain degree of coherence may be found throughout the entire thickness of the cortex only. Even if there are differences in shape, caused by differences of frequency and amplitude, there usually seems to be a correspondence between deep and superficial activity. This assumption cannot be generalized, however. Fig. 6 presents an example of seizure activities recorded epicortically and from a depth of 2500 $\mu$ which seem to have no correlation whatsoever with each other. Neither the potential shape nor the total time sequence of the seizure pattern show any similarities with one another. This finding raises a further question which, however, is beyond the scope of this paper, namely the action of antiepileptic drugs on synchronization. The record was taken half an hour after an ouabaine status was stopped by intravenous application of Clonazepam[R] (Roche) 3.0 mg/kg. It may be that certain drugs have

Fig. 6. Low speed record (1 mm/sec) of reappearing seizure activity of a status epilepticus (locally applied Ouabaine, $10^{-4}$ M) that was temporarily stopped by Clonazepam (Roche) 3.0 mg/kg. 5 longitudinally arranged surface electrodes (1–5, 3 mm interelectrode distances) and the respective intracortical microelectrodes (11–15) at 2500 $\mu$ depth. Note the total independence of epi- and intracortical seizure patterns

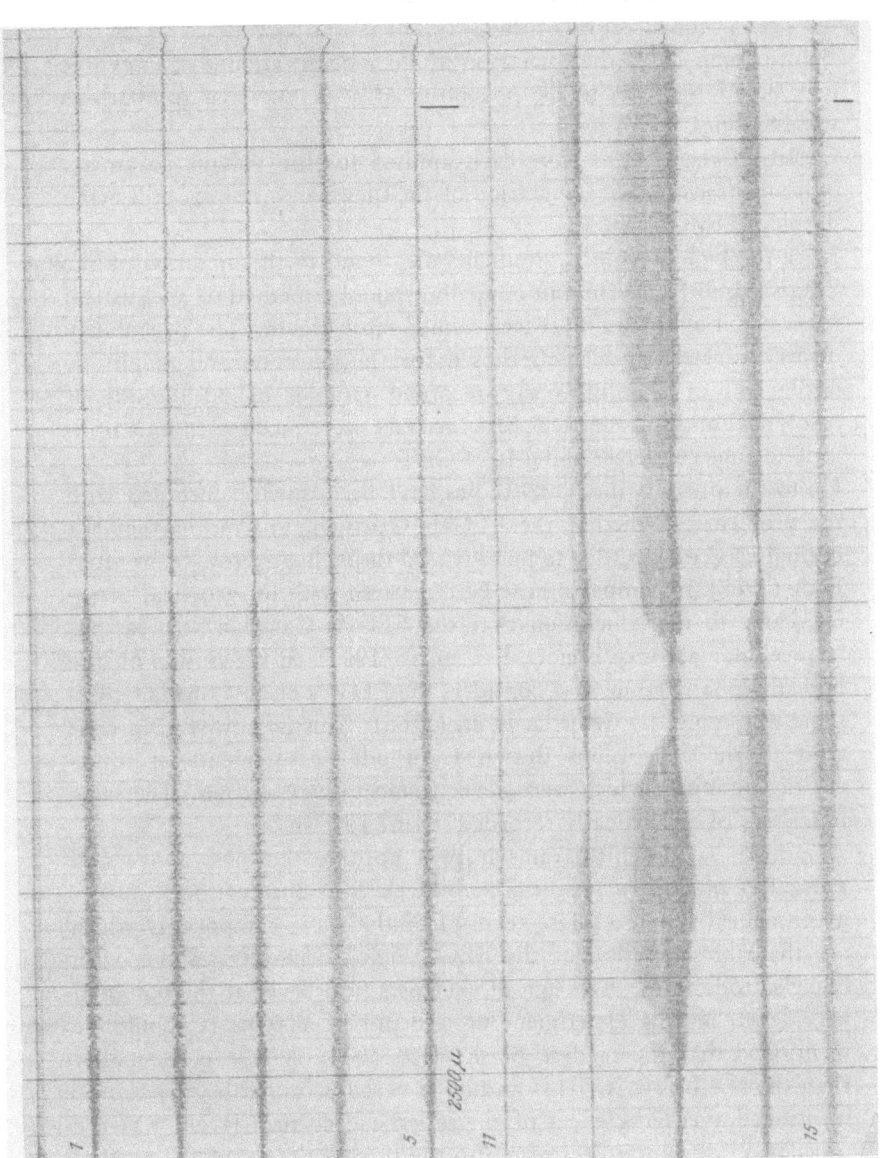

Fig. 6

the effect of restraining synchronization not only in the X-Y-plane but even in the direction of the Z-axis of the cortex.

Before going on to the discussion of these observations, a few words may be advisable as to the possibility of a quantitative approach to these findings.

Quantitative methods have been applied for the voltage parameter only; methods for the evaluation of the time parameter are still being developed by us.

Since the highest voltage gradients are found in the direction of the Z-axis (up to 10 mV/mm in Penicillin spikes) a method of a statistical registration of these differences seemed most urgent. The probability density function (pdf) was chosen for this purpose, and amplitude histograms of simultaneously recorded samples of 30 seconds of epicortical and intracortical EEG activity were made with an amplitude-to-time converter and a 1024 CAT.

Up to the present time, pdf-studies have been chiefly concerned with the problem of whether the EEG is Gaussian, this means that the normal EEG may be due to noise passed through a narrow-band filter. Elul (1967), in comparing the EEG pattern with intracortical activity, came to the conclusion that the EEG is Gaussian whereas the intracellular activity is not. Saunders (1963), in his studies on the human alpha rhythm, also considers the EEG as being Gaussian. The same was found by Walter et al. (1966). This is, however, in contrast to the conclusions drawn from pdf-measurements of intracortical activities which were never found to be Gaussian. The same holds true of epicortically recorded seizure patterns.

As a rule, seizure histograms have a positive skewness above and a negative one below potential zero level. This observation is in fact a consequence of the phase reversal of the wave components which are the largest portion of the EEG. Fig. 7 demonstrates this with one electrode track. The sign of skewness turns over at the potential zero level, in this experiment at a depth of 600 $\mu$. It should be mentioned that the reading of skewness above zero level is smaller than the one below it. This finding is perhaps caused by the above-mentioned averaging effect of the superficial cortical levels. On the other hand, skewness decreases also the further the tip of the recording electrode is away from the generator layer (e.g., in recording points close by the white matter).

The upper curve of Fig. 7 is constructed from statistical estimates of the amplitude ratios depth/surface. These data were calculated from the difference between the 95th and the 5th percentile of histograms and normalized by dividing them through the same difference but calculated from the simultaneous epicortical histograms. The curve demonstrates that—at deep levels of the cortex—voltages up to four

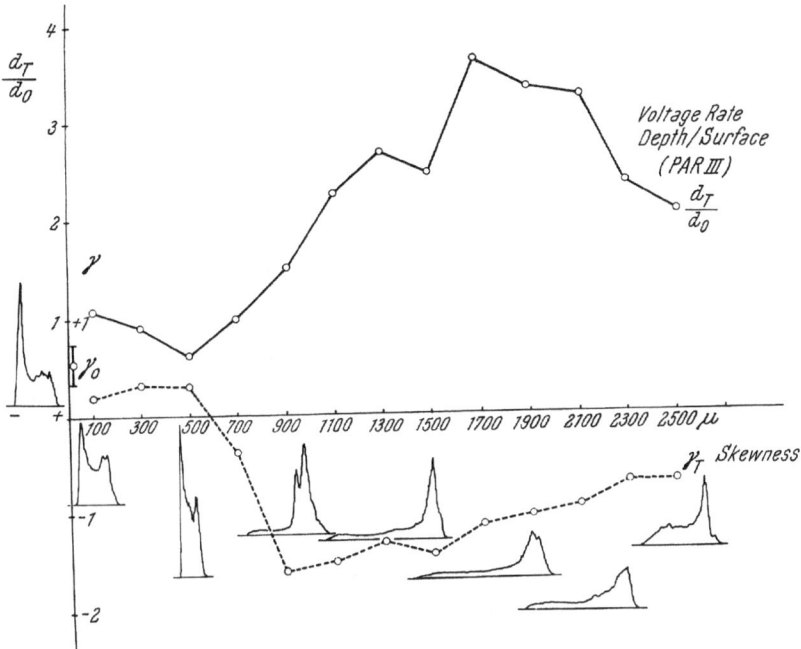

Fig. 7. Voltage ratio depth/surface (upper line), obtained from percentiles of probability density function (see text). Skewness of intracortical activity (dotted line). The skewness of the simultaneous surface record and its standard deviation is indicated at 0.54 ($s = 0.21$) near the Y-axis. Irregular spike-and-wave activity, parietal cortex. The histograms (30 sec) are drawn every 400 $\mu$. Potential zero-zone (at about 600 $\mu$ below surface) corresponds to the turning-over from positive to negative skewness. The histograms show also that the percentage of spike-components is largest at depths greater than 900 $\mu$. On the surface no spikes are seen. Histogram at the left of Y-axis: surface activity (from the shape of the histogram it may be concluded that this activity is almost sinusoidal but with the positive phase more fluctuating than the negative one)

times the surface voltages are measured, whereas at the site of reversing skewness a minimum is attained. At the very zero-level the ordinate ought to become zero; however, a true zero potential over periods of a few seconds is never observed because of the fluctuation of this zone by a few hundred micra as already described.

## Discussion

Based on the above observations, the characteristics of the epi- and intracortical seizure EEG may be outlined in the following tables (Tables 1, 2, 3).

These tables roughly summarize previous findings but are more a programme for future research than a summary of results. This is particularly true of the heading of "coherence". Coherence has not yet been quantified in these experiments; the conclusions on the different degrees of coherence are only derived from the observations on the degrees of similarity of the traces. This question is still to be studied by means of a computer.

One great difficulty in explaining these various observations is that when brain potentials are recorded, there is an interplay between the influences of two basic conditions the proportions of which cannot be clearly differentiated: volume conduction (i.e., conduction in an electrically inactive electrolytic conductor) and active conduction (comparable the one of action potentials in nerves). There is, however, a phenomenon that cannot be sufficiently explained by either of these two qualities, namely the phase-shift of potentials in the X-Y-plane. This phenomenon, which indicates some sort of propagation, has probably something to do with spreading along axons and dendrites plus several additional and complex processes caused by synaptic delays. The only methods exerted up to the present for differentiating the different kinds of conduction involved in this phenomenon are incision to the cortex and its laminar destruction by Ouabaine. These procedures have indeed resulted in some better understanding and evaluation of the relatively low degree of volume conduction in this context, but they do not permit us to estimate whether the proportion of volume conduction is the same in all three parameters of space. It has not been possible to ascertain whether the differences in amplitude gradients epi-, intra- and transcortically are merely a function of the different degrees of volume conduction or of the arrangement of nerve cells. In our opinion, the fairly steep gradients in the Z-axis reflect the very low role of volume conduction (the heavy horizontal fibre bundles of layer II and IV may support

Table 1. *Frequency Distribution of Seizure Waves*

| Recording site | Low frequencies | High frequencies |
|---|---|---|
| Epicortical | +++ | + |
| Intracortical | ++ | +++ |

Epicortically there is a prevalence of low over high frequencies; intracortically, higher frequencies are more pronounced.

Table 2. *Amplitude Distribution of Seizure Waves*
Voltage ratios with respect to the epicortically recorded potentials of low frequency.
Epicortically, the high frequency components ("spikes") are usually lower than "waves".
Zero potential zone: only "spikes" are seen. Intracortically, below zero-zone: spikes are usually larger than waves (the maxima were obtained from Penicillin seizures)

| | Low frequency components ("waves") | High frequency components ("spikes") |
|---|---|---|
| Epicortical | 1 | 0-1 |
| Intracortical (zero potential zone for waves) | 0 | < 7 |
| Intracortical (2000 μ) | < 4 | < 7 |

*Table 3*
Epi — epi: records in the X-Y-plane from surface.
Intra-intra: in the X-Y-plane below zero potential zone.
Epi-intra: transcortical recording (along Z-axis).
For both high and low frequencies the highest degrees of coherence are found in the X-Y-plane of the surface. The highest potential gradients are along the Z-axis. Propagation ("travelling") is most pronounced in the X-Y-plane on the surface, for "spikes" also in the direction of the Z-axis

| Recording (site and direction) | Coherence | | Potential gradient (maximal) | Propagation | |
|---|---|---|---|---|---|
| | "spikes" | "waves" | "spikes" + "waves" | "spikes" | "waves" |
| Epi — epi | +++ | +++ | 2 mV/mm | +++ | +++ |
| Epi — intra | + | ++ | 10 mV/mm | +++ | 0 |
| Intra — intra (below zero-potential zone for "waves") | + | + | 4 mV/mm | + | ++ |

this idea). On the other hand, the smoother gradients in the X-Y-plane may be furthered by the dense horizontal feltwork of fibres, in combination with the parallel arrangement of other parymids and their horizontal connections by which the potential of adjacent regions does not so quickly decline.

The assumed low passive conductivity along the Z-axis may also be responsible for the difference in frequency spectrum epi- and intra-cortically.

As for the amplitude distribution, the structure of the cortex may be responsible too. That the potentials usually are highest in the deep pyramidal layer indicate that here may be the very origin of the main components of a seizure, an assumption that was stressed by other papers presented at this Symposium (Peronnet et al. 1972, Speck-mann et al. 1972, Scherrer et al. 1972). But there are also earlier observations (Adrian 1936, Burns 1958) which emphasize the important role of these deep layers for maintaining and propagating after-discharges.

The cortical structure may also be the reason for the different degrees of coherence between surface and depth. Most obviously, the "filter" effect of the superficial layer is a consequence of 1. the richness of the horizontal fibres and their connections and 2. the poorness of the nerve cells and therefore of the generators contributing to the EEG. Less well explained is the fact that, in the deep layers with their rich horizontal fibre systems, the degree of coherence in the X-Y-plane is much lower. As far as the situation of coherence in the Z-axis is concerned, vertical columns of cells seem to interact in some still obscure way so as to give rise to the most complicated correlations found in vertical direction, as illustrated by the penicillin spikes.

One rather peculiar finding is the basic difference in the origin of "waves" and "spikes" in that the latter propagate from the depth to the surface whereas the first seem to reflect the events on an electrically charged membrane of the character of a dipole layer. That there is a basic difference between the nature of "waves" and that of "spikes" was first shown by Pollen (1964) who found the waves were well correlated with IPSPs and therefore concluded that they are secondarily generated after the EPSPs which seem to be connected to the "spike". The surface large negative wave was considered by Pollen et al. (1964) to "result from the apical dendrites serving as passive sinks for such deep sources". It has not yet been ascertained whether these two kinds of behaviour are representative of two groups of different phenomena and may comprise the total of possible brain potentials. It is certain, however, that a phase-reversal is the rule in slow wave components (that act as a kind of "carrier waves"). Propagating along the Z-axis is usually found in penicillin spikes and the "spikes" of spike-and-wave activity. From our previous findings it may be concluded, however, that the majority of the events called "spikes" in EEG terminology belong to this group.

Moreover, there seems to exist another group of spikes of lower voltage and produced by small generators in certain layers which do not propagate far enough to be detected at all on the surface. Highly synchronized activities other than seizure patterns, e.g., sleep patterns, very often exhibit transients that are seen at a certain depth only. The study of sleep spindles revealed, in many cases, a lacking of coherence between the spindle waves intra- and epicortically. This matter has to be studied more thoroughly, however, if consistent conclusions are to be drawn. The ability of the deep layers to develop self-sustained seizure patterns without the least reflection in super-ficial layers was clearly demonstrated in Fig. 6. This shows that to produce a seizure pattern, the total diameter of the cortex is not needed.

According to these findings, the phenomenon of synchronization which is thought to underlie the EEG becomes very complex and different if studied from the viewpoint of the different directional parameters of the cortex. They show in addition that this problem may only be solved by investigations based on an intimate knowledge of cortical circuitry.

## Summary

The phenomena of synchronization in seizures were studied along the X- and the Z-axis of the rabbit's cortex by means of surface- and intracortical recordings (gross- and semi-microelectrodes).

The most characteristic findings were the greater proportion of higher frequencies of intracortical recordings, its higher amplitudes, a sort of "phase-reversal" of lower-frequency events between surface and deep cortical layers, a minor tendency to a simultaneous occur-rence of potentials in deep than in superficial levels (lower degree of coherence), indicating some sort of filter effect of these latter, and therefore also steeper potential gradients at deep cortical levels than at the surface.

When spike-and-wave activities were studied throughout the cortical diameter, the waves behaved as if caused by a vertical dipole, displaying a phase-reversal, whereas the spikes seemed to propagate from deep levels towards the surface. The latter observation was also confirmed for penicillin spikes which were studied more thoroughly.

Observations during Ouabaine-induced seizures proved that to produce seizure patterns, not the entire diameter of the cortex is needed.

The different possibilities of conduction which may play a role during synchronization are discussed.

## References

Adrian, E. D.: The spread of activity in the cerebral cortex. J. Physiol. *88*, 127—161 (1936).

Burns, B. D.: The mammalian cerebral cortex, p. 119. London: Edward Arnold Ltd. 1958.

Elul, R.: Statistical mechanisms in generation of the EEG. In: Progress in Biomedical Engineering, Vol. I, pp. 131—150. Washington, D.C.: Spartan Books. 1967.

Green, J. D., and H. Petsche: Hippocampal electrical activity. IV: Unitary events and genesis of hippocampal seizures. Electroenceph. clin. Neurophysiol. *13*, 868—879 (1961).

Li, C. L., C. Cullen, and H. H. Jasper: Laminar microelectrode analysis of cortical unspecific responses and spontaneous rhythms. J. Neurophysiol. *19*, 131—134 (1956).

Peronnet, F., M. Sindou, A. Laviron, F. Quoex, and P. Gerin: Human cortical electrogenesis through stratigraphy and spectral analysis. See this book, pp. 235—262.

Petsche, H., und P. Rappelsberger: Der Strophanthinstatus als Modell für die Auswèrtung antikonvulsiver Medikamente. Pharmakopsychiatrie *3*, 151—161 (1970 b).

— — Influence of cortical incisions on synchronization patterns and travelling waves. Electroenceph. clin. Neurophysiol. *28*, 592—600 (1970 a).

— — Spatio-temporal and laminar analysis of self-sustained cortical activity. (In press.)

— — und Zs. Frey: Intrakortikale Mechanismen bei der Entstehung der Penicillin-Spitzen. EEG — EMG. *2*, 176—180 (1971).

— — and R. Trappl: Properties of cortical seizure potential fields. Electroenceph. clin. Neurophysiol. *29*, 567—578 (1970).

— and J. Šterc: The significance of the cortex for the travelling phenomenon of brain waves. Electroenceph. clin. Neurophysiol. *25*, 11—22 (1968).

POLLEN, D. A.: Intracortical studies of cortical neurons during thalamic induced wave and spike. Electroenceph. clin. Neurophysiol. *17*, 398—404 (1964).

— K. H. REID, and P. PEROT: Microelectrode studies of experimental 3/sec wave and spike in the cat. Electroenceph. clin. Neurophysiol. *17*, 57—67 (1964).

SAUNDERS, M. G.: Amplitude probability density studies on alpha and alpha-like patterns. Electroenceph. clin. Neurophysiol. *15*, 761—767 (1963).

SCHERRER, J., and J. CALVET: Normal and epileptic synchronization at cortical level in animal. See this book, pp. 112—132.

SPECKMANN, E. J., H. CASPERS, and R. W. JANZEN: Relations between cortical DC shifts and membrane potential changes of cortical neurons associated with seizure activity. See this book, pp. 93—111.

WALTER, D. O., J. M. RHODES, D. BROWN, and W. R. ADEY: Comprehensive spectral analysis of human EEG generators in posterior cerebral regions. Electroenceph. clin. Neurophysiol. *20*, 224—237 (1966).

WOODBURY, J. W.: Potentials in a volume conductor. In: RUCH, I. C., and J. F. FULTON (eds.), Medical physiology and biophysics, pp. 83—91. Philadelphia: Saunders. 1960.

# Discussion

LEHMANN: You have shown examples of travelling of the maximum value of mapped potential fields. In our map series it is exceptional that a field maximum moves continuously across the recording area. However, we look at larger surfaces (most of the scalp) and at men. Have you found continuous travelling of the maximum value of potential fields over larger recording areas?

PETSCHE: There are several reasons why, in humans, the conditions in which the travelling phenomenon can be observed are much worse than if you record directly from the flat cortex of a rabbit: (1) the layer of generators, *i.e.*, the cortex, is curved in three dimensions, (2) the relatively high conductance of CSF but also of the skull bone and of the skin contribute a great deal to the smoothing and enlarging of EEG potentials. Therefore, in humans, travelling of brain potentials may best be seen in highly synchronized events, such as in generalized patterns of uniform shape.

LEHMANN: Have you observed migration of the maximum value of the potential field over greater distances, *e.g.*, from the occipital pole to the frontal pole?

PETSCHE: Yes. Some of the potentials in the spike-and-wave pattern propagate all over the skull even if, in most cases, there is a delay at the boundary between frontal and parietal cortex.

VERZEANO: When one records, simultaneously, the gross waves and the neuronal activity, with the same microelectrode at the same point, one finds that, in penetrating through the cortical layers, the waveform and polarity of the gross waves are related very closely to the neuronal activity *at the recording* point, whatever its location in the cortex may be. One must conclude that the gross waves in the deep layers are due to the activity of neuronal elements in the deep layers, and the gross waves in superficial layers are due to the activity of elements in the superficial layers. The idea that the development of gross waves is based on the so-called dipoles formed by the apical dendrites and the somata of pyramidal cells may have to be abandoned.

Gloor: In assessing the extent of volume conduction it is necessary to take into consideration the geometry of the cortex. For example, the curved layers of the hippocampus produce powerful field effects which extend over relatively large distances.

Petsche: You are thinking of Woodbury's theorem, aren't you?

Gloor: Yes.

Elul: In contrast to the Gaussian distribution of the EEG in the waking man or in the waking cat, during seizures the EEG is not Gaussian, suggesting interruption of the normal randomness of neuronal synchronization in the cortex.

Petsche: Even an activity as regular as the hippocampal theta rhythm in rabbits is not Gaussian.

Invited Discussion

## Some Results of the Analysis of Epileptic Seizure Patterns by Correlation-Methods

G. Pfurtscheller

Institute of Medical Electronics, Graz, Austria

In this paper I will report on some results of analysis of epileptic seizure activity by correlation methods. The analysis to be reported here relates to one experiment that has been discussed in this symposium by Petsche et al. (1971).

A comprehensive analysis of epileptic seizure patterns by mathematical methods is only possible if the data can be processed with a computer. However it is necessary to record several simultaneously recorded EEGs in binary coded form on an IBM compatible digital tape.

The first figure shows a block diagram of the data acquisition- and data processing equipment which is used.

The EEGs which were simultaneously taken from the cortex surface and the depth of the cortex were amplified and filtered and then recorded with a multi-channel analogue tape recorder. Then the individual analogue signals were transferred from the analogue tape recorder via a multiplexer and an analogue-digital-converter on to an IBM-compatible digital tape. The signals of 4 channels with a sampling frequency of 280 Hz were transferred. 8 bit were chosen for the word length which corresponds to a quantisation accuracy of 1% (Pfurtscheller 1971).

Data processing was carried out with the general purpose computer UNIVAC 494 in the computing center Graz. The autocorrelograms of EEGs with a duration of 5.7 seconds were computed and the spectral power densities were calculated from these values. Fig. 2 shows an example of the distribution of the spectral power densities

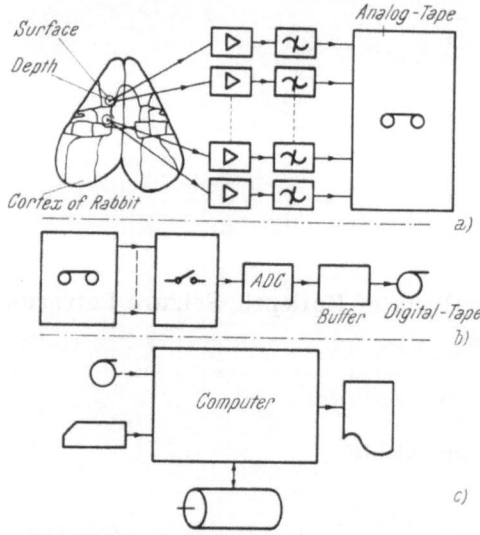

Fig. 1.  Data acquisition and data processing equipment

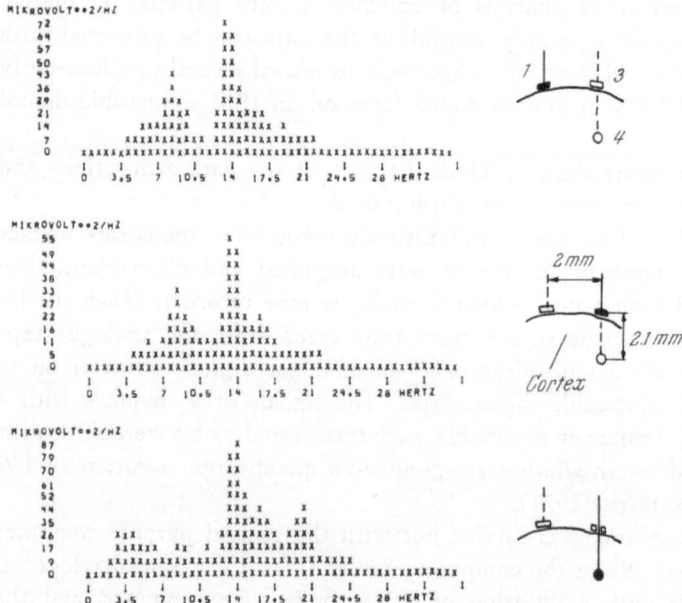

Fig. 2.  Spectral power density distribution of seizure patterns

of two EEGs from the surface and one from the depth of a rabbit's cortex. All EEGs were taken simultaneously.

It is striking that these power densities show approximately the same distribution of surface EEG and depth EEG. The two power density peaks of 8 and 14 Hz probably correspond to the "waves" and the "spikes".

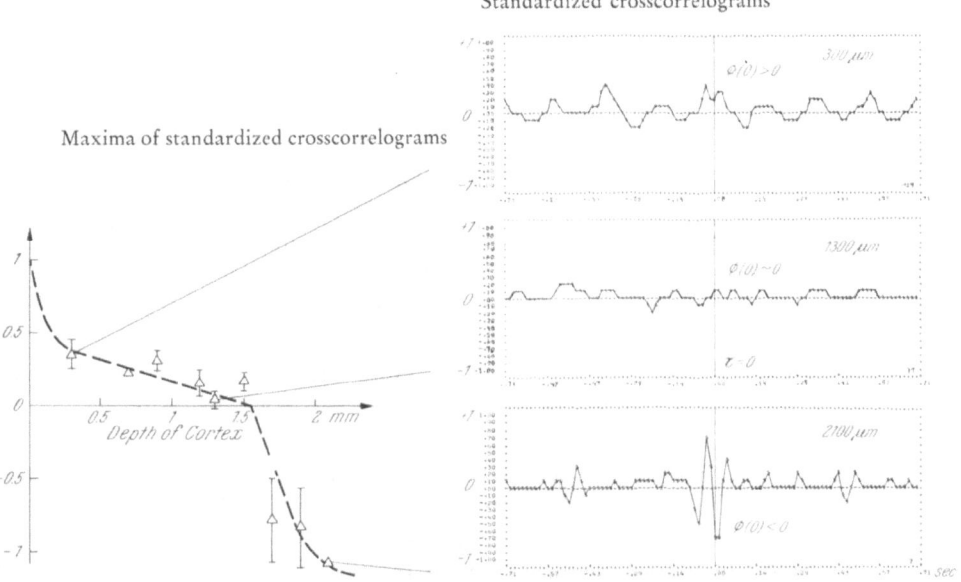

Fig. 3. Crosscorrelogram—maxima at increasing depth and crosscorrelograms

To be able to come to a conclusion concerning the origin of the epileptic patterns, surface- and depth-EEGs, which were taken at the same time, must be compared. To be able also to compare depth EEGs which were obtained at different times, the surface- and depth EEGs are standardized, that is every depth EEG is related to a surface EEG with constant standard deviation. For this purpose the standard deviation of the surface EEG in a certain time range is computed and then the individual voltage values of the surface- as well as of the depth EEG are divided by the standard deviation. This procedure guarantees that the individual surface EEGs which are examined have always the same standard-deviation amounting to 1.

For the purpose of gaining information on the phase and amplitude of the depth epileptic activity in relation to the simultaneously taken

surface epileptic activity, crosscorrelograms (CC) between the standardized surface- and depth EEG were computed and the function
maxima near the zero point on the $\tau$-axis were determined. The
distribution of these maximum values can be seen from Fig. 3.

The CC-maxima show a very characteristic behaviour, that is a
change of polarity in a certain depth. Up to a depth of about
1100 $\mu$m positive maximum values, from a depth of 1700 $\mu$m negative ones can be observed. At a depth of 1300 to 1500 $\mu$m no pronounced maximum values of crosscorrelograms can be observed. The
right side of Fig. 3 shows 3 examples of such CCs computed from
surface- and depth EEG. As can be seen, the highest CC shows a
positive peak and the lowest one a negative peak ($\tau = \emptyset$). The CC
in the middle however shows no pronounced extreme value at
$\tau = \emptyset$.

This peculiar behaviour of the CCs, that is the change of polarity of
the maximum values at a certain depth, perhaps permits the conclusion that there exists a polarized layer of dipole-sources at a depth
of about 1100 to 1700 $\mu$m. Because of the fact that the potential in
the middle of such a polarized layer is approximately zero, the
activities from these layers used for the CC-compution yield no
pronounced maxima in the neighbourhood of the zero point. The
irregularities in the CCs of surface- and depth EEG indicate that
in certain conditions not the whole seizure activity, but only the
greatest part can be explained by a dipole-layer.

We have also studied the behaviour of the square mean roots of the
depth EEGs depending on the derivation depth. For this purpose
we have calculated first the individual square mean roots of several
patterns with a duration of 5.7 seconds derived from the same cortex
depth and then the average value of these square mean roots and
the standard deviation. Fig. 4 shows these mean values plotted
against the derivation depth. The increase from a depth of 1700 $\mu$m
is significant, but a second maximum at 400 $\mu$m and a minimum at
1500 $\mu$m are less significant. It is remarkable that this minimum
coincides with the zero cross of the CC-maxima-distribution shown
in Fig. 3.

If the assumption that a layer of dipole-sources produces an essential
part of the epileptic seizure is correct, the square mean root should
have a minimum in the middle of this layer and a maximum on both
the boundaries of the layer. This behaviour is pointed out in Fig. 4,
but it is not very significant for the first maximum and minimum

and should therefore be further investigated with regard to significance.

The fact that the CC-maximum-distribution in Fig. 3 shows a different increase in the positive and negative range as well as the fact that the second maximum in Fig. 4 is much higher than the first maximum may be due to the difference in the epileptic seizures at derivation from a depth of 700 $\mu$m and 1900 $\mu$m for example, which are not recorded at the same time.

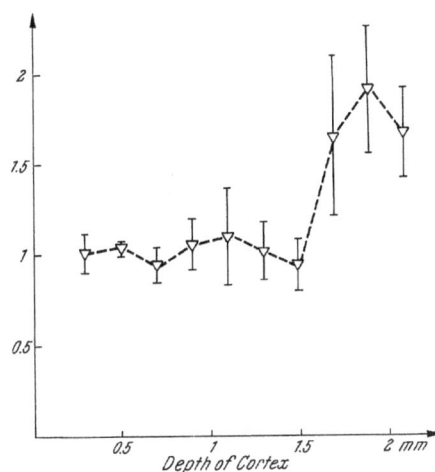

Fig. 4. Square mean roots of standardized depth seizure patterns

As the data we have studied were obtained from only one experiment with one animal, we are not yet able to say whether the different increase of the CC-maxima-distribution up to a depth of about 1500 $\mu$ and from a depth of about 1500 $\mu$, is only accidental, or if it occurs in the same manner in other experiments too and is therefore specific. In any case it is necessary to clear up in further tests and by mathematical analyses if the localization of the electric generators which are responsible for the epileptic seizure activity at a depth of 1300 to 1500 $\mu$m is correct and if the generators have dipole-character.

I hope that I have been able to prove with these examples the importance of mathematical-computer methods in the field of epilepsy research and to show what results can be expected from these methods.

19 Synchronization

## References

Petsche, H., P. Rappelsberger, and Zs. Frey: Intracortical aspects of the synchronization of selfsustained bioelectrical activities. See this book, pp. 263—284.

Pfurtscheller, G.: Die simultane Mehrkanal-EEG-Analyse mit dem Großrechner. Proc. of the IFAC-Symposium of Automatic Control and Computers in the Medical Field, Brüssel (1971).

# Phylogenetic Aspects of Synchronization in the Electrogenesis of Epileptic Phenomena. Thalamo-Cortical Mechanisms in Lower Vertebrates

Z. SERVÍT

Institute of Physiology, Czechoslovak Academy of Sciences

It is generally accepted that excessive synchronization (hypersynchrony) should be one of the essential features of epileptic activity in the nervous tissue (see e.g., JASPER 1969, CREUTZFELDT 1969, and AJMONE MARSAN 1969).

Characteristics, or even the existence of an "epileptic neuron" are still under discussion (AJMONE MARSAN 1969). Nevertheless, an epileptic neuron can manifest its abnormal epileptic reactivity only in an epileptic neuronal aggregate (AJMONE MARSAN 1961) and "the understanding of basic mechanisms of epilepsies can never be achieved by considering properties of single cells alone, for seizures are always consequence of mass discharge of thousands or many millions of cells, usually in excessively synchronized discharge" (JASPER 1969). Both abnormal and normal neurons may be driven in abnormal, excessive synchronization.

On the other hand, synchronization of neuronal activity in a "normal", not excessive form should be a characteristic property observed in and—possibly—inherent to all neuronal aggregates. It has been seen at all levels of vertebrate brain phylogeny and also in the central nervous system of invertebrates (the pioneer observations of ADRIAN (1930, 1931) should be remembered in this connection).

When speaking about synchronization in the brain, two kinds of synchrony may perhaps be discerned, which have special relations to the pathogenesis of epileptic phenomena: the *local synchrony* observed especially in the interictal discharge of a cortical epileptic focus and synchronization of the activity of *distant brain areas* well known, for instance, as the bilateral synchrony of generalized epileptic activity in both forebrain hemispheres.

This is connected with a still open question of central synchronization mechanisms (thalamic pacemaker, thalamo-cortical reverberating circuits, septal pacemaker, centrencephalic system—these are the names indicating the most important problems). The term "synchronization" may be understood in the meaning generally used in electroencephalography, *i.e.*, not in the sense of a strict simultaneity of events. The synchrony can not be absolute because synaptic mechanisms are the most important tool of synchronization. Local synchrony may be considered as a product of activity of a neuronal population in which the electrical transients in individual neurons are only approximately isochronous. In the so-called bilateral synchrony phase differences of the activity in cortical fields of both hemispheres may not be detectable in EEG paper recordings, they thus should not last longer than about 10–20 milliseconds. (It is of course questionable, if the term "synchrony" is appropriate enough in this case—see PETSCHE and RAPPELSBERGER 1970.)

A basic question of epileptology may be the relation of *abnormal* (*i.e.*, excessively synchronized) epileptic activity to *normal* manifestations of synchronized neuronal activity—for instance the alpha activity of relaxed wakefulness, spindle (sigma) and slow wave theta and delta activity of sleep and transitory states between wakefulness and sleep. The relations between behavioral epileptic symptoms and normal states of vigilance (different levels of vigilance), relaxation and sleep, questions of physiological and pathological deafferentation of brain structures may also be significant in this connection.

In this paper I will review the results of some of our comparative electrophysiological studies concerning the mechanisms of local and generalized epileptic hypersynchrony and the relations of these pathological activities to some normal manifestations of synchronized neuronal activity at different levels of brain phylogeny.

The comparative phylogenetic approach makes it possible to find out appropriate experimental models at certain crucial levels of brain evolution, where a definite function which is supposed to play an important role in the mammalian brain, appears first, in a simplified form. The study of such a primitive stage of development may help to understand better the complex phenomena in the mammalian and human brain.

*Focal interictal epileptic activity* could be demonstrated in forebrains of lower vertebrates at all important levels of brain phylogeny. With locally applied penicillin it is possible to evoke an epileptic focus manifesting itself by highly synchronized discharges in primitive laminary organized cortical structures of amphibia and reptiles as well as in the fish forebrain which completely lacks any laminar organization (STREJČKOVÁ 1969, SERVÍT 1970) (Fig. 1).

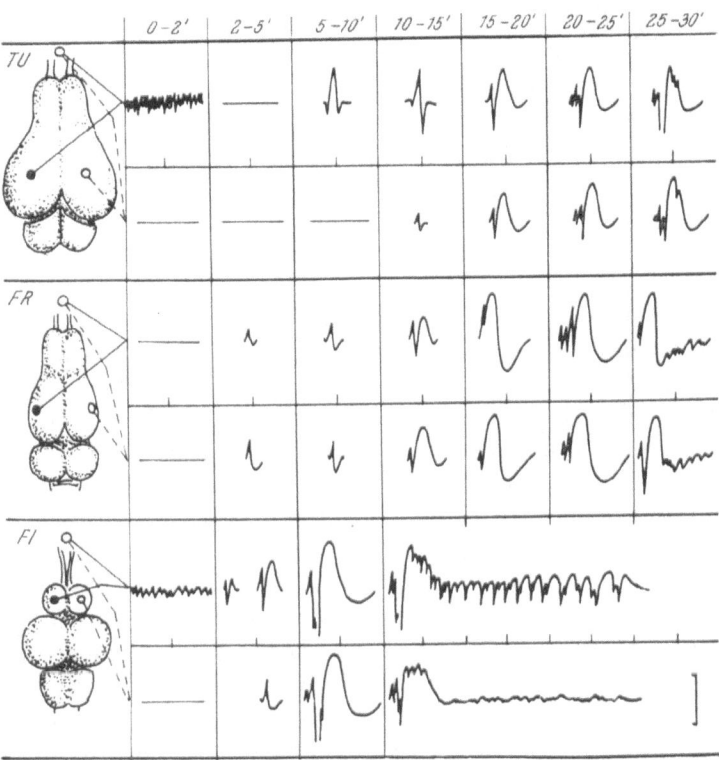

Fig. 1. The development of the pattern of the interictal discharge at the site of applied penicillin and at the contralateral projected focus in turtle (*TU*), frog (*FR*), and fish (*FI*) (semi-schematic drawing, monopolar recording). Left: primary focus; right: secondary focus; time after application of penicillin in min; calibration: 1 mV. (From: SERVÍT 1970)

The primary focus may not only produce projected activity in the contralateral forebrain hemisphere, but also an independent secondary focus both in fish (STREJČKOVÁ 1969, SERVÍT and STREJČKOVÁ 1969) and frogs (SERVÍT and STREJČKOVÁ 1967, WILDER and MORRELL 1967, WILDER et al. 1968, SERVÍT 1970).

The forebrain structures of inframammalian vertebrates are therefore able to generate epileptic neuronal aggregates with highly synchronized interictal activity. A layered cortical structure is not indispensable for this type of synchrony.

In all these animals we can also observe *bilateral synchrony in both forebrain hemispheres.* Electrographic seizures of spike-and-wave activity which may be bilaterally synchronous could be elicited here by different means—e.g., by electroshock (in frogs, lizards, and turtles—Servít et al. 1965, Servít 1965, 1966, Servít and Strejčková 1966, Volanschi and Servít 1969), with metrazol (in toads—Morocutti and Vizioli 1957), or with locally applied penicillin (in fish—Strejčková 1969 and frog—Wilder and Morrell 1967, Servít and Strejčková 1967). Besides spike-and-wave paroxysmal 10 c/sec activity was also observed in frogs (Servít et al. 1965, Servít 1966).

It seemed to us important to identify the *mechanisms engaged in the genesis of these generalized paroxysmal activities.*

From experiments on the mammalian brain two mechanisms are supposed to be active in bilateral synchrony. The first is cortico-subcortical in nature—thalamo-cortical reverberating circuits or—in a broader sense—the centrencephalic system of Penfield and Jasper. The second one may involve commissural connections of cortical fields of both hemispheres (corpus callosum, commissura anterior). The experiments of Marcus and Watson (1966, 1968) and Marcus et al. (1968) have proved that these commissural pathways are sufficient to preserve bilateral synchrony of spike and wave activity independently of thalamic and other brain stem structures. From the extensive work of Petsche and coworkers (Petsche and Šterc 1968, Petsche and Rappelsberger 1970) it follows further that the so-called synchronized activity in large cortical regions is not generally synchronized in the strict sense of the word. Large cortical fields may be successively invaded by travelling waves of synchronization. The synchronization in the mammalian neocortex thus evidently is a rather complex phenomenon, in which local intracortical mechanisms are very important.

In fish, where the functional integration of the brain is very poor, where the connections between forebrain and lower brain structures are limited and thalamo-cortical relations of the mammalian type do not exist, it is possible to elicit very typical paroxysmal spike-and-wave activity, which sometimes is bilaterally synchronous. This is possible both in the intact brain (Strejčková 1969) and in an isolated

forebrain deprived of all connections with diencephalic and mesencephalic centres (SERVÍT and STREJČKOVÁ 1970).

In frogs, paroxysmal spike-and-wave activity of very regular
10 c/sec frequency has also been evoked in the isolated forebrain with
interrupted thalamo-cortical communication. No significant differences concerning the shape and frequency of paroxysmal electrographic activities could be observed here in comparison with the
intact brain (SERVÍT et al. 1966). The paroxysmal 10 c/sec activity
in the frog forebrain has also been compared with human alpha
activity using autocorrelation and spectral density function. No
distinct differences could be detected between both groups of records
(KREKULE et al. 1966).

Evidently, in these species of submammalian vertebrates highly synchronized and sometimes bilaterally synchronous paroxysmal activities can be generated in both forebrain hemispheres independently
of lower diencephalic and mesencephalic structures. Moreover, activity which appears as a normal component of human EEG in relaxed
wakefulness (the alpha-activity) appears in frogs as an electrograph-
ical correlate of a clearly pathological, epileptic state. Synchronized
activity of large neuronal aggregates with the same electrographic
shape and frequency can thus arise under different (physiological and
pathological) conditions in very different brain structures. Thalamo-
cortical synchronization mechanisms are not a necessary condition for
the genesis of these activities at least as concerns their macroelectro-
graphic pattern.

On the other hand, complete transection of forebrain commissures in
fish makes the dissemination of epileptic activities between two forebrain hemispheres impossible (SERVÍT and STREJČKOVÁ 1970). Transverse commissural communication between both hemispheres seems
therefore to be a phylogenetically more primitive tool of bilateral
synchrony.

In the frog and the turtle such a transection of commissural connections does not prevent the spread of epileptic activities to the contralateral hemisphere. Ascending and descending pathways through
diencephalic and mesencephalic structures may play a role here both
in the propagation of focal activity and in the dissemination of the
seizure (SERVÍT 1970).

In amphibia, the first signs of *thalamo-cortical participation* in the
genesis of epileptic electrographic phenomena may be observed. The
formation of a secondary focus in the contralateral hemisphere and

the transition of focal activity into an electrographic seizure after metrazol administration is remarkably lowered in the isolated forebrain of the frog (Servít et al. 1968).

Caspers and Winkel (1952) were able to elicit short episodes of paroxysmal activity in forebrain hemispheres of the frog (sometimes with a behavioral convulsive correlate) by rhythmic electrical stimulation of diencephalic (thalamic) structures.

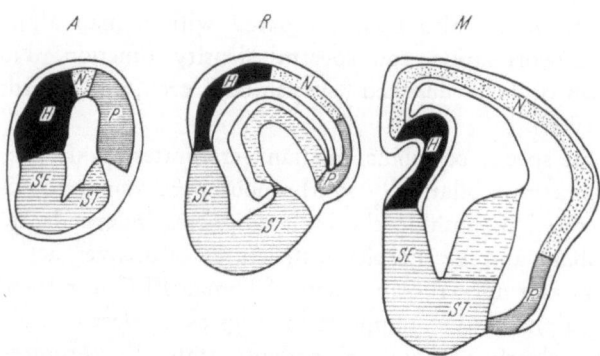

Fig. 2. Schematic outline of phylogenetic development of forebrain structures in amphibia (A), reptiles (R), and mammals (M). H = hippocampus; P = pyriform cortex; SE = septum; ST = striatum; N = mammalian neocortex, general sensory cortex in reptiles, transitory zone between hippocampal and pyriform region in amphibia

In reptiles, thalamo-cortical relations are more extensive and in addition, some new features of thalamo-cortical co-operation probably appear at this level of evolution. This class of vertebrates is situated at an important intersection of evolutionary pathways. A direct phylogenetic line leads to lower mammals, birds being on a side branch of the evolutionary tree. As regards brain structure, reptiles may be divided into two groups—squamata (lizards) and crocodiles which are situated on the side line leading to birds, and chelonia (turtles) being on the direct evolutionary pathway to mammals. In birds, a highly developed striatum plays the role of the leading brain center, thalamo-cortical relations being of secondary order.

A new cortical structure develops in reptiles—the general cortex or general sensory cortex (Elliot Smith 1910, Crosby 1917). This is inserted between primordium hippocampi medially and pyriform cortex laterally (Fig. 2). Afferent optic, acoustic and somatosensory

pathways are terminated here, most of them having a thalamic relay
(ORREGO 1961, VORONIN and GUSELNIKOV 1963, KARAMYAN et al.
1966, GUSELNIKOV 1965, KARAMYAN 1970). This area is also provided
with olfactory afferent fibers. The origin, homology and functional
significance of this cortical region is still under discussion (see KRUGER
and BERKOWITZ 1960, KRUGER 1969, HALL and EBNER 1970), most
probably it may play the role of an associative region, where the first
steps of correlation of different sensory information take place.
Direct thalamo-cortical connections of this region has recently been
proved histologically (HALL and EBNER 1970).
Diverse opinions exist concerning the specific or non-specific nature
of the ascendent pathways terminating in the general cortex (BISHOP
1961, SMIRNOV and MANTEIFEL 1962, GUSELNIKOV 1965, KARAMYAN
et al. 1966, KARAMYAN 1970). The most probable conclusion is that
at this stage of brain phylogeny, the afferent cortical pathways are
not yet differentiated into the specific and non-specific pathways.
They may, however, also operate as the non-specific pathways of the
mammalian brain in modulating the reactivity of forebrain structures
(KARAMYAN et al. 1966, KARAMYAN 1970).
In reptiles, thalamo-cortical reverberating circuits are already engaged in this modulating activity. Several data may be brought forward in favor of this assumption. WINKEL and CASPERS (1953)
elicited electrographic seizures in forebrain cortical regions of lizards
by rhythmic thalamic stimulation. We were able to confirm these
results in experiments on turtles (SERVÍT and STREJČKOVÁ 1971).
Recruiting response could be evoked in the forebrain pallium in both
lizards and turtles (BELEKHOVA 1963, KARAMYAN et al. 1966, SERVÍT
and STREJČKOVÁ 1971).
In 1968, we observed in the corticogram of a curarized, artificially
ventilated turtle electrographic spindles, localized predominantly in
the area of the general cortex (VOLANSCHI and SERVÍT 1969) (Fig. 3).
These spindles were associated in a characteristic way to the interictal
discharges of epileptic foci, localized in the general cortex (VOLANSCHI
and SERVÍT 1969 a, b, SERVÍT et al. 1971).
We assumed that these spindles may be analogous to "sleep spindles"
of the mammalian EEG and performed further studies to investigate
their electrogenesis (SERVÍT and STREJČKOVÁ 1971).
The results of these experiments may be summarized as follows:
The spindles were observed in the resting EEG almost in all experiments when appropriate measures were preserved (curarization,

relative deprivation of acoustic and optic stimuli). They consisted of
the rapid regular activity (usually 15–20 c/sec, extreme variations
10–30 c/sec) superposed on a slow wave component (1–2 c/sec).
Normally they lasted 1.5–3 seconds and appeared three to ten times
per minute. They were localized almost exclusively or predominantly

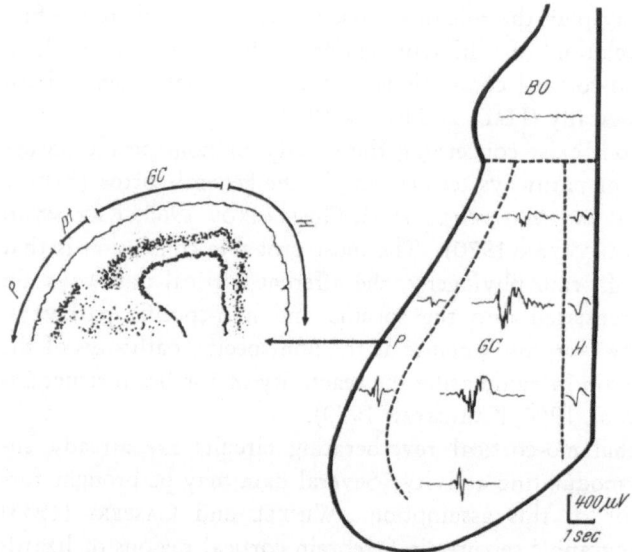

Fig. 3. Right: schematic delimitation of areas of the dorsal forebrain surface of a turtle with different occurrence of electrographic spindles. Left: transverse section approximately at the
level indicated by the arrow. H = primordium hippocampi,
GC = general sensory cortex, P = primordial pyriform cortex
(paleopallium), pt = pallial thickening, BO = bulbus olfactorius.
(From: Servít, Strejčková, and Volanschi 1971)

in the general cortex, especially their rapid component, the slow
transient sometimes irradiated into the neighboring hippocampal
and pyriform regions (Figs. 3 and 4).
The spindles could be elicited (triggered) by sudden short sensory
stimuli (similarly as the K-complex of the human EEG). Optic and
acoustic stimuli as well as electrical stimuli applied to the olfactory
and sciatic nerve were effective, the spindles triggered in this way
were localized in the same cortical region as the spontaneous ones
(Fig. 5).
Similarly as the mammalian electrographic spindles, small doses of

barbiturates (Pentobarbital) raised the number and the length of spindles in the turtle.

The spindles could be recorded simultaneously in the general cortex and in thalamic structures (predominantly in the nucleus rotundus and dorsomedialis anterior). From the same diencephalic area it was possible to trigger the spindles by single electrical stimuli. This was

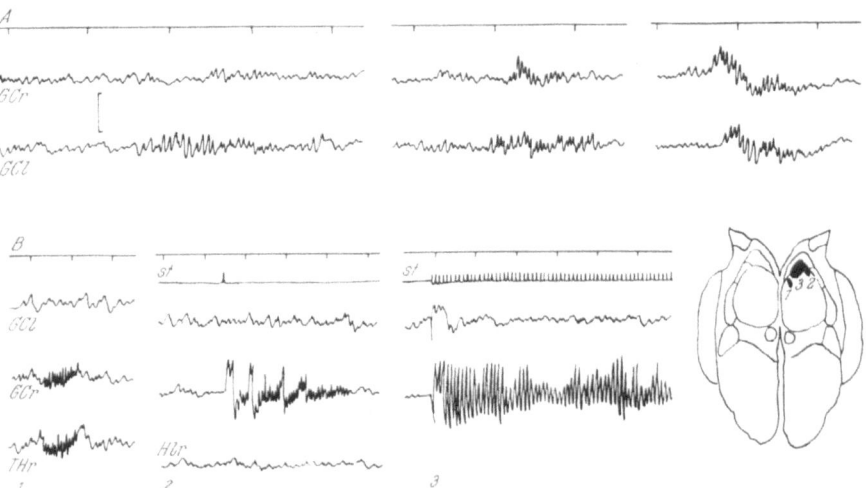

Fig. 4. Participation of thalamus in the genesis of corticographic spindles in the turtle (Servít and Strejčková 1971). A = spontaneous electrographic spindles in the resting EEG of the turtle, $B_1$ = spindle simultaneously recorded in the right general cortex and thalamus, $B_2$ = spindle triggered by single electric stimulus in the homolateral general cortex, $B_3$ = recruiting response elicited in the homolateral general cortex by rhythmic thalamic stimulation. Calibrations: 100 mV, 1 sec. $St$ = stimulus mark. The position of thalamic electrodes marked by small electrolytic lesions. GCl, GCr = general cortex left, right; THr = thalamus right; Hlr = hippocampus right

also the region from which the recruiting response could be elicited (Fig. 4). High frequency stimulation (100 c/sec) of nucleus rotundus suppressed the spindles and resulted in a flattening (desynchronization) of the corticogram.

On the other hand, cortical spindles were not always accompanied by corresponding thalamic activity and thalamic lesions did not always suppress spindles in the corticogram.

It may thus be concluded that many similar features could be demonstrated as concerns the electrogenesis of corticographic spindles in the mammalian and turtle brain. Thalamo-cortical circuits are most probably engaged in their origin. As in mammals, the nature of thalamo-cortical relations in the turtle brain can be interpreted as a

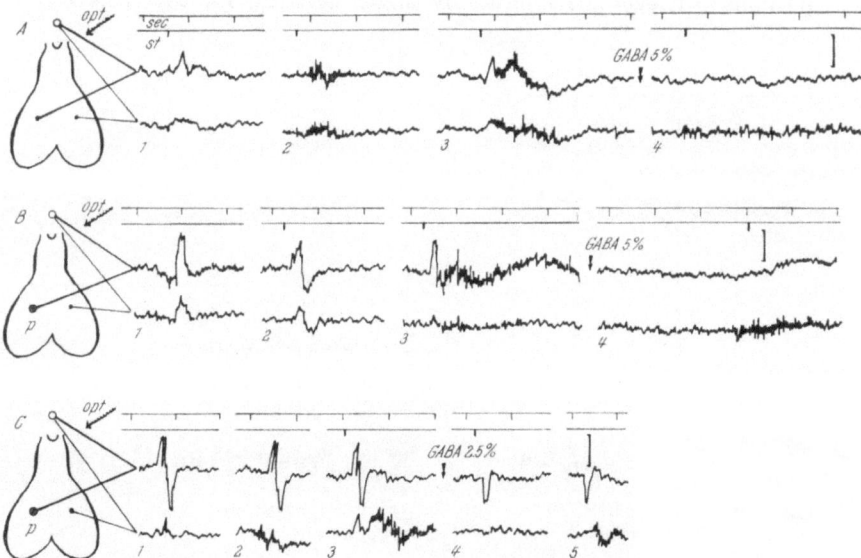

Fig. 5. Evoked potentials, spindles and triggered epileptic discharges in the general cortex of the turtle, elicited by optic stimuli. Influence of local GABA application. Monopolar EEG recordings. $p$ = site of penicillin application, $St$ = stimulus mark. Calibration: 50 $\mu$V for A, 100 $\mu$V for B and C. $A_1$ = evoked potential, $A_2$ = triggered spindle in both hemispheres, $A_3$ = evoked potential, followed by a spindle in the contralateral hemisphere, a spindle in the homolateral one, $A_4$ = local application of 5% GABA solution on the left hemisphere—suppression of spontaneous and triggered spindles and of evoked potentials, $B_1$ = spontaneous discharge of a penicillin focus, $B_2$ = triggered discharge of the focus, $B_3$ = triggered discharge of the focus, followed by a spindle, $B_4$ = local application of 5% GABA solution on the left hemisphere—suppression of both spontaneous and triggered spindles, of the evoked potentials and also of the spontaneous and triggered discharges of the penicillin focus, $C_1$ = spontaneous discharge of a penicillin focus, $C_2$ = pontaneous discharge of a focus in the left hemisphere, spindle in the right one, $C_3$ = triggered discharge of a focus, in the right hemisphere followed by a spindle, $C_{4\,5}$ = local application of GABA solution 2.5% on the left hemisphere—suppression of spindles; the focus still continues to discharge, the discharge pattern being rather modified (the surface negative transients being suppressed or diminished. (From: Servít, Strejčková, and Volanschi 1971)

combination of two relatively independent autorhythmic circuits—
the cortical and the thalamic one (VERZEANO *et al.* 1953).
The question arises if the functioning of the thalamo-cortical complex,
manifesting itself by spindle activity may already be incorporated
into the sleep mechanism at this level of brain phylogeny. This at

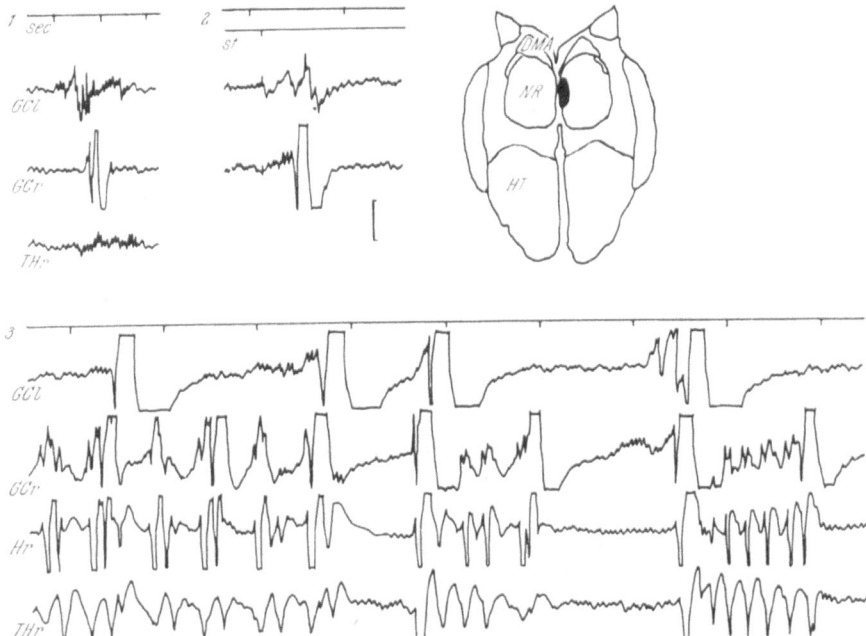

Fig. 6. Thalamo-cortical mechanisms in the genesis of epileptic electrographic
phenomena in the turtle (SERVÍT and STREJČKOVÁ 1971). 1 = electrographic
spindle, recorded simultaneously in the general cortex and thalamus, discharge
of a penicillin focus in the course of the spindle; 2 = a spindle in the general
cortex, triggered by an electric stimulus from the thalamus, discharge of the focus
in the course of the spindle; 3 = spontaneous electrographic seizure following
electrode insertion into the right thalamus. The position of thalamic electrodes
is marked by a small electrolytic lesion. Calibrations: 1 sec, 100 μV; st = stimulus
mark.
GCl, GCr = general cortex left, right; THr = thalamus right; Hr = hippocampus right; NR = nucleus rotundus; DMA = nucleus dorsomedialis ant.;
HT = hypothalamus

present remains an open question. The data demonstrating the
presence of behavioral and electrographic correlates of sleep in reptiles
are still very poor (PEYRETHON 1968, TAUBER *et al.* 1968). It seems

probable that these animals do not have "integrated" sleep similar to that observed in human and higher mammals. They do, however, dispose of thalamo-cortical synchronizing circuits which at subsequent evolutionary stages are included in the mechanisms of sleep.

We were further interested if at this stage of development *thalamo-cortical circuits may be engaged in the genesis of epileptic phenomena.* Several observations justify this assumption.

As it has already been mentioned, interictal discharges of an epileptic cortical focus in the turtle are very often associated with spindles (VOLANSCHI and SERVÍT 1969, 1970, SERVÍT *et al.* 1971, SERVÍT and STREJČKOVÁ 1971) (Figs. 5 and 6). The spindle may immediately precede or follow the discharge of the focus or the interictal discharge of the focus may appear in the course of the spindle. Interictal epileptic discharge in one hemisphere and simultaneous spindle discharge in the contralateral hemisphere were also observed. Similar observations have been reported in mammals (CASPERS 1954 in rats, KILLAM 1969 in baboons).

The discharge of an epileptic focus in the general cortex of the turtle could be triggered by thalamic stimulation. Sometimes, the thalamic stimulus triggered a spindle and the interictal discharge appeared in its course (Fig. 6).

It can be concluded from these observations that in the course of a spindle (which can be initiated in the thalamus) an enhanced disposition to the *interictal* epileptic discharge arises in the general cortex.

Thalamo-cortical activity may also enhance the susceptibility to the generation of *ictal* electrographic phenomena in the forebrain. In one of our studies (SERVÍT and STREJČKOVÁ 1971) in which stimulating and/or recording electrodes were introduced into thalamic structures, we observed episodic ictal activity in cortex and/or thalamus in $^1/_3$ of the experiments. The electrographic seizures followed thalamic stimulation or appeared spontaneously merely after electrode insertion (Fig. 6). As has already been stated above, similar observations were obtained by WINKEL and CASPERS (1953) in lizards.

The emergence of the thalamo-cortical functional complex in brain phylogeny therefore evidently plays an important role in the development of epileptogenetic mechanisms. These mechanisms are not indispensable for the genesis of epileptic phenomena but, once evolved, they become involved in their generation, already at the first primitive stages of their evolution.

In the brain of the turtle we have a model where these first steps of development of thalamo-cortical co-operation can be investigated. The thalamo-cortical complex appears here for the first time as a tool of synchronization of large forebrain areas. Corticographic spindles seem to be the symptom in which thalamo-cortical circuits play a role. This burst of synchronized activity which at higher levels of brain phylogeny will become a part of electrographic phenomena of normal sleep can already be engaged at this stage of evolution in the genesis of pathological epileptic hypersynchrony.

In my report, I have tried to show that the comparative approach following the development of the functional organization of the brain from the simple to the complex may bring some insight into the possible role of normal and pathological mechanisms of synchronization in the genesis of epileptic electrographic phenomena.

## Summary

Excessive synchronization may be one of the essential features of epileptic activity. On the other hand certain modes of synchronized activity may be inherent to all normal neuronal aggregates. The relations between the normal and abnormal (epileptic) synchrony may be of interest for epileptological studies.

The comparative approach can bring useful insight into these problems in following the phylogenesis of synchronization mechanisms of the brain under normal and pathological conditions from simple forms to the complex ones.

In the electrogenesis of epileptic electrographic phenomena two kinds of synchrony may be discerned: the local synchrony taking part in interictal discharges of an epileptic focus and synchronization of distant areas of the brain (e.g., bilaterally synchronized activity of both forebrain hemispheres).

The local synchronization mechanisms are present at all levels of vertebrate brain phylogeny. Neither a layered neuronal structure of a cortical type nor a functional connection with diencephalic and/or mesencephalic structures are indispensable for the generation of interictal epileptic activity in the forebrain structures.

Generalized bilaterally synchronous ictal activities may also be generated in forebrain hemispheres of an isolated forebrain of fish and amphibia. The seizure dissemination is realized here via the transverse commissural pathways which are the more primitive tool of bilateral synchrony.

The thalamo-cortical functional complex is not indispensable for the genesis of generalized electrographic epileptic activities in the forebrain. However, its emergence in the brain phylogeny (in reptiles) brings an important new feature which may already be engaged at this primitive stage of its evolution in the generation of hypersynchronous epileptic phenomena.

Electrographic spindles in the resting EEG of the turtle, the participation of thalamo-cortical circuits in their genesis and in the electrogenesis of epileptic activities have been investigated. The results are discussed in terms of brain phylogeny.

## References

ADRIAN, E. D.: The activity of the nervous system in the caterpillar. J. Physiol. 70, 34—47 (1930).
— Potential changes in the isolated nervous system of the Dytiscus marginalis. J. Physiol. 72, 132—148 (1931).
AJMONE MARSAN, C.: Electrographic aspects of "epileptic" neuronal aggregates. Epilepsia 2, 22—38 (1961).
— Acute effects of topical epileptogenic agents. In: JASPER et al. (eds.), Basic mechanisms of epilepsy, pp. 299—319. Boston: Little, Brown & Co. 1969.
BELEKHOVA, M. G.: Electrical activity of cerebral hemispheres evoked by stimulation of diencephalic structures in Varanus. Fiz. Zh. SSSR im. I. Sechenova 49, 1318—1329 (1963). [In Russian.] Fed. Proc. Suppl. 24, T 159 (1965). [English translation.]
BISHOP, G. H.: The organization of cortex with respect to its afferent supply. Ann. N. Y. Acad. Sci. 94, 559 (1961).
CASPERS, H.: Die Aktivierung kortikaler Krampfstromherde im natürlichen und elektrisch induzierten Schlaf beim Tier. Z. Ges. exp. Med. 124, 176 (1954).
— und K. WINKEL: Untersuchungen über die Bedeutung des Thalamus und Lobus opticus für die Großhirnrhythmik beim Frosch. Pflügers Arch. 255, 391—416 (1952).
CREUTZFELDT, O. D.: Neuronal mechanisms underlying the EEG. In: JASPER et al. (eds.), Basic mechanisms of epilepsy, pp. 397—410. Boston: Little, Brown & Co. 1969.
CROSBY, E. C.: The forebrain of Alligator Mississipiensis. J. comp. Neurol. 27, 325—402 (1917).
GUSELNIKOV, V. I.: Elektrofiziologicheskoe issledovanie analizatornykh sistem v filogeneze pozvonochnykh, p. 266. Moskva: Izdatelstvo Mosk. Univ. 1965.
HALL, W. C., and F. F. EBNER: Thalamoencephalic projections in the turtle (Pseudemys scripta). J. comp. Neurol. 140, 101—122 (1970).
JASPER, H. H.: Mechanisms of propagation: extracellular studies. In: JASPER et al. (eds.), Basic mechanisms of epilepsy, pp. 421—440. Boston: Little, Brown & Co. 1969.
KARAMYAN, A. I.: Funktsionalnaya evolutsia mozga pozvonochnych (Functional evolution of the vertebrate brain), p. 301. Leningrad: Izd. Nauka. 1970.

KARAMYAN, A. I., N. P. VESSELKIN, M. G. BELEKHOVA, and T. M. ZAGORULKO: Electrophysiological characteristics of tectal and thalamo-cortical divisions of the visual system in lower vertebrates. J. comp. Neurol. *127*, 559—576 (1966).

KILLAM, K. F.: Genetic models of epilepsy with special reference to the syndrome of the Papio papio. Epilepsia *10*, 229—238 (1969).

KREKULE, I., T. WEISS, and Z. SERVÍT: Comparison by means of probalistic methods of ten per second quasiperiodic activities recorded in man and frog. In: SERVÍT, Z. (ed.), Comparative and Cellular Pathophysiology of Epilepsy, Proc. Internat. Cong. Ser. No. 124, Liblice 1965, pp. 129—138. Amsterdam: Excerpta Medica. 1966.

KRUGER, L.: Experimental analysis of the nervous system. Ann. N. Y. Acad. Sci. *167*, 102—117 (1969).

— and ELLIS C. BERKOWITZ: The main afferent connections of the reptilian telencephalon as determined by degeneration and electrophysiological methods. J. comp. Neurol. *115*, 125—141 (1960).

MARCUS, M., and C. W. WATSON: Bilateral synchronous spike wave electrographic patterns in the cat. Arch. Neurol. *14*, 601—610 (1966).

— — Symmetrical epileptogenic foci in monkey cerebral cortex. Arch. Neurol. *19*, 99—116 (1968).

— — and S. A. SIMON: An experimental model of some varieties of petit mal epilepsy. Electrical-behavioral correlations of acute bilateral epileptogenic foci in cerebral cortex. Epilepsia *9*, 233—248 (1968).

MOROCUTTI, C., and R. VIZIOLI: Osservazioni sull'attivita elettrica del cervello e sulle crisi convulsive di „Bufo Vulgaris". Riv. neurol. *27*, 669—676 (1957).

ORREGO, F.: The reptilian forebrain. I. The olfactory pathways and cortical areas in the turtle. Arch. ital. Biol. *99*, 425—445 (1961).

PETSCHE, H., und P. RAPPELSBERGER: Die Beeinflussung der Potentialfelder epileptischer Anfälle durch Rindeninzision. Különlenyomat az Ideggyógyászati *2*, 1—12 (1970).

— — Influence of cortical incisions on synchronization pattern and travelling waves. Electroenceph. clin. Neurophysiol. *28*, 592—600 (1970).

— and J. ŠTERC: The significance of the cortex for the travelling phenomenon of brain waves. Electroenceph. clin. Neurophysiol. *25*, 11—22 (1968).

PEYRETHON, J.: Sommeils et évolution. Étude polygraphique des états de sommeil chez les poissons et les reptiles. Thèse de Médecine No. 201, 104, Lyon (1968).

SERVÍT, Z.: Physiopathologie comparée des manifestations électroencéphalographiques de la crise épileptique. J. Physiol. *57*, 731—741 (1965).

— Comparative physiology of the paroxysmal EEG. Pattern and frequency of paroxysmal activity on different levels of brain phylogeny. In: SERVÍT, Z. (ed.), Comparative and Cellular Pathophysiology of Epilepsy, Proc. Internat. Congr. Ser. No. 124, Liblice 1965, pp. 103—111. Amsterdam: Excerpta Medica. 1966.

— Focal epileptic activity and its spread in the brain of lower vertebrates. A comparative electrophysiological study. Epilepsia *11*, 224—240 (1970).

— J. MACHEK, and J. FISCHER: Electrical activity of the frog brain during electrically induced seizures. A comparative study of the spike and wave complex. Electroenceph. clin. Neurophysiol. *19*, 162—171 (1965).

Servít, Z., and A. Strejčková: Comparative pathophysiology of the paroxysmal electroencephalogram. Paroxysmal electroencephalogram (spike and wave activity) in the lizard (Lacerta agilis). Physiol. bohemoslov. *15*, 117—121 (1966).
— — Epileptogenic focus in the frog telencephalon. Seizure irradiation from the focus. Physiol. bohemoslov. *16*, 522—530 (1967).
— — An electrographic focus in the fish forebrain. Conditions and pathways of propagation of focal and paroxysmal activity. Brain Research *17*, 103—113 (1970).
— — unpublished observations (1971).
— — and J. Fischer: Comparative physiology of the thalamic pacemaker of paroxysmal activity. Extirpation of thalamus in the frog. In: Servít, Z. (ed.), Comparative and Cellular Pathophysiology of Epilepsy, Proc. Internat. Congr. Ser. No. 124, Liblice 1965, pp. 270—276. Amsterdam: Excerpta Medica. 1966.
— — and D. Volanschi: An epileptogenic focus in the frog telencephalon. Pathways of propagation of focal activity. Exp. Neurol. *21*, 383—396 (1968).
— — — Epileptic focus in the forebrain of the turtle (Testudo Graeca). Triggering of focal discharges with different sensory stimuli. Physiol. bohemoslov. *20*, 221—228 (1971).
Smirnov, G. D., and J. B. Manteifel: Sravnitelnoe elektrofiziologicheskoe izuchenie mozga v ryadu pozvonochnykh zhivotnykh. Uspechi sovr. biol. *54*, 309—332 (1962).
Smith, Elliot: Some problems relating to the evolution of the brain. Lancet *1*, 147 (1910).
Strejčková, A.: Epileptogenic focus in the forebrain of the fish. Physiol. bohemoslov. *18*, 209—216 (1969).
Tauber, E. S., J. Rojas-Ramírez, and R. Hernández-Peón: Electrophysiological and behavioral correlates of wakefulness and sleep in the lizard Ctenosaura Pectinata. Electroenceph. clin. Neurophysiol. *24*, 424—433 (1968).
Verzeano, M., D. B. Lindsley, and H. W. Magoun: Nature of recruiting response. J. Neurophysiol. *16*, 183—195 (1953).
Volanschi, D., and Z. Servít: Epileptic focus in the forebrain of the turtle. Exp. Neurol. *24*, 137—146 (1969 a).
— — Epileptic focus in the forebrain of the turtle. Pathways of propagation of epileptic activity. Physiol. bohemoslov. *18*, 381—386 (1969 b).
Voronin, L. G., and V. I. Guselnikov: K filogenezu vnutrennykh mekhanizmov analitikosinteticheskoi deyatelnosti golovnogo mozga. Zh. vyssh. nervn. deyat. *13*, 193—206 (1963).
Wilder, B. J., and F. Morrell: Secondary epileptogenesis in the frog forebrain. Neurology *17*, 1041—1051 (1967).
— R. L. King, and R. P. Schmidt: A comparative study of secondary epileptogenesis. Epilepsia *9*, 275—289 (1968).
Winkel, K., und H. Caspers: Untersuchungen an Reptilien über die Beeinflussung der Großhirnrindenrhythmik durch Zwischenhirnreizungen mit besonderer Berücksichtigung des Thalamus. Pflügers Arch. *258*, 22—37 (1953).

# Human Scalp EEG Fields:
# Evoked, Alpha, Sleep, and Spike-Wave Patterns[1]

D. Lehmann

Labor für Neurophysiologie, Neurologische Universitätsklinik,
Zürich, Switzerland

Complete information about the human scalp EEG is available in
the distribution of the electrical field on the scalp, and its change as
a function of time. Any conventional EEG as recorded between unipolar and bipolar electrode combinations can be reconstructed from
complete data of the field distributions. However, the large amount
of data in recordings of field distributions demands considerable data
reduction before evaluation. One strategy of data reduction is particularly challenging: a limited number of intracranial model generators, if possible only one, can be fitted to account for a given field
distribution. A sequence of field distributions could thus be described
by the parameters of the model generator(s) which change as a function of time.

The EEG may be understood as an averaged, smoothed representation
of the general tendency of activity of the neural elements (Geisler
and Gerstein 1961, DeLucchi et al. 1962, Cooper et al. 1965); an
equivalent dipole generator which accounts for the scalp EEG data
will then be a representation of the average tendency of the EEG.
Modelling of EEG field distributions with one or two model generators appears to be a promising strategy to produce small sets of
characteristics for the classification of EEG data obtained in various
functional conditions (see Kavanagh 1972), although it must be
understood that in the first place, the fitted generators are mere
equivalents of the real sources of the EEG scalp fields. On the other
hand, fitting of intracranial generators to the field distributions may

1 Supported in part by USPHS grants NB 06038 and FR 00241, and by a grant
from the Hartmann-Mueller-Stiftung, Zürich.

—given detailed knowledge of the electrical properties of the brain and the surrounding tissue—lead to a useful, direct three-dimensional localization of the source of certain scalp EEG patterns; this is particularly likely in extreme conditions, as in brain tumors and epileptic foci (SCHNEIDER and GERIN 1970). However, we shall see that the careful examination of the field distributions directly allows the establishment of some general characteristics of the distributions. Concise description of the field sequences hopefully will aid in the classification and topographical differentiation of scalp EEG activity during different functional conditions, and result in an improved understanding of conventional EEG recordings. It will be seen that this approach produces new insights into the phenomenology of the EEG, particularly relative to the discussion on phase lags and travelling of EEG waves (WALTER 1947, PETSCHE and MARKO 1959, PETSCHE 1962, 1970, RÉMOND 1968).

Our present apparatus for the collection of simultaneous EEG data from many different points on the human scalp (see also LEHMANN 1971 a) uses 48 pre-amplifiers. The output of each of the 48 channels is sampled 750 times/sec, multiplexed in groups of 8 channels, post-amplified, and recorded as amplitude-modulated pulses on 6 tracks of an analog tape recorder. The seventh track carries control pulses. Further treatment of the data is performed off-line. The demultiplexed and resynthesized EEG recordings are A–D converted, and transformed into series of maps of the scalp field distributions, using general purpose computers. The map series may be produced as semi-schematic potential plots into which equipotential lines may be sketched (Fig. 1), or as maps with computer-interpolated equipotential lines.

When examining the series of equipotential maps we found it most convenient to use as zero reference level the arithmetic mean of all field measurements simultaneously obtained from the equally distributed electrodes at the instant represented by the map. All maps shown in the present paper are constructed in this fashion. Of course, the data are independent from the choice of the reference electrode during recording: the voltage difference between two points in a field is not affected by the reference point from which the voltages are measured.

Viewing of equipotential map series as a movie (ESTRIN and UZGALIS 1969, LEHMANN and FENDER 1969, LEHMANN 1971 b) is esthetically pleasing but permits very little understanding of the sequential development of field distributions. It appears that the unreduced amount of data in the movie sequence is too large for comprehension. Successive inspection of single frames is more fruitful.

## Averaged Evoked EEG Potentials

Fig. 2 illustrates several frames of a map sequence of average, visually evoked scalp EEG responses in man. We note in Fig. 2 that the first frames at 71, 79, and 87 msec after the stimulus are dominated

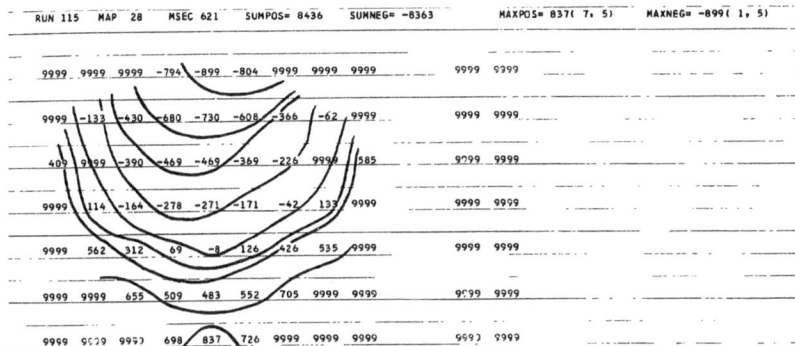

Fig. 1. Computer print of a scalp field distribution, showing the field values which were measured at one instant during a 3/sec spike-wave complex. Semischematic display of the head seen from above; anterior is up. Field values in 1/10 $\mu$V, referred to the average of all instantaneous values. Each figure indicates an electrode position, 9999 means no entry. Equipotentials lines in steps of 20 $\mu$V were sketched in by hand

by a negative field maximum which is situated occipitally in the midline. At 95 msec latency, the field distribution shows two negative maximum values, one over each occipital hemisphere. Mapping of the field obviously allows the detection of the independent activity of the two hemispheres of the brain in response to the light stimulus (LEHMANN et al. 1969 a, 1969 b, BOURNE et al. 1971). Shortly later, at 111 msec after the stimulus, there is a dominating positive maximum value of the field distribution over the midline in the occipital area. The map sequence of Fig. 2 illustrates very clearly the well-known fact that an EEG waveform conventionally recorded from bipolar or so-called unipolar electrode derivations is crucially dependent on the localization of the electrodes. Evidently, a maximal positivity of the occipital electrode in reference to a mastoid electrode

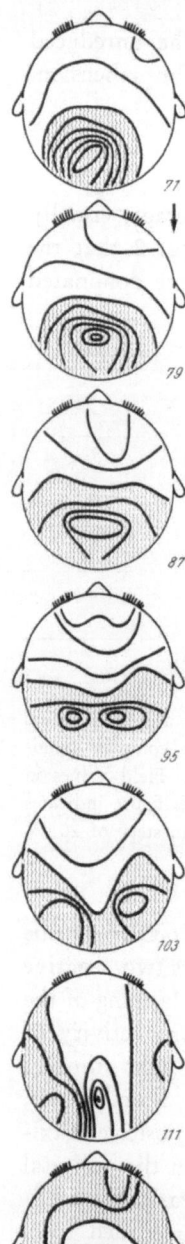

Fig. 2. Sequence of equipotential maps of visually evoked
scalp EEG potentials, averaged during 100 repetitive 3/sec
xenon flash stimuli to a central 18 degree field in the left
eye of a normal subject. Time in msec indicates latency
after the flash. Equipotential lines in steps of 2 $\mu$V. White
positive, grey negative. (From LEHMANN, in preparation)

would have been detected at 111 and 119 msec after the flash, if the occipital electrode had been positioned 3 cm above the inion over the midline; however, if the occipital electrode had been positioned about 3 cm to the left from the midline, only a very minor positivity would have occurred at the same latencies. Further, at msec 95 the lateral electrode would have detected a considerably stronger negativity than the midline electrode. In general, Fig. 2 demonstrates that any electrode combination would have resulted in different waveforms, as it has often been observed. On the other hand, the sequence of field distributions is relatively simple compared with the multitude of different waveforms which can be observed as evoked responses in conventional, averaged EEG traces recorded from different electrode combinations. Finally, the map sequences permit us to make practical conclusions about most suitable electrode positions if, using only a few electrodes, we wish to record dominant features of the field distributions during certain times after the stimulus. This problem had been emphasized by RÉMOND (1968). For instance, the sequence illustrated in Fig. 2 suggests a midline occipital and a lateral occipital electrode site, possibly referred to a mastoid or anterior midline electrode.

## Spontaneous Alpha EEG

The 100 msec-sequence of scalp field distributions illustrated in Fig. 3 a represents approximately one alpha cycle. It was taken from a period of alpha EEG of a normal human (LEHMANN 1971 a). Some of the maps of Fig. 3 a show more equipotential lines than others, indicating that the positive and negative values of the field distributions in these maps are high: in other words, some maps show more relief than others. A measure for the amount of relief of a given distribution is the average, absolute amplitude per electrode, referred to the arithmetic mean of all measurements from the equally distributed electrodes at the moment of measurement. Fig. 3 b shows the changes of the amount of relief during the alpha period from which the maps of Fig. 3 a were taken. The relief curve permits to select most pronounced field distributions. Fig. 3 b demonstrates a typical relief curve during alpha EEG activity, with about twenty peaks/sec. Roughly similar field distributions, but of opposite polarity, are observed at successive, similar phases of the relief curve during alpha. An example of this rule is illustrated by the maps at msec 433 and 481 of Fig. 3 *a* which show the fields at two successive peaks of the relief curve. It is of interest that the periodicity of the relief curve in our example of Fig. 3 *b* is interrupted for one cycle, between msec 350 and 400, without interruption of the regular sequence of field distributions.

The maps of Fig. 3 *a* show relatively simple field distributions, each with one or two positive and negative maximum fields values. The localization of the positive and negative maximum values can thus be used to characterize a field distribution. The general characteristics of a longer map sequence accordingly can be surveyed by charting the localization of the maximum field value of each map during the analysis time (Fig. 4). [A related data reduction tactic for the survey of map sequences is the averaging of maps over time (Lehmann 1971 b), where the standard deviation of the average field value computed for a given electrode position is correlated with the absolute amplitude at this electrode, averaged over time. The topographical distribution of these standard deviations resembles the distribution generated by the mapping of the field maxima during the same analysis time.] The values in Fig. 4 indicate the fraction of the analysis time during which the field maximum was recorded at each electrode site, covering the period of alpha EEG of Fig. 3 *b*. In Fig. 4 *a*, the incidence of positive and negative maximum values at the different electrodes during the analysis time was combined. Figs. 4 *b* and *c* show separately the localization of positive and negative maximum field values at the different electrode sites. We note that positive as well as negative field maxima apparently occurred preferentially in three areas: left occipital, frontal, and right occipital. We also note that the areas of preferential occurrence are separated by zones where the field maxima were found seldom, or never.

Fig. 4 *d* shows similar data obtained from and averaged over five subjects; the observations made above on the data of the first subject

---

Fig. 3. *a*) Sequence of scalp field equipotential maps (in 8 msec intervals) during alpha EEG. White positive; grey negative; equipotential lines in steps of 10 $\mu$V. For each map, upper figure indicates absolute, average amplitude per electrode, lower figure indicates msec, referring to Fig. 3 *b*, of which *a*) illustrates a portion. *b*) Absolute, average amplitude per electrode as a function of time during a period of alpha EEG activity. [From Lehmann (1971 a) by permission of Elsevier Publishing Co.]

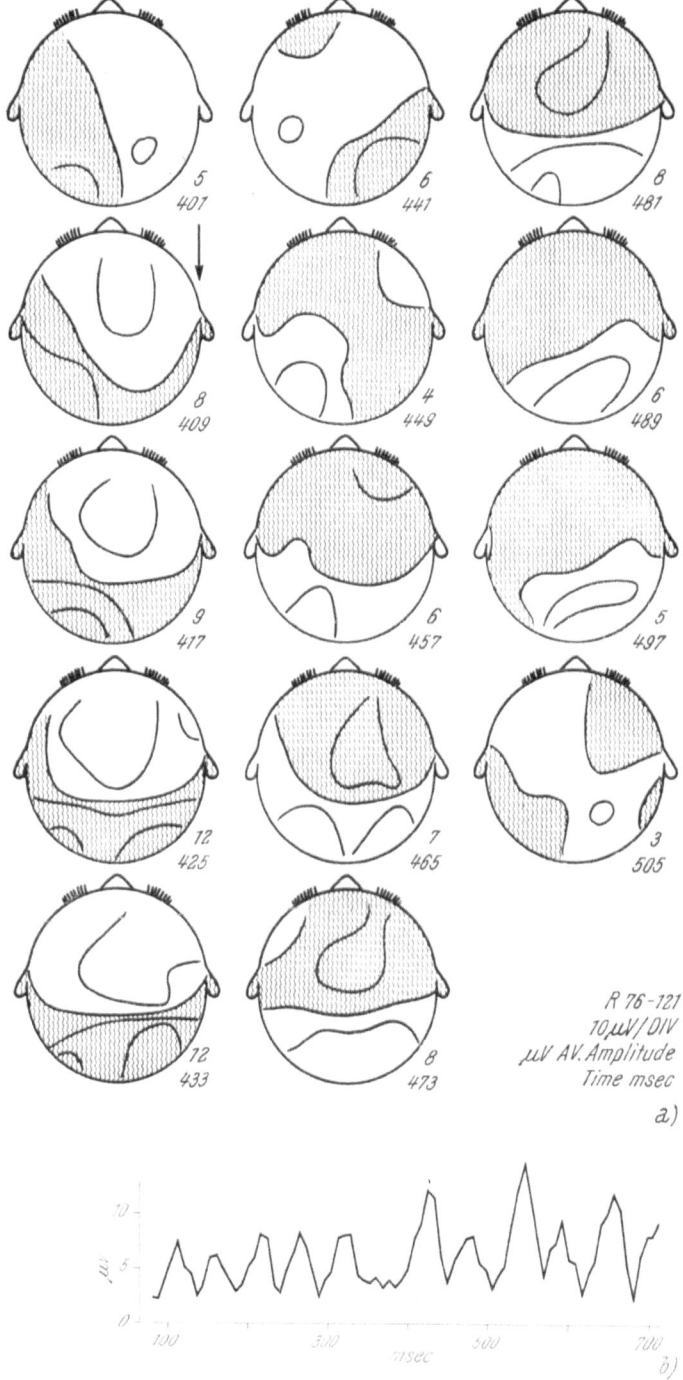

R 76-121
10 μV / DIV
μV AV. Amplitude
Time msec

a)

b)

Fig. 3

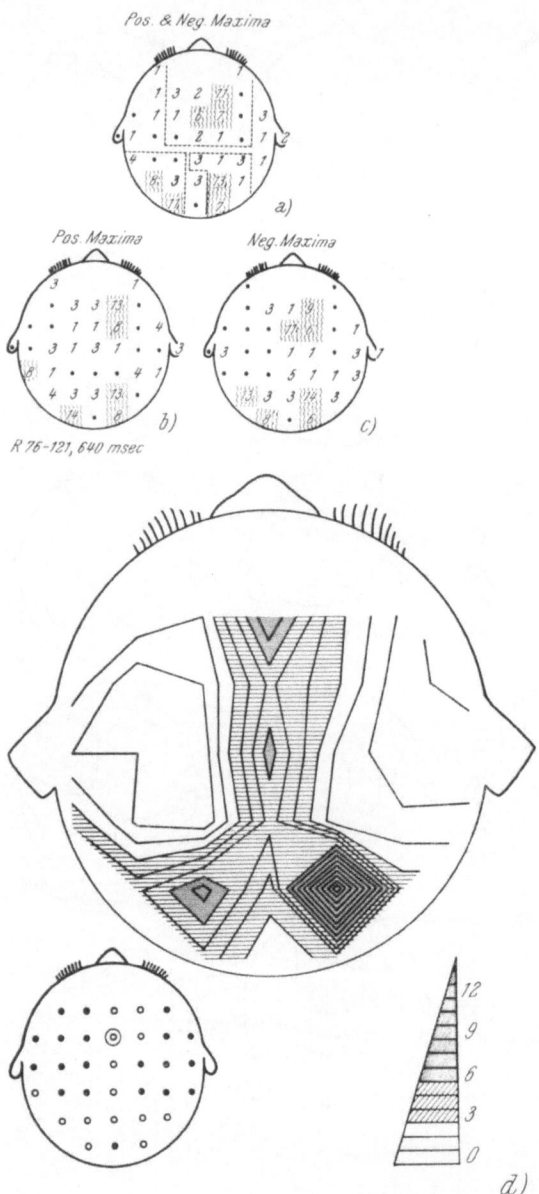

Fig. 4. *a*), *b*), and *c*) Location of positive and negative maximum field values during the 640 msec of alpha activity of Fig. 3. Sampling interval 8 msec. Numbers give the percentage of time that the maximum occurred at indicated electrode. Dots indicate zero%. Rounding errors cause sums to deviate from 100%. *d*) Location of positive and negative maximum field values of alpha EEG, mean of five subjects. Contour lines interpolated linearly between values at electrode positions indicated at lower left; dots mark less than 3% time, open circles more than 3%. Percentage of time indicated by shading, key at lower right. [From LEHMANN (1971 a), by permission of Elsevier Publishing Co.]

(Figs. 4 *a*, *b*, and *c*) thus were confirmed by the data of the other subjects (Lehmann 1971 a).

An examination of the localization of the field maxima as a function of time reveals that there is a systematic sequence in a counter-clockwise (in other cases clockwise) direction. The occurrence of the field maximum first over one, then over the other hemisphere appears as phase lag of the conventionally recorded EEG of the two hemispheres (Garoutte and Aird 1958). However, the spatial progression of the field maximum is not continuous. If the progression were continuous, there would be no preference areas of occurrence of the maxima. In fact, the maximum typically resides in one preference area for some time, then jumps to another preference area, several electrode positions away. In a discussion of generator models which would account for these findings we have proposed three stationary generators which oscillate at similar frequencies, but with different phase angles (Lehmann 1971 a). Rémond (1968), who proposed a four-generator model for averaged, parietal alpha EEG data, pointed out that interaction of stationary generators can result in travelling of conventionally recorded waves.

## Sleep EEG Waves

Sleep slow waves and sleep spindles are patterns of synchronized EEG with a topography which is in general aspects different from alpha topography (Brazier 1949 a, 1949 b). The map sequence in Fig. 5 illustrates a typical sleep slow wave. There is some tendency to bilateral symmetry with field maxima of the same polarity over both anterior areas, contrary to the bilateral-symmetric alpha distributions with two occipital field maxima. In general, the sleep EEG fields during slow waves and sleep spindles tend to distributions which are concentric or symmetric around maxima over anterior areas. Sleep slow waves, contrary to sleep spindles, often show a good amount of irregularity in their fields. The sequence in Fig. 5 demonstrates between msec 17 and msec 129 some migration of the positive maximum value of the field—a migration restricted to the anterior left area (Fig. 6 *b*), with a speed of about 1 m/sec. However, there may still be a problem of spatial resolution. At any case, the migration of the field maximum terminates at msec 129. The change of the field shape at msec 145 and 161 apparently was caused by the emergence of a second positive maximum value, as indicated by the map at msec 177 with its two positive maxima.

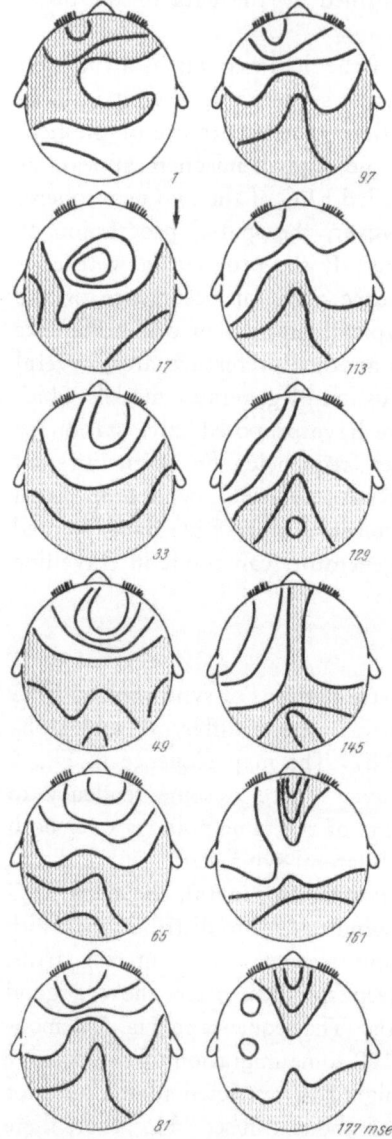

Fig. 5. Sequence of scalp EEG field equipotential maps during sleep slow wave. White positive, grey negative, equipotential lines every 10 $\mu$V, map intervals 16 msec, numbers refer to msec in Fig. 6 *a*

Fig. 5 illustrates how potential field mapping can elucidate conventional bipolar EEG recordings. During the time represented by the map series of Fig. 5, a left anterior temporal *vs* right mastoid EEG derivation would have recorded ½ of a 3/sec wave, whereas a midline frontal *vs* right mastoid derivation would have recorded ¾ of a 4/sec wave—a complicated finding which, however, is explained by the sequence of relatively simple field distributions in Fig. 5. In addition, the map sequence shows once more that in conventional recordings, neighboring electrode combinations may result in different EEG patterns.

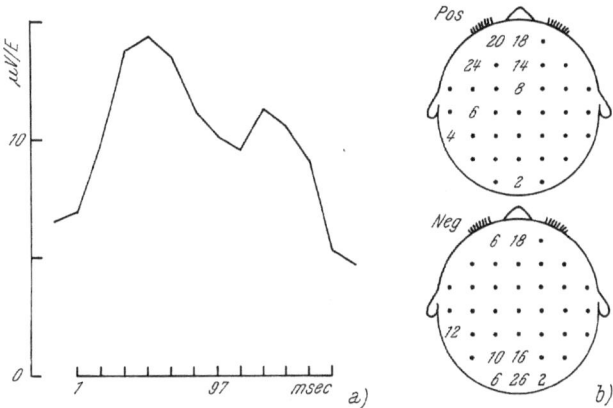

Fig. 6. Sleep slow wave, same as in Fig. 5. *a*) Absolute, average amplitude per electrode as a function of time. *b*) Location of positive (above) and negative (below) maximum values of the fields during the analysis time. Sampling interval 4 msec. For other explanation see Figs. 4 *a*, *b*, and *c*

The relief curve (Fig. 6 *a*) associated with the sleep wave maps of Fig. 5 exhibits a major peak at msec 49 when the frontal maximum is positive, and a secondary peak at msec 129 coincident with the emergence of a frontal field maximum of negative polarity.

## Spike–Wave Complexes

Let us now examine human scalp EEG field distributions during generalized 3/sec spike and wave patterns, one of the most impressive cases of synchronization of brain activity. The various spike-wave complexes which were analyzed in two patients had very similar characteristics. One representative complex will be considered in detail. The scalp field distributions of this spike-wave complex are

513    625    737
529    641    753
545    657    769
561    673    785
577    689    801
593    705    817 msec
609    721    a)     Fig. 7 a

shown as map series in Fig. 7 *a*. The illustrated complex was recorded during a spontaneous "absence" of a 29-year-old patient, who had suffered from frequent absence-type spells without motor phenomena since childhood. The generalized spike-wave trains in his EEG are very uniform. The set of EEG traces of Fig. 8 *a* demonstrates the conventional EEG during the spike-wave complex under analysis, recorded from a longitudinal row of midline electrodes with the left mastoid as common reference. The most frontal electrode was at Fz, and the most posterior electrode was over the inion, as indicated in the schematic of Fig. 7 *b*. The traces of Fig. 8 *a* were reconstructed from the field data, sampling the original recordings every 8 msec.

These EEG traces exhibit the classical sequence (Weir 1965) of spike-wave phenomena: an abortive spike 1, a frontally maximal positive transient, spike 2 (the classical EEG spike), the subsequent positivity, and finally the wave. The EEG traces of Fig. 8 *a* have their peak amplitudes over anterior areas, similar to the five cases of idiopathic petit mal illustrated by Fischgold et al. (1955) in their report on spike-wave patterns (Fig. 10 of Rémond 1955). These graphs are are dealing with a typical spike-wave complex is Rémond's (1955) report of potential distribution graphs (Shaw and Roth 1955) during spike-wave patterns (Fig. 10 of Rémond 1955). These graphs are very similar to the results one obtains from our sample spike-wave complex, if Rémond's graphing technique is used.

The curve of the field relief (Fig. 8 *b*) of the spike-wave complex under study shows three periods during which the field distributions

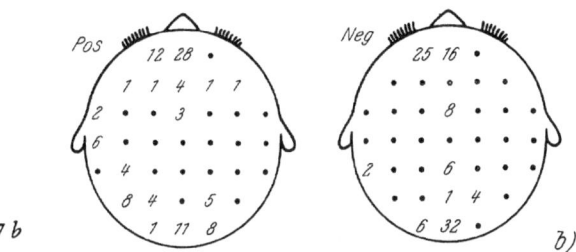

Fig. 7 *b*

Fig. 7. *a*) Sequence of equipotential maps during spike-wave complex. Intervals between successive maps is 16 msec. White positive, grey negative, equipotential lines in 20 μV steps. Numbers indicate msec, referring to Fig. 8. *b*) Location of positive and negative maximum field values during the spike-wave complex illustrated in Figs. 7 *a* and 8. Sampling interval 4 msec. For other explanation see Figs. 4 *a, b,* and *c*

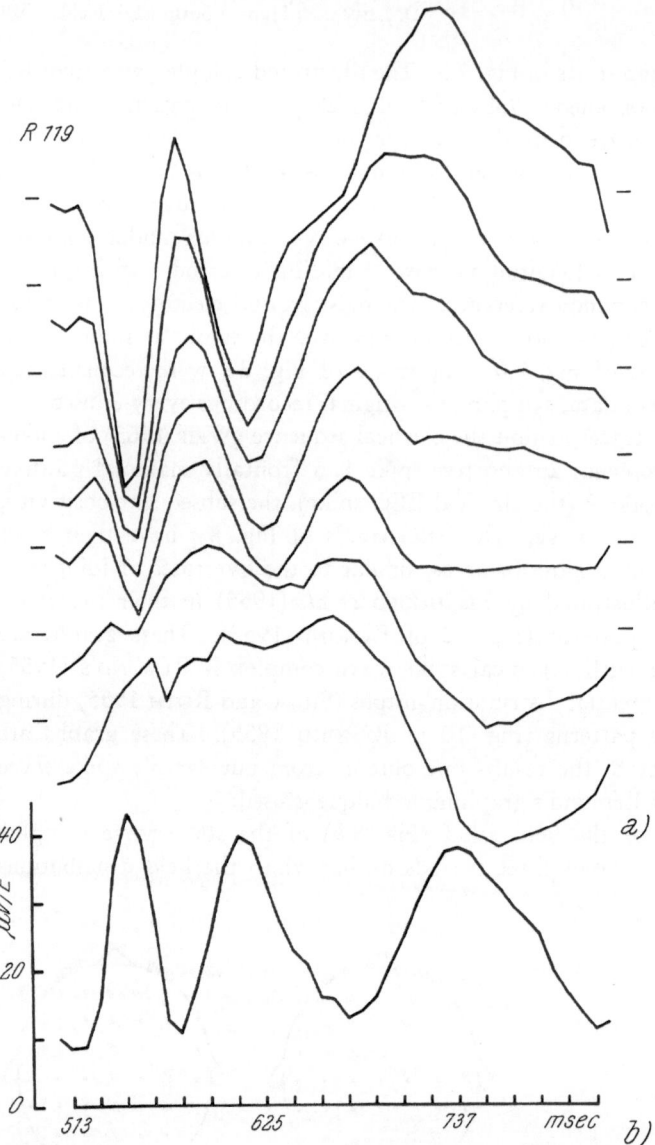

Fig. 8. *a*) EEG traces from an anterior-posterior row of midline scalp electrodes, referred to left mastoid electrode, during the spike-wave complex of Fig. 7.
Most anterior trace on top, most posterior trace on buttom. Reconstruction from
field recordings. Sampling interval 8 msec. Calibration 100 μV. The arithmetic
mean of all data points computed for each trace between msec 521 and 817 is
indicated by horizontal bars (zero level) on both sides. *b*) Absolute, average
amplitude per electrode as a function of time, during the spike-wave complex
of Fig. 7 *a*. Ticmarks indicate time slices at which field distributions were mapped
in Fig. 7 *a*

were most pronounced: these periods coincide with the positive transient, the positivity following spike 2, and the wave (Fig. 8 *a*). We note that the EEG spike 2 thus has a relatively flat field distribution. During the positive transient at msec 545 and 561, the distributions tend to be concentric around a strong anterior positivity, with a high relief. On the other hand, the map at msec 577, at the instant of peak amplitude of spike 2 in the conventional EEG demonstrates a low field relief. There is a persisting, small occipital negativity which had developed during the positive transient. Further, there is now a minor negativity with a small field gradient over anterior areas, replacing the anterior positivity which existed during the positive transient. At msec 577, the instantaneous voltage difference between anterior areas and the left mastoid is slightly lower than between parietal areas and the left mastoid. However, the EEG spike 2 has its maximal amplitude over anterior areas, since its amplitude is measured as voltage difference between a trough and subsequent peak (or vice versa) in the trace—and the preceding trough (the positive transient) was very strong over frontal areas.

The following map at msec 593 reapproaches the distribution of the field during the positive transient at msec 529 and 545, with an increasing degree of relief. The relief reaches its next peak values at msec 609 and 625, concomitant with the EEG positivity after spike 2. This positive peak in the EEG traces of Fig. 8 *a* occurs progressively later as one proceeds from frontal to parietal electrode positions. The phase differences correspond with an apparent travelling of the wave at a speed of about 8.5 m/sec, within the range reported by Petsche and Marko (1959) and Petsche (1962). However, the three field maps at msec 609, 625, and 641 of Fig. 7 *a* which are associated with this travelling peak show a basically stable field distribution without changes of the localization of the maximum field values. The travelling phenomenon is caused by minor changes of the shape of the field, involving also the region of the reference electrode.

While the occipital negativity declines, a second negative field maximum develops centrally during the subsequent trough of the relief curve at msec 657 and 673. This central negativity of frame 673 and 689 has jumped to a frontal position by msec 705, coinciding with a new general buildup of the field relief. The survey of location of the maxima in Fig. 7 *b* shows that at 4 msec sampling intervals, no intermediate position between central and anterior maximum location was detected. Therefore, one may well assume a minor, independent process at the central area, particularly since at msec 769 and 785 the decay of the anterior negativity is advanced at the central location.

On the other hand, the inspection of the EEG traces of Fig. 8 a suggests an occipital generation of the wave which then moves forward over the scalp cresting at frontal areas about 60 msec later than at occipital areas. Using the method of Petsche and Marko (1955) one would have seen an impressive propagation of the end of the wave from occipital to frontal areas as indicated by the zero-level bars of Fig. 8 a. The speed of the propagation would have been about 2–3 m/sec. However, from Fig. 7 a it is evident that the basic aspects of the field distributions do not change between msec 705 and 817. Evidently, there is no continuous movement of the maximum field values over the scalp. The migration of the EEG wave is caused by an interaction of minor changes of the field shape at all electrode sites, including the mastoid. We note here that there is no proven electrically indifferent reference point.

Thus, there is buildup and decay of the spike-wave field around two or three restricted areas where maximum field values are detected. The maximum field values occur in a frontal, central, and occipital preference area, as demonstrated in Fig. 7 b. There is no continuous displacement of the maximum field values across the recording area. However, simultaneous, conventional EEG traces may demonstrate a migration or phase lag of waves defined either by base-line crossings, or by peaks. It appears that, similar to our conclusions in the case of alpha EEC, a stationary generator model could adequately account for the sequence of field distributions which is observed during the spike-wave complex.

## Summary and Conclusions

The technique of multichannel EEG field mapping and strategies of subsequent data reduction are briefly reported. Charting of the orbits of maximum field values over time, and charting of the standard deviation of the field values which were averaged over time at each electrode position may be interpreted to indicate the location of equivalent model generators. Graphing of the absolute amplitude per electrode, averaged over space, reveals periods of pronounced and of flat field relief.

The field mapping technique can detect more than one simultaneous maximum field value of the same polarity: the simultaneous activity of the two brain hemispheres during average evoked potentials is visualized.

The relation between different conventional EEG traces which are recorded simultaneously from different electrode combinations becomes intelligible when the associated sequence of field distributions is taken into consideration (e.g., evoked responses, sleep slow waves).

Knowledge of the field distributions aids in the selection of the best electrode positions for optimal recording of certain wave components, as demonstrated with averaged evoked potentials.

The field distributions during alpha EEG are relatively simple, and the maximum values of the fields are found preferentially in three separate scalp areas. This suggests the modelling of alpha fields by three stationary generators.

A clear indication of a moving generator process would be a continuous displacement of the maximum field value. Such continuous displacement is observed seldom within restricted scalp areas, and very rarely across the whole scalp.

Travelling of EEG waves across the scalp, as determined by baseline-crossing lags and peak lags in conventional EEG traces may be associated with stationarity of the maximum values of the field distributions. This observation is illustrated by an example of 3/sec spike-wave activity. In other words, EEG wave travelling may be accounted for by minor changes of the distribution of the field, whose basic aspects remain stationary. Thus, travelling of conventionally recorded EEG waves does not necessarily indicate a migration of the principal process which produces the EEG waves.

## Acknowledgement

The author thanks Mr. J. M. MADEY for the construction of the multichannel system, and Dr. M. GASSEL for his kind permission to examine one of his patients.

## References

BOURNE, J. R., D. G. CHILDERS, and N. W. PERRY: Topological characteristics of the visual evoked response in man. Electroenceph. clin. Neurophysiol. 30, 423—436 (1971).

BRAZIER, M. A. B.: The electrical fields at the surface of the head during sleep. Electroenceph. clin. Neurophysiol. 1, 195—204 (1949 a).

— A study of the electrical fields on the surface of the head. Electroenceph. clin. Neurophysiol. Suppl. 2, 38—52 (1949 b).

21*

Cooper, R., A. L. Winter, H. J. Crow, and W. G. Walter: Comparison of subcortical, cortical and scalp activity using chronically indwelling electrodes in man. Electroenceph. clin. Neurophysiol. *18*, 217—228 (1965).

DeLucchi, M. R., B. Garoutte, and R. B. Aird: The scalp as an electroencephalographic averager. Electroenceph. clin. Neurophysiol. *14*, 191—196 (1962).

Estrin, T., and R. Uzgalis: A moving topological display of the EEG. Electroenceph. clin. Neurophysiol. *27*, 658—659 (1969).

Fischgold, H., H. Torubia et P. Mathis: Le champ électrique du complexe pointe-onde. Rev. Neurol. *93*, 468—474 (1955).

Garoutte, B., and R. B. Aird: Studies on the cortical pacemaker: Synchrony and asynchrony of bilaterally recorded alpha and beta activity. Electroenceph. clin. Neurophysiol. *10*, 259—268 (1958).

Geisler, C. D., and G. L. Gerstein: The surface EEG in relation to its sources. Electroenceph. clin. Neurophysiol. *13*, 927—934 (1961).

Kavanagh, R. N.: Localization of sources of human evoked responses. Thesis, Calif. Inst. of Technology, Pasadena, Calif., 1972.

Lehmann, D.: Multichannel topography of human alpha EEG fields. Electroenceph. clin. Neurophysiol. *31*, 439—449 (1971 a).

— Topographische Erfassung des EEG. Zschr. EEG EMG *2*, 146—147 (1971 b).

— and D. H. Fender: Multichannel analysis of electrical fields of averaged evoked potentials. Electroenceph. clin. Neurophysiol. *27*, 671 (1969).

— R. N. Kavanagh, and D. H. Fender: Field studies of averaged visually evoked EEG potentials in a patient with split chiasm. Electroenceph. clin. Neurophysiol. *26*, 193—199 (1969 a).

— J. M. Madey, M. Koukkou, and D. H. Fender: Mapping of visually evoked EEG responses on the human scalp. Invest. Ophthal *8*, 651 (1969 b).

Petsche, H.: Pathophysiologie und Klinik des Petit Mal. Wien. Z. Nervenheilk. *19*, 345—442 (1962).

— Quantitative analysis of EEG data. In: Schadé, J. P., and J. Smith (eds.), Progress in brain research, Vol. 33, Brain mechanisms, pp. 63—86. 1970.

— und A. Marko: Toposkopische Untersuchungen zur Ausbreitung des Alpharhythmus. Wien. Z. Nervenheilk. *12*, 87—100 (1955).

— — Zur dreidimensionalen Darstellung des Spike-Wave-Feldes. Wien. Z. Nervenheilk. *16*, 427—435 (1959).

Rémond, A.: Orientations et tendances des méthodes topographiques. Rev. Neurol. *93*, 399—432 (1955).

— The importance of topographic data in EEG phenomena, and an electrical model to reproduce them. Electroenceph. clin. Neurophysiol. Suppl. *27*, 29—49 (1968).

Schneider, M., et P. Gerin: Une méthode de localisation des dipôles cérébraux. Electroenceph. clin. Neurophysiol. *28*, 69—78 (1970).

Shaw, J. C., and M. Roth: Potential distribution analysis. I. A new technique for the analysis of electrophysiological phenomena. Electroenceph. clin. Neurophysiol. *7*, 273—284 (1955).

WALTER, W. G.: Analytical means of studying the nature and origin of epileptic disturbances. Res. Publ. Ass. Nerv. Ment. Dis. (N. Y.) 26, 237—251 (1947).

WEIR, B.: The morphology of the spike-wave complex. Electroenceph. clin. Neurophysiol. 19, 284—290 (1965).

# Discussion

GLOOR: I wish to comment on the modelling of a dipole from EEG data, especially with regard to the location of this dipole in the head. This is a formal mathematical exercise, but the postulated dipole and its location has an anatomical reality. The real physiological generators are cortical and extremely close to the surface; the larger the area of synchronized cortical activity, the deeper the theoretically derived dipole representing this activity appears to be displaced in the brain. This localization derived from the theoretical treatment of the problem on the basis of the single dipole hypothesis is erroneous. Application of the solid angle theorem to fields created by dipole layers is much more useful and closer to the physiological and anatomical reality. It also indicates that one has to be careful with these purely formal mathematical treatments which assume the existence of a single dipole.

LEHMANN: Generally, I agree. However, I think that the realistic localization of generators is only one goal and that at present, a very useful aspect of modelling of the fields is the compression of the EEG data into small sets of numbers, which can be used for classification. In this application, systematic distortions of the localization parameters are not important; distortion even might enhance differences between different functional states.

PETSCHE: As far as our 56 patients with absences and spike-and-wave patterns are concerned, we studied them with 8 electrodes put along our parallel to the midline and across the skull. The most obvious travelling pattern was seen in the most regular EEG patterns. Particularly noticable was a discontinuity in spreading which was mostly seen at the boundary between frontal and parietal cortex. This is also the place where two cortices of quite different thickness and structure come together. This observation led us to the idea that the morphology of the cortex may play an importent role for travelling; this was ascertained by further observations in rabbits.

Lehmann: I think that it is very interesting that the travelling characteristics change at the border between parietal and frontal areas. This reminds me of the preferred localization of the field maxima: frontal and parietal-occipital. One could hypothesize that the overlapping fields produced by the two "generators" (principal generating processes) undergo minor shape changes of different time course. Hence, EEG waves would travel at a fairly uniform rate within the area dominated by one generator, and at a different rate (or even, with a different direction) in the area primarily influenced by the other generator.

# Mathematical Simulations of Alpha Rhythms Recorded on the Scalp

J. P. Joseph [1], H. Rieger, N. Lesèvre [2], and A. Rémond [3]

Laboratoire d'Electroencéphalographie et de Neurophysiologie appliquée (L.E.N.A.), Hôpital de la Salpêtrière, Paris, France

## I. Introduction

The purpose of this work is the description of the structure of brain potentials based on EEG rhythms, recorded on human scalp, using bipolar derivations. Although this description is essentially concerned with the study of alpha rhythms, still the methods used herein could be transferred to studies of other types of rhythms and the conclusions about alpha rhythms might well clarify underlying neurophysiological mechanisms and also aid us in our efforts to discover the cause of the synchronization of cortical activities and that of the so-called "travelling waves".

The analysis and description of the gradient structure of alpha rhythms is but one step in the analysis of its potential structure. The next step is, using the gradients as a basis, to find the phase and frequency relations among the different potentials which appear at varying places on the scalp. Experience shows that the latter is difficult. It does not suffice to integrate the topogram of the gradients to get an answer to questions about the phase and frequency relations involved since integration adds nothing to this kind of information and, in fact, the potential topogram thus obtained masks the elements essential to its own description (extrema, points of inflexion), these elements showing up much more clearly on the topogram of gradients. Besides, such an integration of gradients would necessitate a "reference" level and the choice of this reference would be as difficult

1 Attaché de Recherche au C.N.R.S.
2 Maître de Recherche à l'I.N.S.E.R.M.
3 Directeur de Recherche au C.N.R.S.

—or as impossible—as that of the common lead reference made in the case of the so-called "monopolar" montages.

To get around the difficulties of this situation and to understand the potential structure as well as to clarify the problem of its origin, mathematical simulations of EEG rhythms were made. A potential field was simulated whose properties very closely approximated those of the cortical potential field recorded on the scalp, the similarity between real and simulated activity being measured by the similarity between recorded experimental gradients and simulated gradients.

## II. Method and Experimental Data

Alpha rhythms were recorded using bipolar derivations consisting of 9 electrodes in line, 2 cm apart; the montages were transversally placed crossing the midline at different levels going from the inion to 16 cm ahead, so as to record simultaneously the activity of both hemispheres, and longitudinally placed along the midline in the parieto-occipital regions. The EEG rhythms were digitized one line and processed by a BGE M 40 computer. The digitalized data were then filtered in the frequency band 7–14 Hz (Joseph et al. 1970) and plotted as spatio-temporal maps (Rémond 1960, Rémond et al. 1969).

The rhythms were studied in the form of filtered, non-averaged alpha or of filtered average alpha using the method developed by Rémond and his co-workers (Lairy et al. 1966, Rémond et al. 1969).

The alpha rhythms of approximately 150 young normal adults of both sexes were studied. The accumulated data allow a cataloguing of the many possible chronotopographical structures of these rhythms. The experimental data collected show that the alpha potential on the scalp most often possesses the following characteristics:

a) the *amplitude* of the potentials, such as they appear from the gradient recordings along the transverse sections, seem to be bell-shaped curves. The position of the extremum and the spread of these curves vary as a function of time; these parameters also vary with the section observed as well as with the subject studied. Along a median posterior longitudinal section the distribution of the potential amplitudes appears to have the shape of a $chi^2$ function whose maximum is reached 4 to 6 cm above the inion;

b) as far as *phase relations* among the different potentials that can be inferred from the experimental gradients are concerned, it seems that in certain cases it is possible to characterize the potential oscillating separately on each hemisphere as being in phase and in other cases as being out of phase; it seems that the propagation of a "wave of activity" could explain the case of out-of-phase potentials.

To make the foregoing statements more concrete and particularly to try to measure the possible phase shift and wave-lengths and in order to attempt to locate the barycentres of activity, certain phenomena suggested by this analysis of the experimental gradients have been simulated giving rise to models of alpha activity.

## III. Models

In a preceding study devoted to average alpha rhythm and its analysis, a simulation of potential was suggested using a physico-mathematical model of EEG activity (JOSEPH *et al.* 1969); this model was composed of two charges each connected with a hemisphere, which pulse sinusoidally and may be out of phase. The simulations that correspond to this model fitted in rather well with certain features of experimental alpha rhythm, particularly with those of what has been called by RÉMOND and his co-workers "typical alpha average" (Fig. 1, RÉMOND *et al.* 1969).

This model suggested that with a small number of parameters it was possible to describe the potential of some of the bipolar recordings. The essential thing in this model was the surface potential structure; to suppose that two charges exist is the same thing as to assert that in certain cases—namely that of the "typical alpha average"—the synchronization of cortical cells is sufficiently good to make plausible the idea that the electrical image of the cortical potential is a single charge for each hemisphere.

The recordings made on 150 normal subjects since this first work was published have shown that what RÉMOND *et al.* call the "typical alpha average"—which was characterized on transverse montages by a fixed extremum situated on the midline of potential (Fig. 1) was, in fact, relatively rare. This first model was a sort of oversimplification. The experimental data obtained since then—the analysis of which has been summed up in the preceding paragraph—show a great variety of alpha rhythms structures. These various chronotopographical structures have suggested the following simulations which make no assumptions about the origin of the potentials on the scalp.

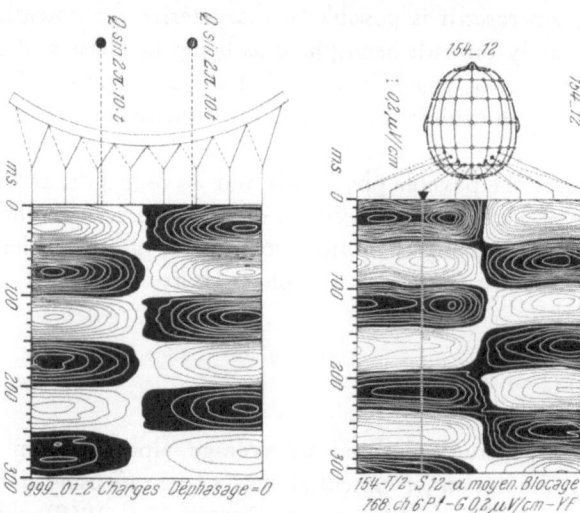

Fig. 1. Two spatio-temporal maps of alpha activity obtained with a montage 16 cm long (bipolar derivations);
*on the right*: experimental "typical alpha average" activity
obtained on a transverse montage by triggering the averaging process from the left hemisphere (from channel 6,
indicated by an arrow); *on the left*: simulated transverse
alpha activity obtained from 2 charges beating at a mean
frequency of 10 c/s with a standard deviation of 1 c/s. In
these maps and those of succeeding figures (except Figs. 3
and 4) zones of negative gradients are in black, zones of
positive gradients in white; the amplitude is given in the
form of contour lines of potential gradients with respect to
space. On these maps and on following maps (except those
of Fig. 7), time is in ordinate and space in abcissa, the
values between the 8 bipolar derivations being obtained by
instantaneous parabolic interpolation. In the experimental
map one sees an image of what has been called "typical
alpha average": each alpha wave is represented by the
succession of 2 instantaneous doublets of gradients of
opposite polarity, thus showing a phase difference of $\pi$ rd
between both hemispheres. The reversal of polarity of the
gradients—which corresponds to an extremum of potential—
is located near the mid-line. The simulated alpha activity
seen in the left map shows a very similar organization to
that of the experimental "typical alpha average"

These simulations are characterized as follows:

a) the 2 hemispheres are replaced by 2 contiguous plane surfaces on which the potentials appear;

b) at each point of a transverse section, the amplitudes of the potentials are proportional to 2 Gaussian distribution curves centered separately on each hemisphere. The respective positions, spreads, and amplitudes of these curves vary continuously from one transverse section to another, one moment to another, and from one subject to another;

c) at every point of a longitudinal section these amplitudes are equal to the corresponding values of a $chi^2$ curve;

d) the phase relations of the gradients, obtained from studying the experimental results, are such as to hinder, for the phase relations among the potentials, the construction of a unique model that takes all the recorded phenomena into account. This is why we shall study two models in turn, each simulating different phenomena.

For the first model—model 1—the potentials on each hemisphere are in phase but between hemispheres can be out of phase. For the second model—model 2—the appearance of potential fields is connected with the existence of a wave of activity that gradually spreads over the surface of the cortex, causing a phase shift between potentials which depends on the wave-length. (Note that a continuous transition from model 1 to model 2 can be carried out if the assumption is made that the direction of propagation of the wave is perpendicular to the montage;)

e) the oscillation frequencies are the same at all points. This is obviously a simplification of the phenomena being studied (alpha rhythm), which, if not filtered into a narrow band, can show notable frequency differences from one region to another.

Experience has shown that the measurement of activity arising from these simulated potential fields, obtained by the method of bipolar recordings, provides pictures of gradients very similar to those obtained from certain experimental curves. Both spontaneous and average alpha rhythms can be interpreted in numerous cases as the manifestation of a single, unique phase displacement between 2 hemispheres. The structure of the potentials could in this case be referred to model 1. In other cases, the structure is more complex and can fit in with model 2. We thus shall study in turn model 1 and model 2 and with each, the structure of the gradients measured using transverse and longitudinal montages.

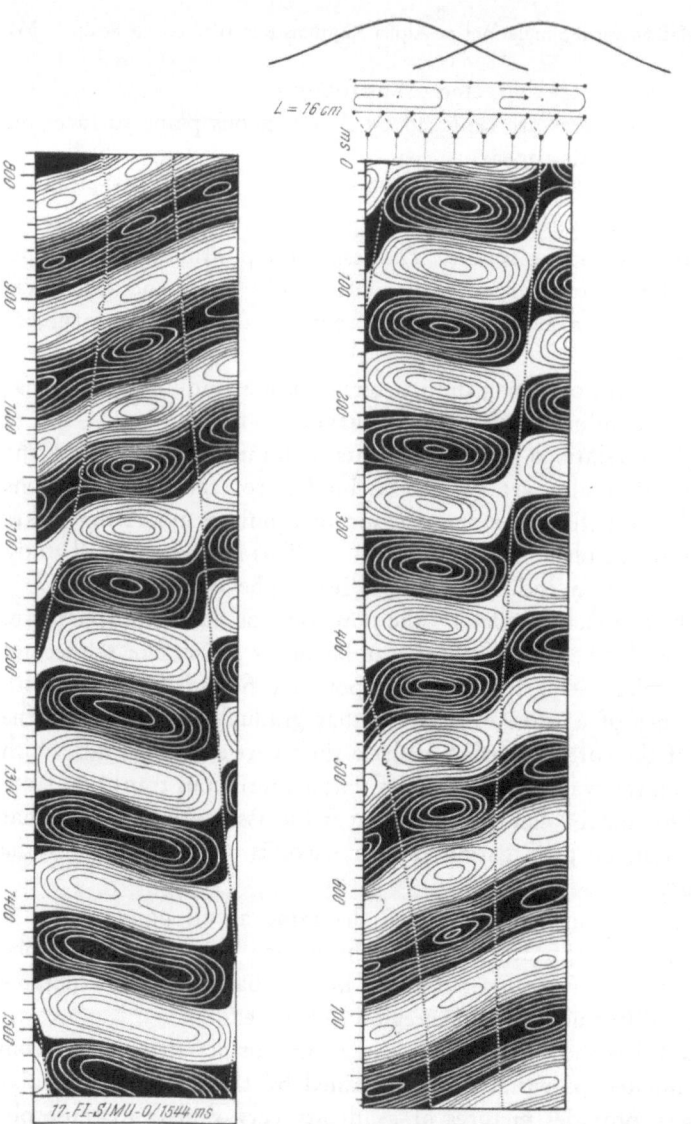

Fig. 2. Two spatio-temporal maps of simulated transverse alpha rhythms (gradients) obtained, with a montage 16 cm long, by summing up 2 bells of potential. The amplitude of both potentials oscillate sinusoidally at 10 per second and the maxima are represented by the two bells above the right map. The oscillation of the right potential is in advance of $\pi/2$ rd. The maxima of the bells oscillate in space separately on each hemisphere around channel 2 (right hemisphere) and channel 7 (left hemisphere). The different positions of these maxima at each moment are marked by 2 dotted lines. The amplitude of the oscillation is 2.5 cm. According to the topography of both maxima of potential one sees how the structure of the gradients changes and how the direction of the apparent travelling wave is reversed

## First Model: In-Phase Oscillations on Each Hemisphere

### A. Structure of the Simulated Rhythms Recorded by Transvere Bipolar Montages

The relations between the apparent phase shift of the gradients and the phase shift of the potentials can be studied qualitatively from Fig. 2, which shows that the direction of the apparent flow (or travel-

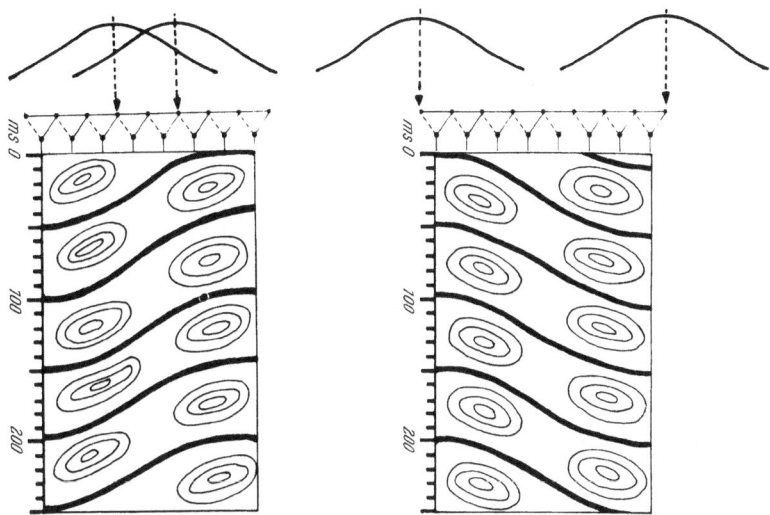

Fig. 3. Two spatio-temporal maps of simulated transverse alpha rhythms (gradients) obtained by summing 2 bells of potential oscillating at 10 per second. The right bell is in advance of $\pi/4$ rd. On this figure and Fig. 4, the polarity has not been differentiated, positive and negative gradients being in white; the thick black lines represent the zero of gradient, each successive fine black line represents the amplitude in the form of isogradient curves. On the *left map*: the 2 bells are close together so that the summed potential is maximum in the middle of the montage, on the *right map*: the 2 bells are far a part so that the summed potential is minimum in the middle of the montage. The direction of the apparent flow changes from one map to the other; this change is only due to differences in topography of the bells

ling wave) on maps of gradients is dependent not only on the chronographic characteristics—frequencies and phase shifts, which are fixed quantities on this figure—but also on the topographic characteristics. When the maxima of activity are close together and the centre of the montage is a zone of maximum activity, the phase shift of the rhythms has a different direction from that shown when the maxima of activity are far apart and the centre of the montage is a region of minimum activity.

The main features of Fig. 2 can be summed up by the maps of Fig. 3 which depict the relations of amplitudes and phases among alpha rhythms. From the right maps of Fig. 3 we may deduce a phase advance on the left side if the tops of the bell-curves are brought closer together (relative maximum at the centre) and a phase delay if their tops are placed farther apart (relative minimum at the centre).

The simulations made by modifying some of the parameters of the model (phase shifts between hemispheres and spacing of the bell shapes) give the relations between the pictures of gradients and the corresponding potentials. As there is a large number of these pictures, we shall assume for purposes of simplification that the 2 bell-shaped potentials have the same amplitude and the same standard deviation (Fig. 4).

A close look at the spatio-temporal maps of Fig. 4 leads to two observations:

a) as we have already noted, the direction of the travelling wave changes according to whether the centre region of the maps is a region of minimum or of maximum potential;

b) the phase shift of the gradients can increase when the phase shift of the potentials decreases.

B. Structure of Rhythms Recorded by Longitudinal Bipolar Montages

Assuming that the potentials oscillate in phase on either hemisphere alone but are out of phase from one to the other, the longitudinal recordings provide but little information on the electrical activity of

---

Fig. 4. Six spatio-temporal maps of simulated transverse alpha rhythms (gradients) obtained by summing up 2 bells of potential oscillating at 10 per second. On the 3 left maps, the right potential is in advance of $\pi/2$ rd. On the 3 right maps the advance of the right potential is of $\pi/4$ rd. The 2 upper maps are obtained from a situation where the maxima of the potentials of each hemisphere are close together. The two lower maps are on the contrary obtained from a situation where the 2 potentials are far apart. The 2 maps in between are obtained from an intermediate situation. One sees how different structures appear in changing only 2 parameters (phase shift and spacing of the bells)

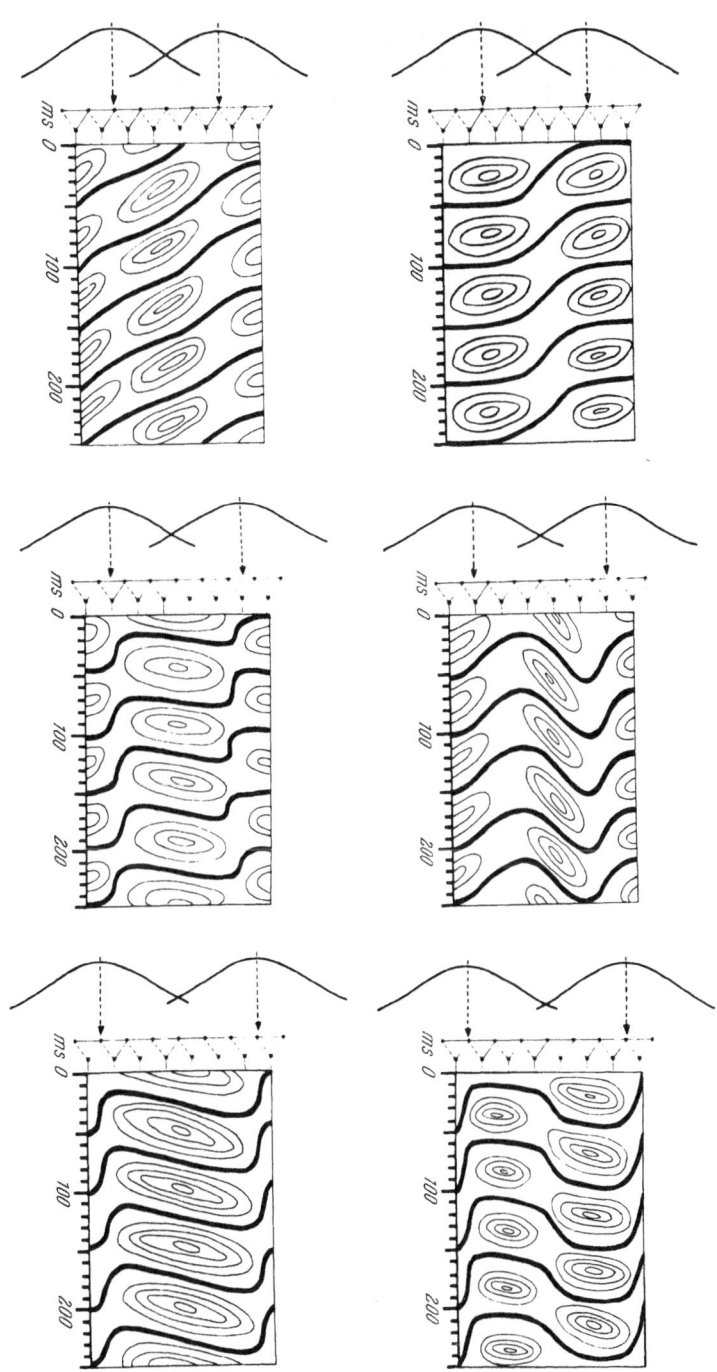

Fig. 4

the cortex. In this respect, the apparent phase shift of the gradients of potential on the longitudinal montage will have, for an equivalent phase shift of potential, a different direction depending on the respective positions on this montage of the maximum of activity coming from the left and from the right hemisphere. This is the reason why, once the direction of phase shift between hemispheres has been determined by a transverse montage, the longitudinal montage will then supply information only on these respective positions. (Note that if these positions are merged statistically, the gradients obtained on the longitudinal montage are statistically in phase, whatever the transverse phase shifts may be.)

### Second Model: Potentials Linked with Activity Wave Trains

As has been already stated in the introduction, the assumption that the potentials of each hemisphere are in phase does not incorporate all the aspects revealed by the experimental data; frequently there can be observed phase differences between potentials on one and the same hemisphere that can be related to the propagation of a wave of activity.

A wave could be represented by an equation of the form:

$$P = A \cos 2\pi \left( \frac{t}{T} - \frac{X}{L} \right)$$

This equation means that:
every point $X$ undergoes an oscillation of period $T$;
the points $X$ are out of phase. At each moment $t$, the amplitude of different points is represented by a sinusoidal wave of spatial period $L$ (also called the wave length); the wave of activity can be interpreted as a sinusoidal wave advancing in space with speed $V$,

$$V = \frac{L}{T}$$

On the longitudinal montages, it is possible that the wave trains coming from the 2 hemispheres are superimposed; in this case, a new system of waves is created, the amplitudes of which are modulated by nodes and anti-nodes. Note that these nodes and anti-nodes do not represent regions of lesser or greater activity; they are but an artifact of measurement. In the case of 2 waves $P_1$ and $P_2$, the resultant wave length is equal to:

$$2\,\frac{L1 \cdot L2}{L1 + L2}$$

and the resultant period is equal to:

$$2\,\frac{T1 \cdot T2}{T1 + T2}$$

When $T1 = T2$, the resultant activity is unstable, and the nodes and anti-nodes of activity are displaced at the speed:

$$\frac{L1 \cdot L2}{L1 - L2} \,\Big|\, \frac{T1 \cdot T2}{T1 - T2}$$

The slow translations of topographic characteristics that can be associated with such superpositions appear often in the activity recorded longitudinally.

One must note here that the wave-lengths read off the montages are in all cases longer than the real lengths of the waves, when the direction of the montage is different from that of the propagation of the wave.

## A. Structure of Rhythms Recorded by Bipolar Longitudinal Montages

Suppose that we have a potential field linked with activity wave trains. We assume that the envelope of these potentials has, as stated before, the form of a chi$^2$ curve with a maximum and a point of inflexion (Fig. 5). If this potential field is recorded by the usual montage made of bipolar derivations in line, one has, depending on the values of $L$, different topographic structures (Fig. 5). One sees that the maximum of potential can be characterized just as well by a minimum as by a maximum of gradients, and that the smaller the wave length, the greater the phase shift between adjacent regions becomes.

## B. Structure of Rhythms Recorded by Bipolar Transverse Montages

When the 2 potential fields of the hemispheres linked with these waves are recorded, we obtain different structures depending on the location of the maxima of the potential curves, just as in the case when the potentials were oscillating in phase.

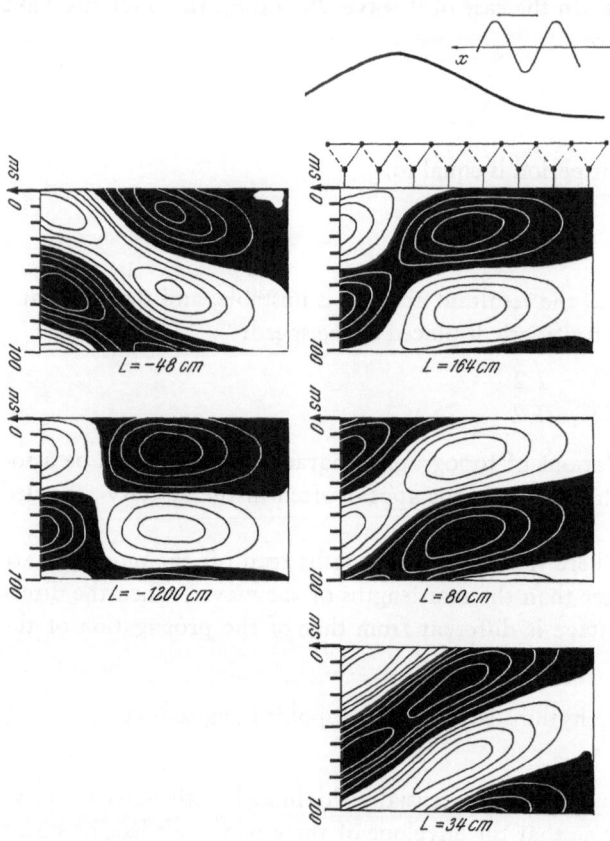

Fig. 5. Five spatio-temporal maps of simulated longitudinal alpha rhythms (gradients) obtained from potentials linked with waves of activity of various lengths and of period $T = 100$ ms. The length of the montage is 16 cm long. The curve represented above the right maps is a chi$^2$ curve which is the envelope of the potentials. The sinusoïdal curve above represents the wave of activity, the arrow showing the direction of its propagation. The wave-length $L$ has successively 164, 80, 34, $-48$ and $-1,200$ cm. One sees that the general chronotopographical organization changes according to the different wave-lengths; particularly the zero of gradient—which usually corresponds to an extremum of potential—shows a tendency to be replaced by an extremum of gradient as the wave-length gets shorter

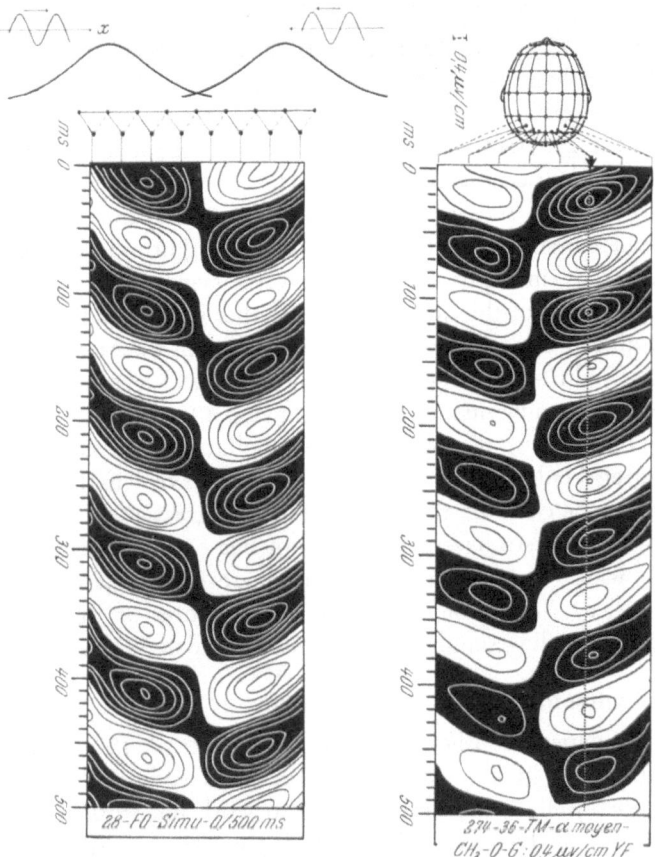

Fig. 6. Two spatio-temporal maps of experimental and simulated alpha rhythms (gradients). *On the right*: map of experimental alpha average obtained from EEG recorded on the scalp with a bipolar transverse montage, 16 cm long, crossing the mid-line 8 cm above the inion; the alpha activity of channel 3 on the right hemisphere has been utilized to trigger the averaging process. *On the left*: map of simulated non-averaged transverse alpha rhythms, obtained with a montage 16 cm long. The field of potential is obtained by summing up 2 potentials linked with 2 centripetal waves of activity. The period of both waves is 100 ms; the wave length is 40 cm for the right wave and 56 cm for the left one. The 2 bells drawn above this map represent the envelope of the 2 potentials separately linked with 2 waves of activity. The sinusoïdal curves above represent the 2 waves, the arrows showing the centripetal direction of their propagation. The chronotopographical structure of both simulated and experimental maps are very similar

In the simulated part of Fig. 6, we shall assume that:

a) the wave lengths are, on both sides, about three or four times longer than the length of the montage;

b) the activity wave trains are centripetal;

c) the 2 potential curves have identical shape and amplitude alignment, their centers being far apart;

d) the right side is in advance; the phase advance of the right hemisphere means that the regions on the right side of the scalp have their maximum of activity before the symmetric regions on the left. The source of this advance can lie either in a real advance of the wave trains or in a difference of wave lengths; in the case of an advance of the right side due to a difference in wave-length, the wave-length would then be smaller on the right (Fig. 6).

As in the case of potential fields oscillating in phase, simulations can be effected under other conditions of wave length and phase shift and when one of the two fields has a greater amplitude having its origin in a stronger polarization of one of the hemispheres. The case in which the maxima of potential are close together has also been studied. The chronotopographical structures of the rhythms of the corresponding simulations are found in numerous experimental maps.

## Discussion

In a paper published elsewhere (Joseph et al. 1971) we have shown that most of the experimental material collected fits these models rather well; it has been possible to work out statistics for the topography of the barycentres of activity of each hemisphere and for their average phase relationships during periods of the order of one minute. These models have also enabled us to distinguish travelling waves due to the waves of activity and those due to phase shifts occurring between spatially fixed "generators". It has likewise been possible to describe the characteristics of waves of activity when these were clearly present.

Filtered alpha rhythms show structures that during a short period fit all the simulations successively. The structure of alpha rhythm can thus be characterized by a succession of periods in which activity is apparently in phase separately on each hemisphere, followed by periods in which activity is related to the presence of waves of activity (Fig. 7).

But the assumption in the study of these two models, which is that the frequencies of different regions of the same hemisphere are equal, is an oversimplification of the reality. On the contrary, the tracings show that there is often not one, but several alpha rhythms, appearing simultaneously and having different but close frequencies which are probably connected to different alpha domains (COOPER and MUNDY-CASTLE 1960), whose spreads and frequencies themselves develop in time. The phenomenon of spontaneous alpha "beats", the different alpha wave lengths on different derivations during one minute or on one derivation during several minutes studied with the aid of interval histograms, all show clearly these differences.

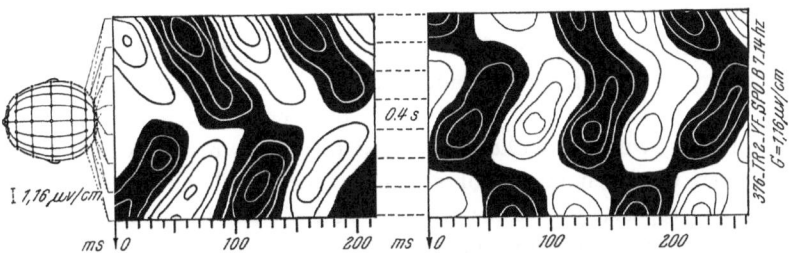

Fig. 7. Two spatio-temporal maps of non-averaged experimental alpha rhythms (gradients) filtered in the 7–14 Hz frequency band and recorded with a transverse montage 16 cm long crossing the midline 2 cm above the inion. The 2 maps show the organization of the alpha rhythm during 210 ms (left map) and 260 ms (right map), 400 ms separating each map. Contrary to the maps of other figures, in this map time is in abcissa and space in ordinate. *On the left map,* the alpha activity is centripetal, linked with two centripetal waves of activity, and shows a slight advance of the left hemisphere. Both waves stop at the level of the midline. *On the right map,* the centripetal activity has disappeared and is replaced by potentials apparently in phase separately on each hemisphere, with an advance of the right hemisphere of approximately π/2 rd (*c.f.,* Fig. 4 and intermediate maps of Fig. 6 where one sees quite similar pictures obtained by simulated gradients)

This problem of the different frequencies recorded at various points on the scalp, often on the same hemisphere, is linked to the problem of synchronization—of its measure and its origin—which we shall consider now.

In the framework of the "alpha average" method (JOSEPH *et al.* 1969) we suggested measuring the degree of synchronization of right and left hemispheres using an index calculated from amplitudes values of two symmetric right and left derivations.

The criticism of this alpha average method and the analysis of its artefacts (JOSEPH 1971) has shown that the above-mentioned index measured not only possible phase or frequency uncertainties between hemispheres but also topographic uncertainties. In other words, this criticism showed that a faulty index of synchronization was not necessarily an indication of a bad synchronization but might depend as well on a topographic instability of the potentials, that is, displacements of the center of polarization maxima. Consequently, this index is an ambiguous quantity and—generalizing somewhat our own conclusions—it seems that the problem of the measurement of the synchronization of cortical activities is a complex one whose study cannot be carried through thoroughly on the basis of gradient measurements on the scalp or of parameters derived from these measurements.

But the measurement of the synchronization of activities is perhaps not a necessary factor in understanding its origin. While it seems difficult to explain this phenomenon of synchronization from EEG recordings, we would like nonetheless to set down some correlatives between the phenomenon of synchronization and that of waves of activity, especially as some workers in the field have noted a link between synchronization and travelling waves (PETSCHE and ŠTERC 1968, PETSCHE and RAPPELSBERGER 1970, PETSCHE, RAPPELSBERGER, and TRAPPL 1970).

First it is important to discriminate, in travelling waves phenomena, those which characterize phase shifts between stationary bells of potentials and those which come from waves of activity. This distinction becomes possible once the amplitudes of gradients providing information on the topography of electrical activity barycentres are taken into account, as is suggested by several simulations. Thus a study of the results collected over several years leads to the following conclusions:

Transhemispherical apparent flow (travelling waves) is probably due to the phase shifts of each hemisphere potentials and not to transhemispherical waves of activity (Fig. 7).

When waves of activity are present, the barycentres of alpha activity, that is, the regions of maximum potential, do not constitute the regions from which the wave of activity arises. Put in another way, the wave does not propagate radially from the regions of maxima of activity.

In general, the waves of activity (when it is possible to detect them), come from anterior and lateral regions and gradually spread through the central and occipital regions which are those of alpha potential maxima (Fig. 7).

In respect to the links between synchronization and waves of activity, some experiments show that:

*On the transverse montages.*

The better the various cortical regions seem synchronized, the better the wave of activity shows up. However, this wave does not always appear when there is a good frequency synchronization between various cortical regions. In this case, the absence of the wave could be explained by assuming a direction of wave propagation per- pendicular to the montages used (Fig. 7).

*On the longitudinal montages.*

The wave of activity appears very frequently whether the rhythms on transverse recordings are synchronized or not. Generally, the direction of its propagation is antero-posterior. The measurement of its characteristics is a delicate if not impossible operation, because the apparent phase shifts of the gradients depend not only on the wave-length but also on the shape of the envelope of cortical potentials.

These observations are insufficient either to affirm or to invalidate the assumption that the wave of activity synchronizes the cortical cells, especially as certain maps obtained from longitudinal posterior montages and analysed in the average alpha form, show the presence of this wave even when the frequencies of rhythms from one region to another are statistically different, that is, when there is manifest frequency asynchronism. In this case, if we retained the assumption of the synchronization of cells by the wave, it would then be neces- sary to suppose either that the wave acts only as an impulse and that the alpha domains pulse with their own frequency or that the wave properties (frequency and wave length) are modified while crossing the different alpha domains.

## Summary

The present study was undertaken to describe the structure of brain potentials based on alpha rhythms recorded on human scalp and to clarify the underlying mechanisms.

The experimental data consisted of bipolar recordings obtained with many transverse and longitudinal montages from 150 normal adult subjects.

The alpha rhythms were studied after having been digitized and filtered in the frequency band 7–14 Hz and plotted as spatio-temporal maps.

This chronotopographical analysis was achieved by simulations of alpha rhythms. These simulations were based on two different models which take in account two different aspects of experimental maps: 1. in-phase oscillations on each separate hemisphere; 2. out-of-phase oscillations on each separate hemisphere linked with the propagation of waves of activity.

The following parameters are varied: topography of the barycenters of activity and phase relations between hemispheres for model 1 and 2, and wave lengths and direction of propagation for model 2.

In these different cases, the chronotopographical structure of the simulated maps is very similar to that of many experimental maps.

It was possible with these models to discriminate in travelling waves phenomena those which result from phase shifts between stationary bells of potential and those which come from the propagation of waves of activity.

The link between synchronization of cortical potentials and propagation of waves of activity has been discussed.

## Acknowledgement

We are grateful to J. C. Bourzeix and J. Furet for their technical assistance.

## References

Cooper, R., and A. C. Mundy-Castle: Spatial and temporal characteristics of the alpha rhythm. A toposcopic analysis. Electroenceph. clin. Neurophysiol. 12, 153—165 (1960).

Joseph, J. P.: Critique de la méthode d'alpha moyen. Trace. 5, 177—181 (1971).

— Simulations mathematiques des potentiels corticaux. Trace. 5, 211—224 (1971).

— et J. F. Baillon: Analyse de fréquence du signal EEG. Rev. Neurol. 122, 480—482 (1970).

— A. Rémond, H. Rieger, and N. Lesèvre: The alpha average. II. Quantitative study and the proposition of a theoretical model. Electroenceph. clin. Neurophysiol. 26, 350—360 (1969).

Lairy, G. C., A. Rémond, N. Lesèvre et H. Rieger: Introduction à l'étude EEG des fonctions visuo-motrices chez l'enfant. Rev. Neurol 115, 15—59 (1966).

Petsche, H., and J. Šterc: The significance of the cortex for the travelling phenomenon of brain waves. Electroenceph. clin. Neurophysiol. 25, 11—22 (1968).

— and P. Rappelsberger: Influence of cortical incisions on synchronization pattern and travelling waves. Electroenceph. clin. Neurophysiol. 28, 592—600 (1970).

— — and R. Trappl: Properties of cortical seizure potential fields. Electroenceph. clin. Neurophysiol. 29, 567—578 (1970).

Rémond, A.: Recherche des renseignements significatifs dans les enregistrements électrophysiologiques et mécanisation possible. Actualités neurophysiologiques, 2ème série, pp. 167—210. Paris: Masson. 1960.
— Integrated and topographical analysis of the EEG. Electroenceph. clin. Neurophysiol. Suppl. 20, 64—67 (1961).
— The importance of topographic data in EEG phenomena, and an electrical model to reproduce them. Electroenceph. clin. Neurophysiol. Suppl. 27, 29—49 (1969).
— N. Lesèvre, J. P. Joseph, H. Rieger, and G. C. Lairy: The alpha average. I. Methodology and description. Electroenceph. clin. Neurophysiol. 26, 245—265 (1969).

# Discussion

Shaw: Why do you use bipolar derivations since Cooper has shown that this introduces errors due to the differences at different reading sites?

Rieger: This is an old question in clinical electroencephalography, but without any importance in our case since potential distribution may be calculated by computer. Mathematically, one can easily transform potential distribution maps into maps of potential gradients, and vice versa, by differentiation or integration. Monopolar and bipolar recordings contain the same information.

Shaw: But if you attempt to interpret your bipolar derived topograms in terms of some physiological processes you are not getting a true account of the time relations.

Rieger: I don't think that there is an error in the time domain. The resolution of measurements is in the order of 1 to 4 msec. It is true, as Rémond pointed out, that bipolar recordings often present more "contrast" than monopolar tracings, but I cannot see here any distortion.

Firneis: Could not a generalized nonlinear generator-system, as e.g., described by Van der Pol differential equations, explain the observed phenomena without the necessity of having to resort to the travelling wave hypothesis? The induced constants would give additional degrees of freedom which could be used to identify the system in some optimal way.

Rieger: I believe that you are right. But it would be preferable to discuss this with Joseph, who is a better mathematician than I am.

Lehmann: Wouldn't three dipole generators suffice to account for your data? In our field studies it appears that this number of model generators will be sufficient for human alpha.

RIEGER: We thought we would need four oscillators to explain the potential distribution in both the longitudinal and the transverse direction. But there are some effects of overlapping between the activities of the right and the left hemisphere, and I think that the simpler model we proposed here may be sufficient.

SHAW: I would like to make a comment on the question of using dipole models. Considering the potential distribution at a particular time slice, there are many source distributions that can fit this particular distribution of potential. If you take several time slices so that you have a changing potential configuration, can you say that the number of source distributions which will fit this picture are reduced, or can you still have many possible source distributions? The model must fit the physiological situation and physiologically we have to think of a change of potential on the surface as a change in activity of a large spatial distribution of cells. What advantage is there in reducing this to a few dipole sources?

RIEGER: This is the problem of average rhythms. Their spatio-temporal configuration is generally rather simple and stable in time; we searched for the simplest mathematical model to "explain" and to reproduce them. But much work has to be done to point out the exact physiological significance of these rhythms. This simplicity has to be paid for by some loss of information, the irregularities being averaged out.

LEHMANN: Dr. KAVANAGH (Localization of sources of human evoked potentials, Thesis, California Institute of Technology, Pasadena, Calif. 1972) from Cal Tech has designed a computer program which fits two model generators to a given scalp EEG field distribution. He has used the program to treat some of our evoked potential field data. The program finds unique solutions for the location of the two equivalent dipoles.

# Photically-Induced Epilepsy in Papio Papio: The Initiation of Discharges and the Role of the Frontal Cortex and of the Corpus Callosum

R. Naquet, C. Menini, and J. Catier

Département de Neurophysiologie appliquée de l'Institut de Neurophysiologie et de Psychophysiologie du C.N.R.S., Marseille, France

Until recently the experimental study of photosensitive epilepsy was obstructed by a major difficulty, namely the absence in the experimental animal of a syndrome comparable to that in man; indeed such a syndrome could be provoked only by the injection of convulsant drugs.

The abnormal photosensitivity of a high percentage of Papio papio coming from the Casamance region of Senegal (described in 1966 a by KILLAM et al.) has proved to be so similar to photosensitive epilepsy in man that these authors (KILLAM et al. 1967 a, NAQUET et al. 1968) have proposed it as a true experimental model.

In 60–80% of baboons coming from the Casamance region, intermittent light stimulation at 25 c/sec provokes myoclonic movements which involves the eyelids initially and which can spread to the face, trunk and limbs. In general these myoclonic movements stop at the same time as the photic stimulation stops, but in the most photosensitive animals, self-sustained discharges can continue after the end of photic stimulation and develop into a tonic-clonic seizure of "grand mal" type.

Until now most of the studies undertaken on baboons have been aimed at defining the nature of the photosensitivity or elucidating its mechanisms, particularly the role played by certain structures. Thus somewhat paradoxically, the frontal cortex has been found to have unusual properties, which undoubtedly contribute to the manifestations of photosensitivity even if they do not explain them. Also, quite paradoxically, the occipital cortex at first appeared to play a merely subordinate not to say negligible role in this form of epilepsy; its involvement in the photosensitive responses has been indicated by some recent experiments.

In this presentation we shall emphasize especially the mechanism of initiation of the paroxysms and the role of different cortical areas in their development; finally we shall consider the role of the corpus callosum in the synchronization of the interictal paroxysmal discharges, and in the morphology and evolution of the self-sustained seizure discharges provoked by photic stimulation.

## I. Frontal Origin of EEG Paroxysmal Discharges

### A. Cortical Recordings

In all photosensitive animals the resting EEG record shows high amplitude paroxysmal discharges of spikes-and-waves, which are specifically located in the fronto-rolandic area. These abnormalities are always bilateral, synchronous and symmetrical, and appear only when the animal has his eyes partially or fully closed. They are immediately blocked by eye opening or any alerting reaction. They became more prominent during drowsiness, and persist during stage I and II of sleep; they diminish during stage III and are absent during REM-sleep (Balzano 1968). Even when they are seen in other regions they always have a pre-rolandic point of origin.

During intermittent photic stimulation at 25 c/sec, after a variable latent period, paroxysmal discharges appear in the frontal regions. They behave like the spontaneous discharges in that they can invade other territories secondarily but the point of origin is always frontal.

After the end of stimulation when self-sustaining ictal activity is seen, it too always begin in the frontal cortex. It occurs synchronously in the two frontal regions and progressively invades all the cortical areas, finally reaching the occipital lobe. At that time it is fully generalized. In the intact animal, and in the absence of certain drug injections (such as inhibitors of the synthesis of GABA (Meldrum et al. 1970) seizures originating in the occipital cortex are not seen.

### B. Recordings from Deep Structures

In the absence of photic stimulation, paroxysmal activity is not observed in deep structures, except when it occurs also at the cortical

level. During stimulation paroxysmal discharges invade deep structures especially when generalized myoclonus occurs; their point of
origin is always the frontal cortex. The specific visual pathways
(optic tract and lateral geniculate body) are only rarely invaded by
spike-and-wave discharges, even when other subcortical structures
are also involved. After the end of stimulation, when a self-sustaining
discharge develops, it always begins at the cortical level and only
secondarily involves deep structures; but it can fail to involve certain
other deep structures such as the rhinencephalon (FISCHER-WILLIAMS
et al. 1968). However, this clear contrast between cortical and subcortical structures during seizures produced by photic stimulation has
not been found by WADA et al. (1971) (but their techniques of pre-
paring and recording from the animals were somewhat different).

## C. Activity of Single Units in the Frontal Cortex during Paroxysmal Discharges

Extracellular recording of single units in various cortical territories
have been made with tungsten microelectrodes. As the animals used
also had chronic cortical macroelectrodes it was possible to compare
EEG data with simultaneous microelectrode recordings.

In the absence of cortical paroxysmal activities normal behaviour is
seen in neurones in all cortical regions and in all baboons, whether
photosensitive or not, and with and without photic stimulation. With
the onset of paroxysmal EEG activity certain neurones in the fronto-
rolandic cortex show a firing pattern very similar to that seen in man
or in animal with a focal epileptogenic lesion or after the local application of a convulsant drug. However traditional histological
techniques have not produced any evidence for a lesion or anomaly in
this region (RICHE et al. 1970).

This close correlation between the firing of frontal neurones and the
EEG paroxysmal discharges has not been found either with other
sensory modalities or in other cortical zones. Although a cortical
paroxysm is always accompanied by a burst of single unit discharges
in the frontal cortex, it is nevertheless possible to observe a burst at
the single cell level in the absence of a paroxysmal EEG discharge
(Fig. 1). We therefore suggested that the frontal paroxysmal activities
originate in this region and are the statistical result of the activity
of the underlying neurones (MENINI et al. 1968 a, MORRELL et al.
1969). However in a recent work JAMI (1972) observed similar
relationships after the systemic injection of metrazol in the cat. This

observation does not tend to disprove the local origin of the paroxysmal EEG manifestations and does not suggest that this pattern of activity does not necessarily indicate the existence of a localized "epileptogenic focus" but may simply result from an abnormal reactivity of the cortex.

Lastly, it has been observed that intermittent photic stimulation at 25 c/sec tends to produce or enhance a synchronization of the single unit discharges in the fronto-rolandic cortex.

## II. Afferent Visual Pathways to the Frontal Cortex

The evidence for visual afferent systems in the frontal cortex comes from single unit or evoked potential records during photic stimulation.

### A. Visual Evoked Potentials

As a general rule, high amplitude potentials evoked by photic stimulation are limited to the occipital cortex. Nevertheless, in the precentral region a low amplitude, short latency (50 m/sec) evoked response is seen; this is always largest in the fronto-rolandic territory (area 6) or more precisely around the superior frontal sulcus and the caudal branch of the arcuate gyrus. This early evoked response is seen in all the baboons.

In the most photosensitive baboons, the responses are of higher voltage and extend over a larger area than in the other animals (KILLAM et al. 1966 b, 1967 b). Moreover, a rhythmic after-discharge, localized to area 6, appears and seems greater in amplitude and duration. It is seen most often when the animal has his eyes closed and not after stimulation in modalities other than visual (MENINI et al. 1968 b, c, 1970). However it must be noted that recently WADA et al. (1972) failed to find such a difference between photosensitive and non-photosensitive animals, although there appeared to be a marked difference between Papio papio and the Rhesus monkey.

---

Fig. 1. Recording from a single unit in the fronto-rolandic cortex; in each section can be seen: above, the artefact from photic stimulation; then a frontal derivation, an occipital derivation and below the single unit record. Spontaneous activity shows few bursts of unit discharges (left, above). The cell responds to each flash of light (left, below). On the right: frontal spikes triggered by photic stimulation. A spike is always synchronous with a burst of unit discharges; but there are bursts seen in the absence of spikes

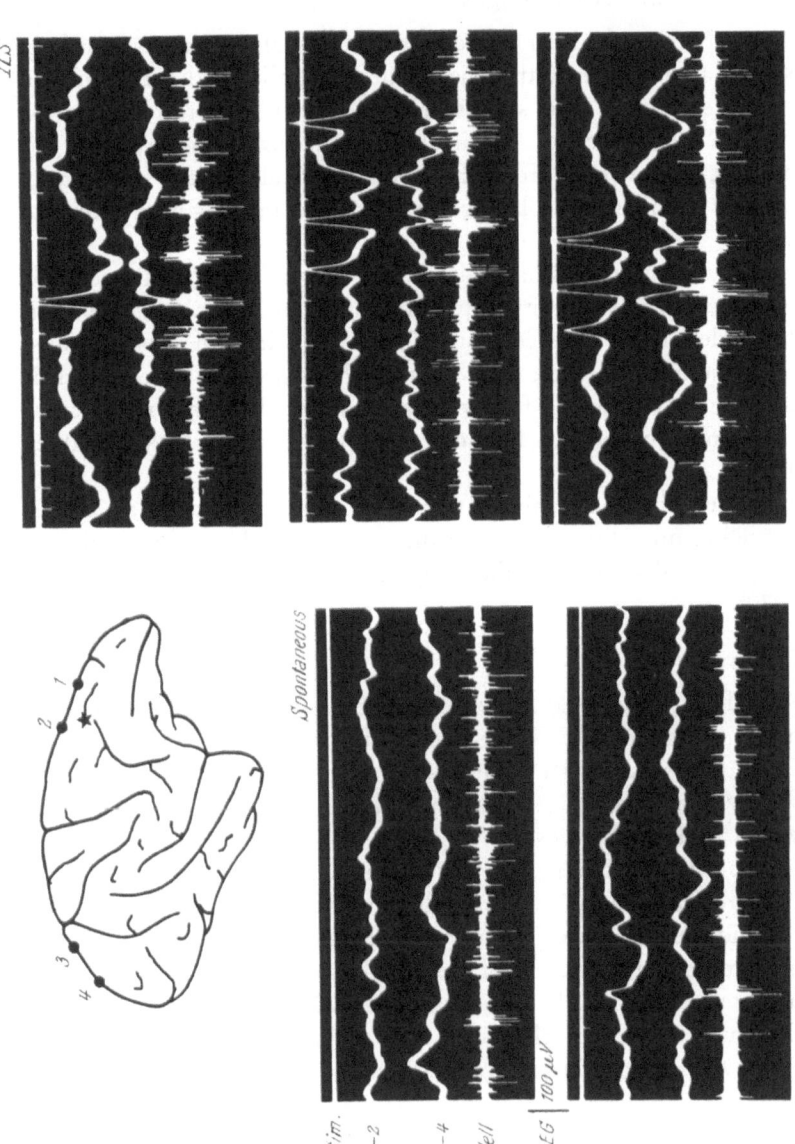

Fig. 1

## B. Single Unit Evoked Responses

The existence of visual afferents to the visual cortex has also been shown at the level of the single unit, especially in the more photo-sensitive animals. The shortest latencies (20 msec) are found in the deepest cortical layers. Their shortness seems to indicate that the afferent pathway is from the lateral geniculate body and passes directly to the fronto-rolandic cortex without relaying in the occipital cortex.

Single unit recordings have also shown activation of fronto-rolandic neurones during intermittent photic stimulation. Certain units, practi-cally silent in the absence of stimulation start firing in a tonic fashion after the initial flashes; other units without notably increasing their discharge rate, show a following response at the same frequency as the stimulation or at a harmonic of it (Menini and Rostain 1969, 1970).

It is interesting to note that intermittent photic stimulation at 25 c/sec as well as provoking changes in the fronto-rolandic cortex, facilitates single unit discharges in the occipital cortex (Menini and Rostain 1970).

## C. Recovery Cycle

A study of the cortical recovery cycle for visual stimulation has been made. This shows that the excitability of the occipital cortex of Papio papio varies after stimulation in a manner corresponding to the classical findings in man; in addition the time course does not vary with the photosensitivity.

In the fronto-rolandic territory, a marked increase in excitability (300–400%) appears at a stimulus interval of 40 msec, which cor-responds to the most epileptogenic frequency (25 c/sec). This increase involves the later elements of the visual evoked response. In the most photosensitive animals it reaches 600–700% and produces a late rhythmic activity of the same type as is seen after single shocks (Dimov et al. 1969, Menini et al. 1970).

## III. Somatosensory Afferents to the Frontal Cortex

Whereas the importance of the visual input in photosensitive epilepsy seems obvious, the part played by the somatosensory system (and in particular the periorbital somatosensory afferents) is actually studied after a series of findings (Naquet et al., to be published):

immobilization of the animal with gallamine considerably reduces the frequence of occurrence of paroxysmal discharges during photic stimulation;

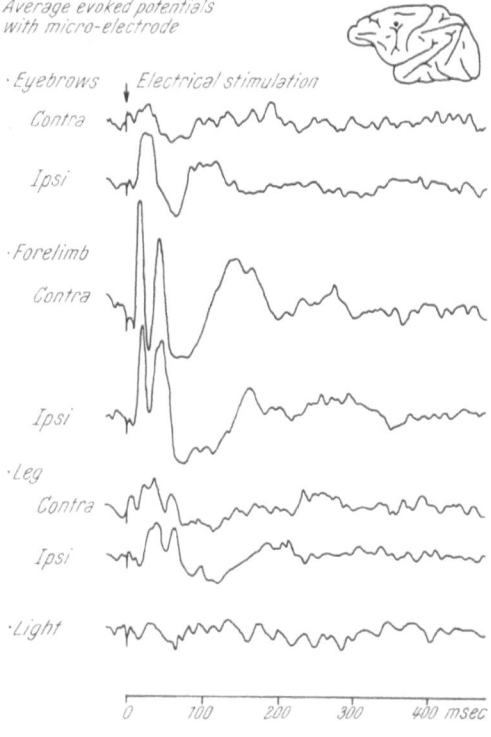

Fig. 2. Somatosensory and visual evoked potentials recorded intracortically with an extracellular micro-electrode. Responses are obtained at the same point for all the somatosensory stimulations employed; the response to stimulation of the homolateral eye-brow is larger than for contralateral stimulation; the inverse is true for stimulation of the limbs

tactile stimulation of the eyebrow region facilitates the occurrence of paroxysmal discharges during intermittent light stimulation;

in contrast, local anaesthesia applied in this region reduces EEG paroxysmal discharges and myoclonic movements of the limbs and trunk.

For these reasons, we have particularly studied the cortical projections

from the periorbital skin, and as a control the projections of the fore-
and hind-limbs in photosensitive and non-photosensitive animals
(CATIER *et al.* 1971).

## A. Evoked Potentials Recorded from the Cortical Surface

With either tactile or electrical stimulation, in all animals an early
positive wave is evoked in the contralateral post-rolandic cortex in
a zone corresponding to the specific projection area of the territory
stimulated. Its latency is 12 msec for the eyebrow and 14 msec for
the forepaw.
All the cutaneous territories stimulated also give rise to evoked
responses in the premotor cortex (area 6).
There is at this level a convergence of somatic afferents, without an
exact somatotopic arrangement. It is perhaps significant that the
spatial distribution of the responses to stimulation of the eyebrows
overlaps that for the forelimbs (Fig. 2).
The shortest latencies (14 msec for the eyebrows and 17 msec for the
forepaw) and the most marked convergence in the frontal lobe are
found near the superior frontal sulcus; in this region response are
obtained for homolateral as well as for controlateral stimulation.
Lastly with stimulation of the eyebrows the responses evoked in the
frontal cortex are remarkable in that the homolateral responses are
larger than the contralateral (200–300%); this is not found either
in other cortical territories or with other forms of stimulation (MENINI
*et al.* 1971).

## B. Evoked Potentials Recorded Trans-cortically

These provide an argument for the local origin of the frontal somatosensory evoked responses; indeed, bipolar transcortical recordings
show that in area 6 the focus of evoked activity is intracortical. In
the neighbouring territories, and especially in area 4, the responses
to stimulation of the face are very superficial in origin and appear to
be due to diffusion from the post-rolandic cortex.

## C. Single Unit Recordings

The existence of somatosensory afferents to the frontal cortex has
been confirmed at the single unit level: numerous frontal cortical units

are fired with a short latency after various stimulations, both electrical and tactile.

An exact somatotopic study has not been made, but it seems that certain units possess some specificity; some of them are fired only by stimulation of the hind-limbs, other by stimulation of the fore-limbs, and others by stimulation of the face. In all cases the receptor field is usually vast and bilateral.

The other distinctive features observed in evoked potential studies were also found in microelectrode studies:

some units showed a heterotopic convergence;

with stimulation of the eyebrows, it was possible to confirm that single unit responses to homolateral stimulation have often a larger duration, and a higher discharge frequency than responses to contralateral stimulation. No interpretation of this observation has yet been found.

## IV. Effect of Various Cortical Lesions

Different kinds of experimental lesions have been made in order to determine the role of the various cortical territories in the development of the seizures. There are recent experiments and certain results (such as those with ablations of the frontal lobe) still need to be confirmed.

### A. Chronic Epileptogenic Occipital and Frontal Lesions

Irritative occipital lesions were produced by injecting beneath the pia mater an epileptogenic agent comprising a mixture of alumina cream and pure cobalt powder (DIMOV and LANOIR 1969).

We shall not give details of this occipital focus, but it is significant that in none of the animals did the lesion lower the threshold for the induction of fronto-rolandic paroxysmal discharges by photic stimulation. It could be considered that the chronic occipital epileptogenic focus had, during its evolution, clearly diminished the photosensitivity of all the animals because intermittent photic stimulation was without effect even in the animals that were initially the most sensitive (in one case only, intermittent photic stimulation triggered a seizure originating in the occipital cortex but it was in the hemisphere contralateral to the lesion and the seizure remained localized in the posterior territory).

Thus an evolving lesion, whether irritative or not, of a single occipital cortex is capable of blocking photosensitivity; this observation, like some findings at the single cell level, allows us to speculate that the occipital cortex can play a role in the initiation of the paroxysmal discharges by photic stimulation.

On the other hand, when the animal has an irritative lesion of the same type in the fronto-rolandic cortex of one hemisphere, the induction of paroxysmal discharges by photic stimulation is facilitated during the developing phase of the focus (DIMOV and LANOIR, to be published).

## B. Bilateral Ablations of Various Cortical Territories

A number of photosensitive baboons have undergone complete bilateral ablations either of the occipital lobes or of the temporal lobes or of the frontal lobes (except area 4); in the latter case the findings concern only one animal whose post-operative survival (3 weeks) does not allow us to give definitive results for this situation.

Post-operatively the photosensitivity of each animal was determined (WADA et al. 1972). Some of the experiments are still in progress but it can be concluded at this time that:

Animals without occipital lobes show a very rapid loss of sensitivity to intermittent photic stimulation. A transitory reduction in photosensitivity normally follows any major neurosurgery; this does not usually last more than a month. So that a complete absence of photosensitivity during the 3–5 months following occipital lobe ablation appears highly significant.

Animals having undergone ablation of their frontal or temporal lobes show no change in their photosensitivity (although the reservations given above must be born in mind).

## V. Section of the Corpus Callosum

Recordings have been made for many months (up to 15 months) in Papio papio after complete callosal section. Except for the paroxysmal discharges, the EEG activity is bilaterally synchronous and symmetrical and computer frequency analyses (power spectra) of homologous derivations are identical. These results confirm the work of BATINI et al. 1967.

In contrast, spontaneous paroxysmal abnormalities in the fronto-rolandic region generally appear asynchronously. When they are induced by photic stimulation they appear either asynchronously or synchronously in the two hemispheres (Fig. 3). They can be predominant either homolateral or contralateral to the side of the surgical approach to the corpus callosum. Similarly discharges occurring after the end of intermittent photic stimulation can begin pre-

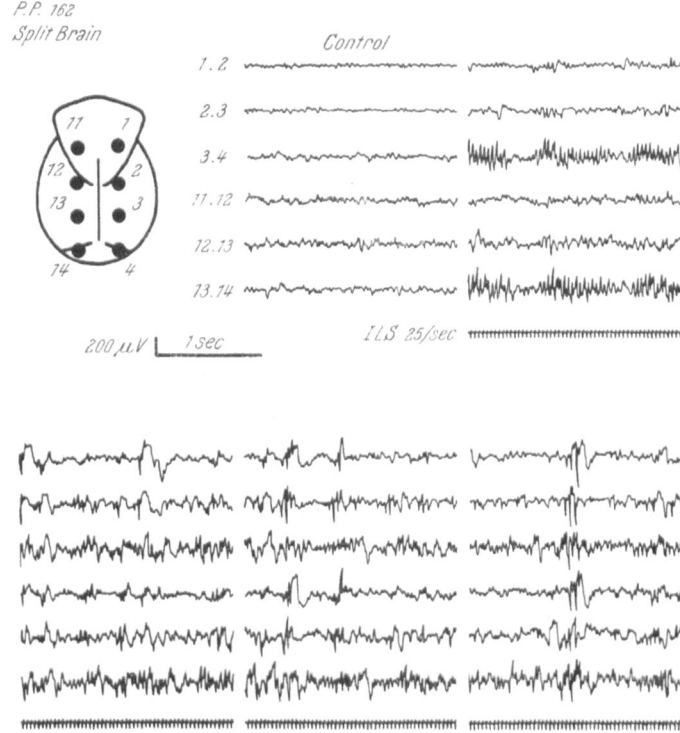

Fig. 3. Induction of paroxysmal discharges during ILS after section of the corpus callosum (right-sided surgical approach). *Upper records*: on the left: control; on the right: occipital following responses during intermittent photic stimulation are bilaterally symmetrical. *Lower records*: 3 examples of paroxysmal discharges induced by photic stimulation, appearing variably on the left or on the right or bilaterally

dominantly in one hemisphere. They can for example in the same animal remain confined to one hemisphere during one experiment, and diffuse transiently to the other hemisphere during another experiment (Figs. 4 and 5).

Clinically the epilepsy is no longer manifest as a generalized tonic-clonic seizure, but by a seizure which may be unilateral or bilateral but which always predominates on the side of the body opposite to the cortical hemisphere in which the maximal seizure discharge is recorded.

When for whatever reason, the sectioning of the corpus callosum led to a significant lesion of the fronto-rolandic cortex on one side, there is a more marked asynchrony between the two hemispheres: on the EEG the photically induced seizure occupies only the "healthy" hemisphere, and, clinically, it remains localized to the contralateral half of the body; it is followed by electrical silence in the cortex with the epileptic discharge and by a transitory paresis (Catier et al. 1970).

From this series of experiments it is apparent that section of the corpus callosum:

a) diminishes the synchrony of both the spontaneous and the photically induced discharges and,

b) converts what would otherwise be generalized seizures into focal or rather hemispheric seizures. These experiments tend to confirm that the fronto-rolandic cortex plays a role in the genesis of the seizures because animals which have undergone some destruction of this area in association with section of the corpus callosum only give unilateral seizures in the cortex opposite to the damaged side.

## Discussion

The fronto-rolandic cortex of Papio papio is distinctive by reason of certain unusual properties which can be summarized thus:

it is the point of origin for the spontaneous paroxysmal EEG discharges as well as those induced by intermittent photic stimulation;

EEG paroxysmal discharges are not found in other cortical or deep territories unless thay are also present in the frontal cortex;

---

Fig. 4. Induction of self-sustaining epileptic discharges continuing unilaterally after the end of photic stimulation, in an animal with transection of the corpus callosum (right-sided surgical approach) (PP 166). *Above*: on the left, control record; in the centre, ILS induces paroxysmal discharges greater on the left than on the right; on the right, the paroxysmal discharges become bilaterally synchronous, then they are continuing on the right when a seizure discharge begins in the left hemisphere anteriorly. *Below*: development of the seizure in the left hemisphere while the animal shows a deviation of the head to the right and clonic movements of the right half of the body. After the end of the seizure a right hemiparesis persists for several minutes

P.P. 166
Split Brain
Control
1.2
2.3
3.4
4.5
5.6
11.12
12.13
13.14
14.15
15.16
100 µV 1 sec
50 µV
ILS 25/sec
Deviation of the head and eyes to the right

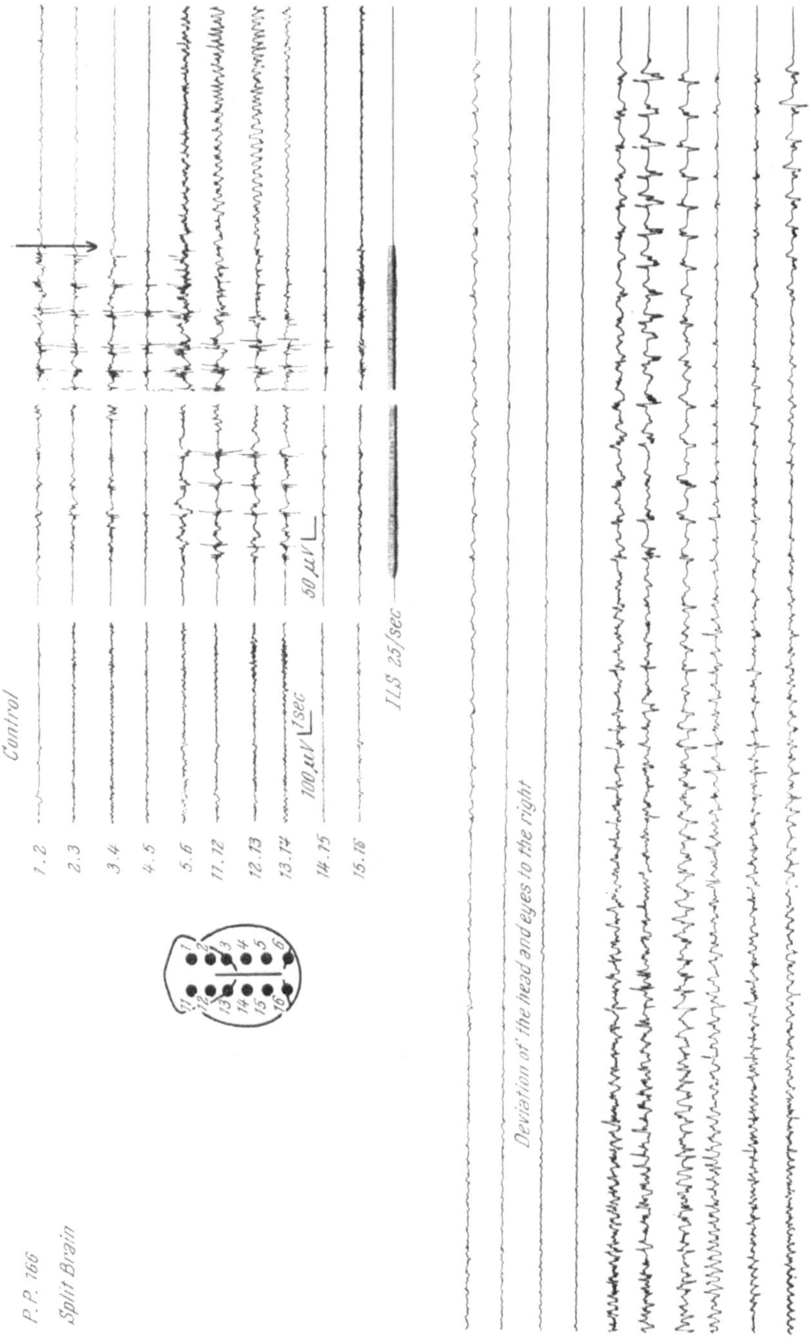

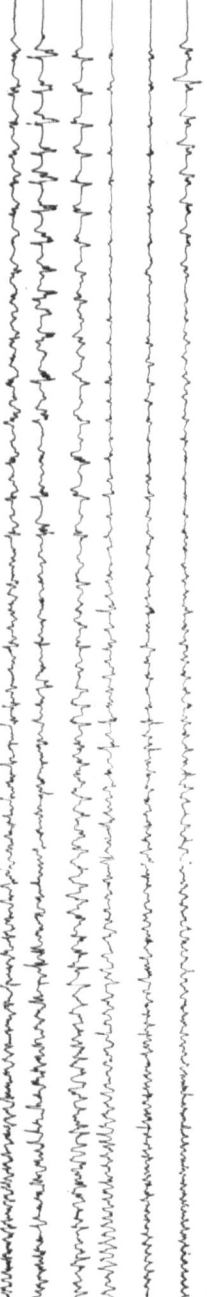

Fig. 4

it shows a significant hyperexcitability during photic stimulation at 25 c/sec;

burst firing of frontal neurones, spontaneously or during photic stimulation, is synchronous with the EEG paroxysmal discharges, which supports the idea that the paroxysmal discharges originate in this territory;

paroxysmal discharges in the fronto-rolandic cortex triggered by photic stimulation appear bilaterally synchronous and symmetrical;

the frontal cortex possesses multisensory afferents, including particularly some visual inputs whose nature is not yet defined, and some major somatosensory afferents from the facial region.

The fact that there is heterosensory convergence exactly in the cortical area where the paroxysmal discharges originate does not permit the conclusion that the photosensitivity of certain baboons is due to the existence or exaggeration of visual afferents to the fronto-rolandic cortex (NAQUET 1969, NAQUET and MENINI 1972).

However the magnitude of the periorbital somatosensory projections to the frontal cortex supports the proposition that in the most photosensitive baboons and in the absence of any drug, photic stimulation induces by a feed-back mechanism an excessive multisensory input to the frontal cortex. The periorbital afferents are immediately brought in because the clinical paroxysmal signs always begin with myoclonus of the eyelids. It must be admitted that the fronto-rolandic cortex of photosensitive Papio papio is not necessarily to be considered as a natural epileptogenic focus, which is confirmed by the absence of definite histological lesions at this level. The cortex overall is considered hyperexcitable, and it is known that this cortical region has a very low convulsive threshold (FRENCH et al. 1956, GREEN and NAQUET 1957). The cortex discharges the more easily

---

Fig. 5. Induction of self-sustaining seizure discharges continuing after the end of ILS after section of the corpus callosum (surgical approach on the right side, same animal as in the previous illustration but several weeks later). *Above* on the left: control records, no specific abnormalities; in the centre: control record with a smaller amplitude (as in following records); on the right: ILS induces discharges of spikes and waves which are more marked on the right than on the left; the end of photic stimulation is followed by a seizure discharge originating anteriorly bilaterally and then becomes generalized. *Below*: Development of a seizure discharge, which is initially synchronous, then continues in the left hemisphere; the animal then shows clonic movements on the right side; this seizure was followed by a transient hemiparesis of the right side accompanied by an abnormal EEG on the left side

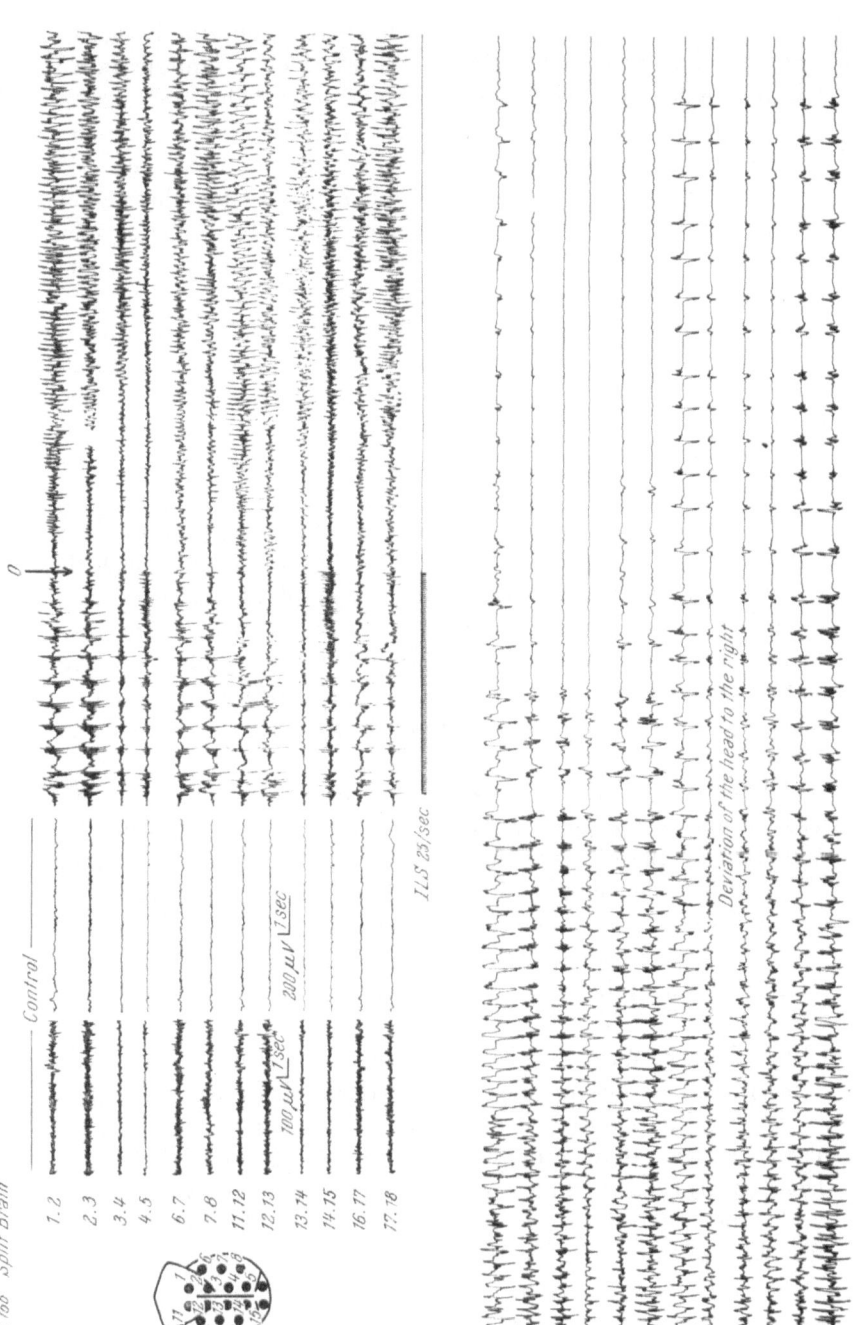

Fig. 5

because it is subjected to an excess of afferents due to the hetero-sensory convergence. This epilepsy, according to this interpretation, will be not uniquely photosensitive but will be rather multisensory with a predominant sensibility to photic stimulation. But it is neces-sary to perform experiments to confirm or refute this hypothesis.

The experiments with cortical lesions and ablations, whose inter-pretation is difficult, and the experiments with callosal section, pose the probleme of the respective roles of the frontal and occipital cortex in the initiation of the EEG paroxysmal discharges. The mod-ification of the paroxysmal signs following callosal section with its associated damage to the fronto-rolandic cortex argues for the signifi-cance of the fronto-rolandic cortex in photic epilepsy in Papio papio. On the contrary, the lesion or ablation experiments involving the occipital cortex indicate that it plays an important role. Among the possible hypotheses let us consider two:

1. The visual input involved in photosensitive epilepsy requires the integrity of the occipital cortex either because it passes through the visual cortex, or because the cortex is controlling synaptic trans-mission at certain subcortical relays (Buser et al. 1963). If the input passes via the occipital cortex it remains to be shown how it arrives at the frontal cortex: is it a cortico-thalamo-cortical or a cortico-cortical pathway? Experiments now in progress may provide an answer.

2. All the unusual features observed in the frontal lobe, and sum-marized above, are only an "epiphenomenon". This view may be supported by the observation in one animal that bilateral ablation of area 6 left the photosensitivity intact. But we have seen in the evoked potential and single unit studies that it is often difficult to dissociate areas 6 and 4. It is necessary to perform experiments to see if ablation of both area 4 and area 6 fails to modify photo-sensitivity.

Lastly, the experiments we have described show that the corpus cal-losum plays a significant role in the interhemispheric synchronization at least for the paroxysmal discharges. This point of view is inter-mediate between those of Bremer (1958) and Batini et al. (1967).

These experiments confirm earlier findings in the cat (Bach-y-Rita et al. 1966, 1969) but are more definitive because the earlier authors provoked seizures with metrazol which causes a propagation of seizure discharges (Machek et al. 1963) very different from that found in an animal in the absence of convulsant drugs.

# Summary

Sixty to eighty per cent of baboons (Papio papio) from Casamance, Senegal, show epileptic manifestations induced by intermittent photic stimulation at 25 c/s.

Electrical recordings from the cortex and deep structures showed that paroxysmal discharges always appeared first in the frontal cortex. At this level, unit recordings demonstrated synchronous neuronal activation of paroxysmal discharges; the occurrence of such activation is similar to that observed in animals with experimental focal epileptic lesions, and in those without lesions who have received an injection of metrazol. Furthermore, histological techniques did not reveal any abnormalities in the frontal cortex of the photosensitive baboon.

The existence of visual afferents (sometimes of very short latency) and of somesthetic afferents from the whole organism, and more particularly from the face, were shown at the level of the frontal cortex; this latter, therefore, appears to be a center of multisensory convergence. A study of excitability cycles during photic stimulation was performed; in the most photosensitive monkeys, a marked hyperexcitability localized in the frontal lobe was observed for the latter part of the response and for a 40 msec time period.

Chronic epileptogenic cortical lesions were induced; focal lesions located in the occipital region elevated the photosensitive threshold of the baboons during the whole period of their evolution, in contrast to the effect of frontal lesions, which facilitated the appearance of paroxysmal discharges. Some of the animals were subjected to bilateral ablation of the occipital lobes; all demonstrated a rapid and complete desensitization to intermittent photic stimulation. After bilateral ablation of the frontal lobes, the mortality rate was such that, at the present time, no significant conclusion may be drawn regarding the effect of such intervention on photosensitivity.

Finally, transection of the corpus callosum, in photosensitive animals, decreased the synchrony of spontaneous paroxysmal discharges or those induced by intermittent light stimulation; furthermore, under these circumstances, generalized seizures were changed into focal seizures. When this preparation was accompanied by a unilateral destruction of the frontal cortex (pathway of approach of the corpus callosum), the animal demonstrated only unilateral seizures localized in the opposite hemisphere.

These results suggest that the frontal cortex plays an important role in the genesis of photosensitive epilepsy; however, the role of the occipital cortex should not be neglected at the present time. Since some of these results seem contradictory, it is difficult for the time being to provide a definite explanation for the mechanisms of photosensitive epilepsy in the Papio papio.

## References

Bach-Y-Rita, G., M. Poncet, and R. Naquet: Morphology and spatio-temporal evolution of ictal discharges induced by cardiazol in the presence of various corticodiencephalic lesions in the cat. In: Gastaut, H., H. H. Jasper, J. Bancaud, and A. Waltregny (eds.), The physiopathogenesis of the epilepsies, pp. 256—267. Springfield: Thomas. 1969.

— C. Trevarthen, and M. Poncet: Effe:s des lésions thalamiques unilatérales associées à une section interhémisphérique sur le comportement et l'activité électroencéphalographique au cours de la veille et du sommeil après injection de pentetrazol chez le chat. J. Physiol. (Paris) 58, 452—453 (1966).

Balzano, E.: Etude polygraphique du sommeil nocturne du Papio papio, babouin du Sénégal. Marseille. 1968.

Batini, C., M. Radulovacki, R. T. Kado, and W. R. Adey: Effect of interhemispheric transection on the EEG patterns in sleep and wakefulness in monkeys. Electroenceph. clin. Neurophysiol. 22, 101—112 (1967).

Bremer, F.: Physiology of the corpus callosum. Ass. Res. nerv. Dis. Proc. 36, 425—448 (1958).

Buser, P., P. Ascher, J. Brunner, D. Jassik-Gerschenfeld, and R. Sindberg: Aspects of sensorimotor reverberation to acoustic and visual stimuli. The role of primary specific cortical areas. In: Moruzzi et al. (ed.), Progress in Brain Research, Vol. 1, Brain mechanism, pp. 294—322. Amsterdam: Elsevier. 1963.

Catier, J., M. Choux, J. P. Cordeau, S. Dimov, D. Riche, A. Eberhard, and R. Naquet: Résultats préliminaires des effets électrographiques de la section du corps calleux chez le Papio papio photosensible. Rev. neurol. 122, 521—522 (1970).

— C. Menini, G. Charmasson, and E. Carlier: Mise en évidence de projections corticales somesthésiques an niveau du lobe frontal chez le babouin. J. Physiol. (Paris) 63, 121—122 (1971).

Dimov, S., and J. Lanoir: Effets de lésions epileptogénes chroniques occipitales (cobalt-alumine) chez le Papio papio. Rev. neurol. 120, 480 (1969).

— C. Menini, and R. Naquet: Cortical recovery cycles to light stimulation in Papio papio and their relation to the photomyoclonic manifestations. Communication au symposium "Visual information processing and control of motor activity", Sofia, 23–26 juillet 1969.

Fischer-Williams, M., M. Poncet, D. Riche, and R. Naquet: Light-induced epilepsy in the baboon Papio papio: cortical and depth recordings. Electroenceph. clin. Neurophysiol. 25, 557—569 (1968).

FRENCH, J. E., B. E. GERNANDT, and R. B. LIVINGSTON: Regional differences in seizure susceptibility in monkey cortex. A.M.A. Arch. Neurol. Psychiat. *75*, 260—274 (1956).

GREEN, J. D., and R. NAQUET: Etude de la propagation locale et à distance des décharges épileptiques. Acta Medica Belgica *57*, 226—249 (1957).

KILLAM, K. F., E. K. KILLAM, and R. NAQUET: Mise en évidence chez certains singes d'un syndrome photomyoclonique. C.R. Acad. Sci. (Paris) *262*, 1010—1012 (1966 a).

— — — Etudes pharmacologiques réalisées chez des singes présentant une activité E.E.G. paroxystique particulière à la stimulation lumineuse intermittente. J. Physiol. *58*, 543—544 (1966 b).

— — — An animal model of light sensitive epilepsy. Electroenceph. clin. Neurophysiol. *22*, 497—513 (1967 a).

— — — Evoked potential studies in response to light in the baboon (Papio papio). Electroenceph. clin. Neurophysiol. Suppl. *26*, 108—113 (1967 b).

MACHEK, J., J. SAIER, D. DESPLAN, and R. NAQUET: Etude de la crise cardiazolique après thalamectomie. J. Physiol. (Paris) *55*, 290—291 (1963).

MELDRUM, B. S., E. BALZANO, M. GADEA, and R. NAQUET: Photic and drug induced epilepsy in the baboon (Papio papio). The effects of isoniazid, thiosemicarbazide, pyridoxine and amino-oxyacetic acid. Electroenceph. clin. Neurophysiol. *29*, 333—347 (1970).

MENINI, C., J. CATIER, G. CHARMASSON, and E. CARLIER: Projections corticales des afférences péri-oculaires chez le Papio papio. Communication à la Sté d'EEG et de Neurophysiologie clinique, Paris, Oct. 1971. Rev. EEG Neurophysiol. *1*, 432—433 (1971).

— S. DIMOV, G. VUILLON-CACCIUTTOLO, and R. NAQUET: Réponses corticales évoquées par la stimulation lumineuse chez le Papio papio. Electroenceph. clin. Neurophysiol. *29*, 233—245 (1970).

— F. MORRELL, and R. NAQUET: Enregistrements corticaux au moyen de microélectrodes chez le Papio papio photosensible. J. Physiol. *60*, 498—499 (1968 a).

— and J. C. ROSTAIN: In: NAQUET, R. (ed.), Enregistrements unitaires dans le cortex fronto-rolandique du Papio papio photosensible. J. Physiol. *61*, Suppl. 2, 352 (1969).

— — Activités unitaires évoquées par la stimulation lumineuse dans différents territoires corticaux chez le Papio papio. J. Physiol. *62*, Suppl. 3, 414—415 (1970).

— G. VUILLON-CACCIUTTOLO, and N. LESÈVRE: Chronologie et topographie des réponses corticales évoquées par différents types de stimulations chez le Papio papio. J. Physiol. *60*, 277—278 (1968 c).

— — and R. NAQUET: Etude morphologique, chronologique et topographique des potentiels évoquées visuels d'un cercopithecinae. Rev. Neurol. *118*, 474—475 (1968 b).

MORRELL, F., R. NAQUET, and C. MENINI: Microphysiology of cortical single neurons in Papio papio. Electroenceph. clin. Neurophysiol. *27*, 708—709 (1969).

Naquet, R.: Discussion. Photogenic seizures in baboon. In: Jasper, H. H., A. A. Ward, and A. Pope (eds.), Basic Mechanisms of the Epilepsies, pp. 566—573. Boston: Little, Brown & Co. 1969.

— F. Ames, E. Carlier, G. Charmasson, J. Catier, and C. Menini: Afférences périoculaires et photosensibilité du Papio papio. Etude clinique et électroencéphalographique. Rev. EEG Neurophysiol. 1, 430—431 (1971).

— K. F. Killam, E. K. Killam, J. Bimar, and M. Poncet: Un nouveau « modèle animal » pour l'étude de l'épilepsie : le Papio papio. In: Actualités Neurophysiologiques, 8ème série, pp. 213—230. Paris: Masson. 1968.

— and C. Menini: La photosensibilité excessive du Papio papio. Approches neurophysiologiques et pharmacologiques de ses mécanismes. Supplement Electroenceph. clin. Neurophysiol. (In press, 1972.)

Riche, D., D. Gambarelli-Dubois, M. Dam, and R. Naquet: Repeated seizures and cerebral lesions in photosensitive baboons (Papio papio). A preliminary report. In: Brierley, J. B., and B. S. Meldrum (eds.), Brain Hypoxia, pp. 297—301. London: Spactics Intern. Med. Publ. 1971.

Wada, J., J. Catier, G. Charmasson, C. Menini, and R. Naquet: Elimination of photogenic seizure susceptibility by bilateral occipital lobectomy in photosensitive epileptic baboon "Papio papio". Science. (In press, 1972.)

— A. Terao, and H. E. Booker: Longitudinal correlative analysis of epileptic baboon, Papio papio. (In press, 1972.)

## Discussion

Elul: Did you try monocular stimulation?

Naquet: It is not easy to induce seizures by monocular stimulation in a normal animal because the amount of the light reaching the brain is fairly small. We did not use monocular stimulation on animals after section of the corpus callosum; nor had these animals their chiasmas severed.

Elul: Another question: have you tried to stimulate other afferents, e.g., through the sciatic nerve?

Naquet: Yes, we tried some electrical stimulation at high frequency. But if the stimulation is painful the paroxysmal discharge is blocked. By comparison, non-painful stimulation of the peri-ocular region facilitated the appearance of a clonus of the eyebrows; for this reason we tried to block this effect by injection of procaine in this region or by cutting the two facial nerves.

Gloor: Does facial nerve section affect the EEG, or only the motor manifestation?

NAQUET: The section of the facial nerves does not affect the EEG; it blocks for some days the clonus of the eyebrows, but permits some clonus of the head to appear. The effect of procaine is more complicated to interpret: with small concentrations we blocked the clonus of the eyebrows without affecting the EEG. With higher concentrations we blocked the clonus of the eyebrows and the EEG paroxysmal discharges diminished. It is possible that this effect is the consequence of a diffusion of this compound in the general circulation.

# Mechanisms of Cortical Discharges in "Generalized" Epilepsies in Man

J. BANCAUD

Unité de Recherches I. N. S. E. R. M., Paris, France

## I. Introduction

The stereotaxic EEG investigations in man, performed with macro-multilead electrodes (macro-SEEG) and intended to facilitate possible surgical intervention in recalcitrant cases of epilepsy, are unable to provide decisive information concerning the mechanisms of discharge synchronization. Nevertheless, the simultaneous investigation of a large number of cortical and subcortical structures implicated in the epileptic process can furnish some indications as to the site of origin and mode of propagation of critical discharges. We shall thus present several of the findings which have emerged from such studies regarding the highly controversial issue of whether the bilateral synchrony of the so-called "generalized" epilepsies is primary or secondary.

In order to clarify our presentation, we shall first briefly recall the often contradictory findings which have been published since PENFIELD (1954) proposed his model of the "centrencephalic" origin of these discharges, without however discussing the symptomatologic, EEG and etiopathogenic evidence which has also helped to bolster this hypothesis.

It is well known that experimental evidences concerning the mechanism of production and the origin of critical discharges in the "generalized epilepsies" are based on the electrical stimulation of various cortical and subcortical structures, on the creation of epileptogenic foci by the application of "irritant" substances, and on the action of convulsants introduced into certain parts of the cerebrovascular system.

However, all the findings adduced in favor of the primary role of the subcortical formations in triggering seizures and producing their electrical correlates (bilaterally synchronous and symmetrical spikes and waves, or discharges of diffuse spikes) run counter to experiments which stress, on the contrary, the secondary role of these structures and the major if not exclusive importance of the cerebral cortex. To the results of Sager et al. (1957), Pollen (1963), and Weir (1964) we may therefore oppose those of Kaada (1951), Lennox and Robinson (1951), Ingvar (1955, 1959), Echlin (1959), Crain (1969), Ralston (1956, 1961), Marcus (1964, 1969), and Naquet (1968). The primary role of various thalamic nuclei after the creation of epileptogenic foci as demonstrated by Kopeloff (1950), Sasaki (1955), Ralston and Ajmone Marsan (1956), and Guerrero-Figueroa et al. (1963) has been questioned if not contradicted outright by many authors, particularly by Marcus et al. (1964, 1969), who propose an experimental model in which the cortex and its direct interhemispheric connections play a major role.

As for convulsant agents administered via different cerebrovascular pathways, their effects on cortical and subcortical structures are interpreted differently by the supporters of the "centrencephalic" hypothesis and by those who doubt its applicability to animals (Shimizu et al. 1952, Starzl et al. 1953) and to man (Bennett and Gibbs 1952).

## II. Material and Methods

Rather than review the methodology underlying SEEG investigations in epilepsy, which has been described on a number of occasions (Bancaud et al. 1963, 1965, 1969, 1971), we should like to make three points which indicate the capabilities and particularly the limitations of this type of study.

Firstly, it should be stressed that our series, which derives from a neurosurgery service devoted primarily to epilepsy, contains a limited number of cases (500 SEEG studies performed in approximately 400 patients). These were due for the most part to a localized or diffuse lesional process and thus include only a part of the "generalized" epilepsies: 60 subjects had either petit mal seizures (of various types), "tonic axial" grand mal seizures (to use the terminology of Gastaut 1967) or psychomotor automatisms (to use the terminology of Penfield and Jasper 1954).

It is also obvious that the SEEG studies as carried out in our service cannot follow a protocol comparable to that used in experimental neurophysiology and that we are often obliged to omit the control tests that might be considered desirable. However, these studies do perhaps offer the advantage of being focused on the epileptic process itself in man and that they sometimes point to a surgical therapy by which the value of the interpretation may be gauged more accurately.

Lastly, this type of investigation of course does not permit the recording of the intercritical and critical electrical activity of all the subcortical structures, in particular the lower mesencephalon. Thus, it may always be objected that a discharge of bulbar or cerebellar origin may well go unrecorded by our electrodes. However, within the framework of this objection one would still have to explain why the thalamic-reticular system shows no evidence of such discharge, even though it is likely that the SEEG technique makes it possible to record from and stimulate all the nuclei of the thalamus and the basal ganglia.

## III. Results

In this presentation we shall not discuss the interpretation of the clinical and EEG data obtained during stereotaxic EEG investigations of the "generalized" epilepsies (BANCAUD 1971). We shall limit ourselves to a consideration of the role of dysfunction, firstly, of subcortical structures, particularly the thalamus, and, secondly, of various formations of the cerebral cortex in the production of spontaneous or electrically induced ictal discharges.

### 1. Role of the Thalamus

One of the basic pieces of experimental evidence advanced in support of the centrencephalic origin of epilepsies which are "generalized" from the start is that stimulation of the thalamus in the animal, especially the intralaminar nuclei, causes the appearance of diffuse spikes and waves whereas the cerebral cortex is believed to be either secondary or completely passive in the production of these discharges.

In man, in our experience, there seems to be no doubt but that thalamic stimulation can trigger bilaterally "synchronous" and symmetrical spikes and waves with a broad central hemispheric extension. However, these EEG patterns only appear in response to stimulation of the dorsomedial nucleus, the central median nucleus, the anterior pole of the thalamus and the region of the internal medullary lamina. Stimulation of the lateral nuclei, in contrast, triggers spikes and waves of varied location but which are first lateralized before becoming generalized. Nevertheless, even in subjects with petit mal epilepsy a very high stimulation voltage is required to produce spikes and waves.

Above all, it should be noted that the responses to thalamic stimulation may show marked variations which are not closely related to the stimulation parameters. For example, one may sometimes obtain 3 c/s spikes and waves with a stimulation frequency ranging from 1 to 300/sec; the discharges may also cease before or persist after the end of stimulation. In addition, identical stimulation may either give rise to spikes and waves or on the contrary interrupt a spontaneous paroxysm.

The demonstration of spike-and-wave complexes by stimulation of a portion of the "centrencephalon" cannot by itself demonstrate that this region is the site of origin of the epileptic discharges responsible for petit mal or grand mal seizures. There must also be evidence of a behavioral symptomatology which in our opinion should reflect the basic mechanisms involved by the electrical paroxysm.

A certain number of findings, which we shall only summarize briefly, run counter to the hypothesis of a primary or exclusive role of the thalamus in producing such seizures.

Regardless of the nuclear formation stimulated, we have been unable to induce from the start of stimulation both spikes and waves at the same frequency as the shocks delivered, and clinical manifestations recognized by the patient as well as the observer as being characteristic of his usual attacks. Generally speaking, the disturbances induced occur only after an appreciable latency period and disappear during stimulation or fade away gradually only after stimulation has ceased. Thus, the cessation of the clinical manifestations does not coincide with the end of the spike-and-wave paroxysm.

It would appear that the paroxysmal clinical disturbances induced by thalamic stimulation represent one of the modalities of expression of, for example, a petit mal absence but not its fundamental essence. In other words, it seems probable that under certain conditions the thalamus may be affected by the critical discharge during a spontaneous episode but that its involvement is only secondary, regardless of the origin of the discharge.

## 2. Role of the Cerebral Cortex

The major argument advanced by the Montreal group in favor of the centrencephalic concept is "the impossibility for a cortical epileptic lesion or for stimulation delivered to any site of the cortex to be at the origin of a petit mal or grand mal seizure and to reproduce the electrographic pattern attributed to these attacks".

However, SEEG studies of spontaneous seizures possessing all the characteristics of the electroclinical paroxysms defining absences, "generalized" tonic-clonic seizures or tonic axial seizures, show in many cases that the discharges may in fact be initially partial, originally starting from the cerebral cortex in the vicinity of a verified lesion.

These seizures may also be reproduced, at very low stimulation voltage and for different stimulation frequencies, by electrical stimulation of the same cortical epileptogenic zone.

Fig. 1 shows that slow rhythmic stimulation of the internal surface of the frontal lobe causes the immediate appearance of bilaterally synchronous and symmetrical spikes and waves with a central predominance; these discharges are clearly visible at the level of the scalp and are accompanied from the start by clinical manifestations in every way suggestive of a petit mal absence and lasting as long as stimulation.

It should be noted that while the two Ammon's horns are not affected by the spike-and-wave discharges, the dorsomedial nucleus does become involved after a very short delay.

In the same patient (Fig. 2), an increase in stimulation frequency at the same site induces a generalized tonic-clonic seizure whose different phases are distinctly recognizable. Fig. 3 provides an example of a typical "tonic axial" seizure recorded from the scalp. Here again, the apparently diffuse discharge involves the two hemispheres in a synchronous and symmetrical fashion.

Our SEEG studies (Fig. 4) show unequivocally that there is in fact a primary involvement of the left medial frontal cortex. In this case, the spontaneous discharge spread very quickly not only to the contra-

Fig. 1. 1, 2: (L) amygdala; 3, 4: (R) amygdala; 5, 6: (R) suppl. motor area; 7, 8: (L) suppl. motor area; 9, 10: (R) gyri orbitalis; 11, 12: (R) dorso-medialis; 13 (EEG): (R) post-frontal-occipital; 14 (EEG): (L) post-frontal-occipital; 15: EKG. Calibration in all figures: 50 $\mu$V and 1 sec

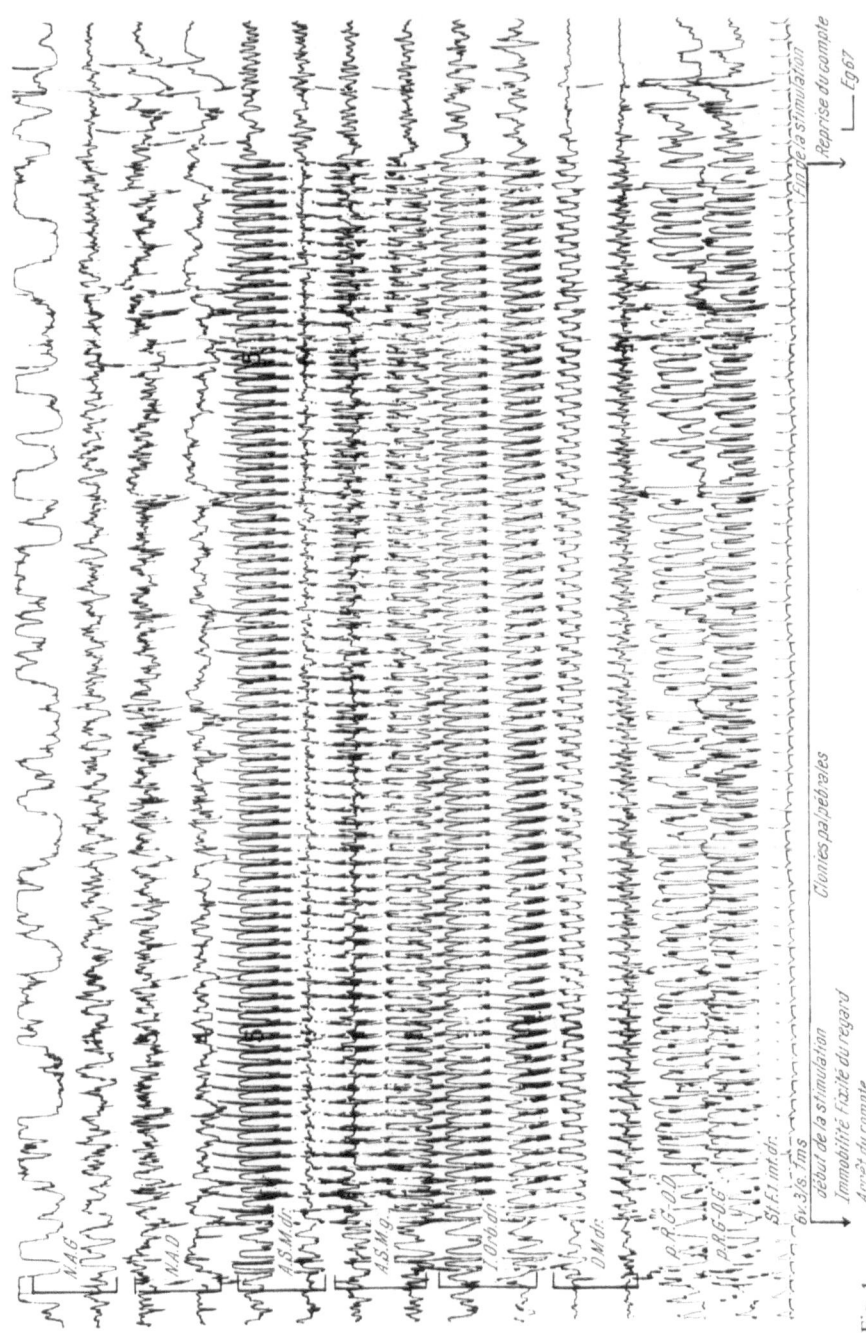

Fig. 1

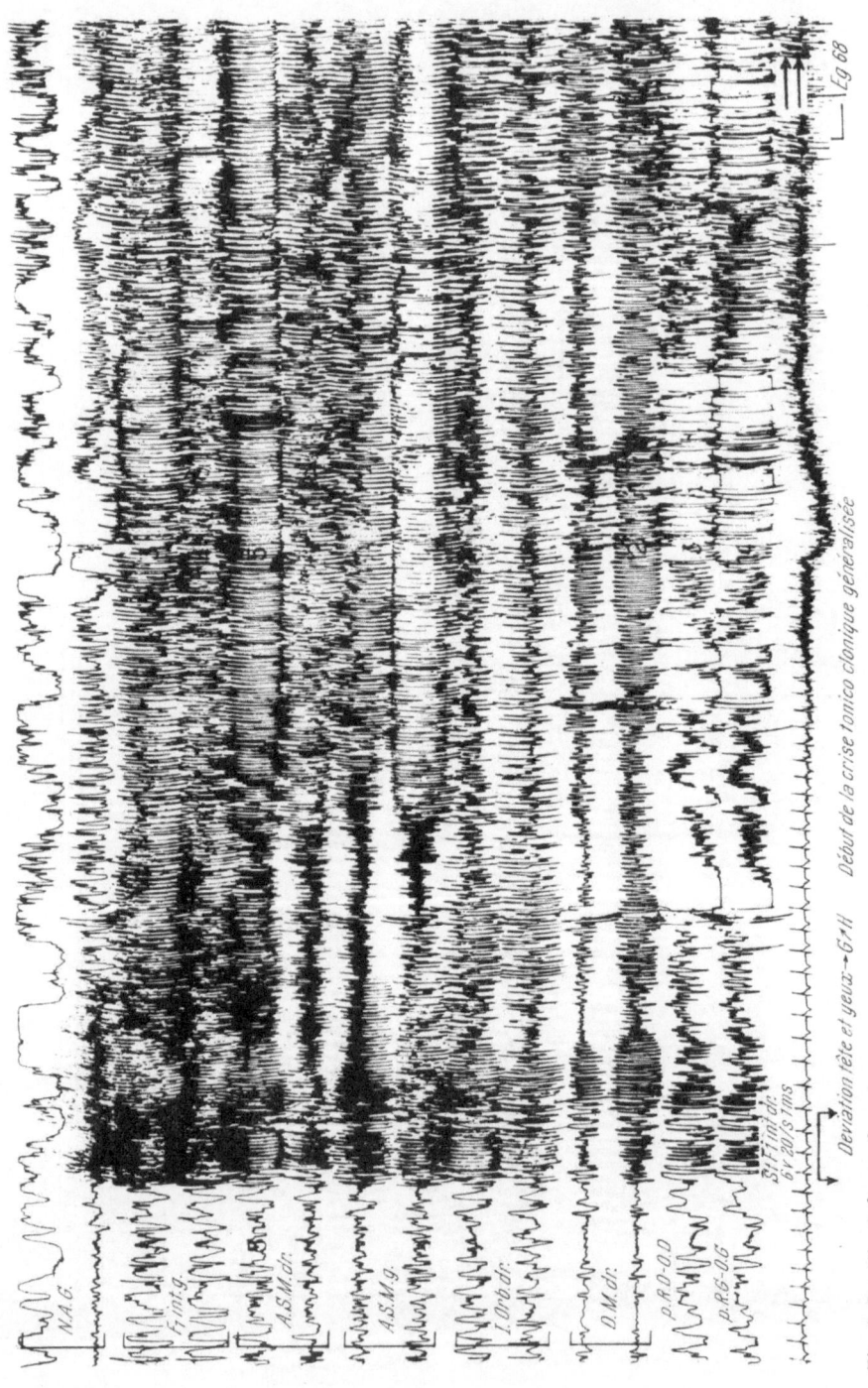

Fig. 2. See Fig. 1 for explanation

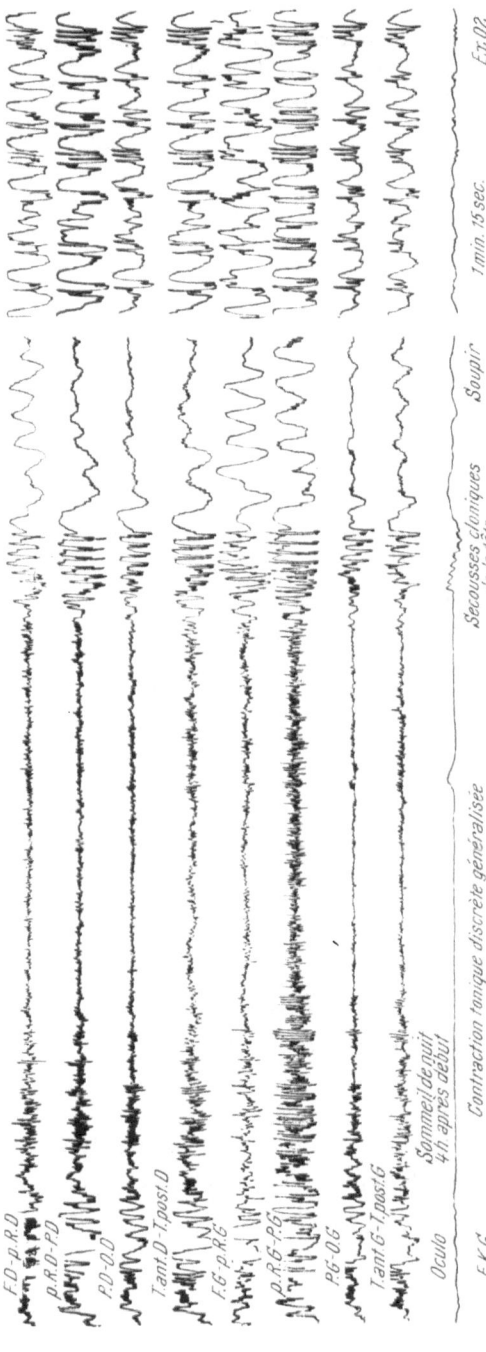

Fig. 3. EEG; 1 = (R) ant-frontal—post-frontal; 2 = (R) post-frontal-parietal; 3 = (R) parieto-occipital; 4 = (R) ant-temporal—post-temporal; 5, 6, 7, 8 = same as L

lateral frontal region but also to the two Ammon's horns and especially to the anterior portion of the thalamus.

In this paper we shall not attempt to define the role of the different cortical structures whose critical involvement seems to us to be responsible for the various "generalized" paroxysmal manifestations. Let us simply note that the site of origin of the discharges is undoubtedly the frontal lobes and in particular their internal surface at the level of the supplementary motor area, or more especially in a further anterior direction up to the orbital face.

Most often, the incriminated region consists of an extensive territory (sometimes encompassing the cingulate gyrus). Above all, it is indisputable that the epileptogenic zone sometimes includes broad portions of both frontal regions, just as in cases of bilateral multifocal cortical epilepsy. This is consistent with the findings of Marcus and Watson (1964) in his experimental model and those of Petsche et al. (1957—1970) in their field studies.

## IV. Conclusions

The "centrencephalic" concept of epilepsies which are "generalized from the start", as proposed by Penfield and Jasper (1954), corresponds to an explanatory schema which has long seemed sufficiently relevant to account for the pathophysiologic mechanisms involved. However, the evidence advanced in favor of this concept, particularly at the experimental level, has in our view been met by so many negative control studies that it can no longer be accepted without discussion.

While we have not proved the meso-diencephalic origin of petit mal, grand mal and tonic axial seizures in man on the basis of our stereotaxic method, we have been able to adduce a certain amount of evidence in favor of the primary participation of the cerebral cortex in these discharges. In any event, the classic criteria underlying the centrencephalic hypothesis appear to us at the present time to be inadequate or dubious.

However, the fact that critical activity primarily affects the cerebral cortex by no means implies that it does not involve the lower for-

---

Fig. 4. 1, 2 = (R) frontal mesial cortex; 3, 4 = (L) frontal mesial cortex; 5, 6 = (R) frontal mesial cortex; 7, 8 = (R) cornu Ammonis; 9, 10 = (L) cornu Ammonis; 11, 12 = (R) ant-thalamus; 13 (EEG) = (R) post-frontal-occipital; 14 (EEG) = (L) post-frontal-occipital; 15 = EKG

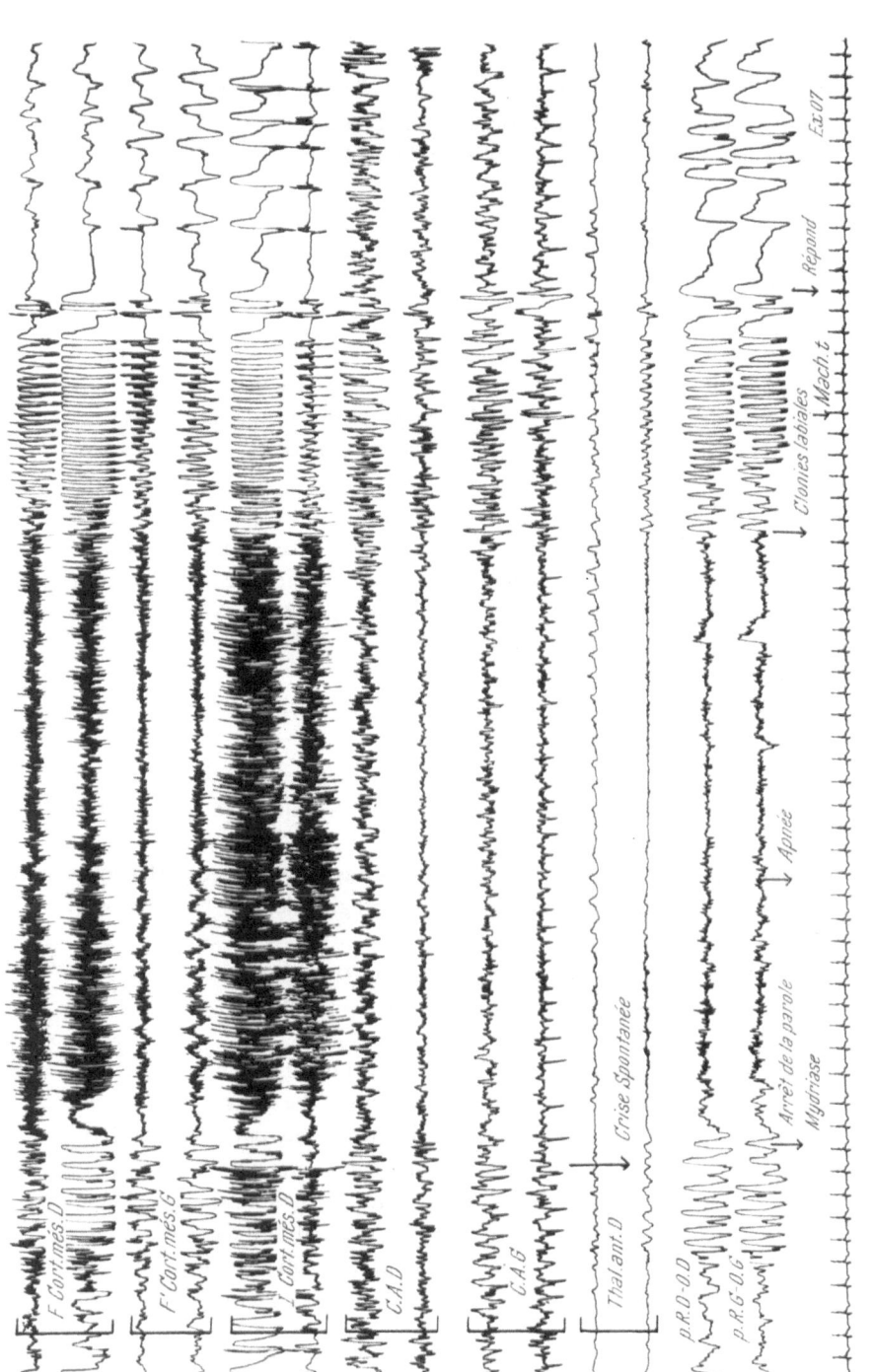

Fig. 4

mations. On the contrary, SEEG studies show that the initial discharge follows a multidirectional cortico-subcortical trajectory in most if not all partial or "generalized" epilepsies.

Moreover, it is not at all certain whether the dichotomy between diffuse epilepsy and focal epilepsy should be stressed or even maintained, now that we are beginning to understand more clearly their capacity for interconversion.

We have virtually no information about the mechanisms which trigger a seizure and which control to a large extent the very nature of the epileptogenic zone. However, it seems to us that in some cases this zone covers a limited territory whereas in others it apparently reflects a massive lowering of the convulsion threshold of very extensive portions of one or both hemispheres.

The probably biochemical factors which facilitate the intracortical extensions and subcortical spread of a discharge are unknown to us.

We therefore cannot offer a new electrophysiologic schema to account for the enormously complex mechanisms involved in the various forms of "generalized" epilepsy.

## Summary

The "centrencephalic" concept of epilepsies which are "generalized from the start", as proposed by PENFIELD and JASPER (1954), corresponds to an explanatory schema which has long seemed sufficiently relevant to account for the pathophysiologic mechanisms involved. However, the evidence advanced in favor of this concept, particularly at the experimental level, has in our view been met by so many contrary findings that it can no longer be accepted without discussion.

While we have not proved the meso-diencephalic origin of petit mal, grand mal and tonic axial seizures in man on the basis of our stereotaxic EEG investigations of approximately 400 patients, we have been able to adduce a certain amount of evidence in support of the primary participation of the cerebral cortex in these discharges.

However, the fact that critical activity primarily affects the cerebral cortex by no means implies that it does not involve the lower formations. On the contrary, SEEG studies show that the initial discharge always follows a multidirectional trajectory in most if not all partial and generalized epilepsies.

# References

BANCAUD, J.: Physiopathogenesis of generalized epilepsies of organic nature. In: GASTAUT, H., et al. (eds.), The physiopathogenesis of the epilepsies, pp. 158—187. Springfield: Thomas. 1969.

— Rôle du cortex cérébral dans les épilepsies « généralisées » d'origine organique. Apport des investigations stéréoélectroencéphalographiques (S.E.E.G.) à la discussion de la conception « centrencéphalique ». Presse Méd. 79, 669—673 (1971).

— et J. TALAIRACH: La stéréo-électro-encéphalographie dans l'épilepsie. Paris: Masson. 1965.

— — A. BONIS, C. SCHAUB, G. SZIKLA, and P. MOREL: Information neuro-physiopathologique apportée par l'investigation fonctionelle stéréotaxique dans les épilepsies. Rev. Neurol. 100, 81—86 (1963).

BENNETT, F. E., and F. A. GIBBS: Intracarotid and intravertebral metrazol in petit mal epilepsy. Electroenceph. clin. Neurophysiol. 4, 382 (1952).

CRAIN, S. M.: Electrical activity of brain tissue developing in culture. In: JASPER, H., et al. (eds.), Basic mechanisms of the epilepsies. Boston: Little, Brown & Co. 1969.

ECHLIN, F. A.: The supersensitivity of chronically isolated cerebral cortex as a mechanism in focal epilepsy. Electroenceph. clin. Neurophysiol. 11, 697—722 (1959).

GASTAUT, H., H. JASPER, J. BANCAUD, and A. WALTREGNY: The physiopatho-genesis of the epilepsy. Springfield: Thomas. 1969.

— and N. PINSARD: Séméiologie clinique des crises épileptiques chez l'enfant. Vie méd. 48, 865—872 (1967).

— — and J. ROGER: Séméiologie électro-encéphalographique des crises épileptiques chez l'enfant en fonction de l'âge. Vie méd. 48, 873—878 (1967).

GUERRERO-FIGUEROA, R., and A. DARROS: Experimental petit mal in kittens. Arch. Neurol. 9, 297—306 (1963).

INGVAR, D. H.: Reproduction of the 3/sec spike and wave EEG pattern by subcortical stimulation in cats. Acta physiol. scand. 33, 137—150 (1955).

— On the pathophysiology of the 3/sec spike and wave epilepsy. Electroenceph. clin. Neurophysiol. 11, 187 (1959).

JASPER, H. H., A. A. WARD, and A. POPE (eds.), Basic mechanisms of the epilepsies. Boston: Little, Brown & Co. 1969.

KAADA, B. R.: Somatomotor automatisms and electroencephalographic response to electrical stimulation of rhinencephalon or other structure in primate, cat and dog. Acta physiol. scand. 24, Suppl. 83 (1951).

KOPELOFF, N., J. R. WAITIN, B. L. PACELLA, and L. KOPELOFF: The epileptogenic effect of subcortical alumina cream in the rhesus monkey. Electroenceph. clin. Neurophysiol. 1, 163—168 (1950).

LENNOX, M. A., and F. ROBINSON: Cingulate-cerebellar mechanisms in the phy-siological pathogenesis of epilepsy. Electroenceph. clin. Neurophysiol. 3, 197—206 (1951).

MARCUS, E. M., and S. JACOBSON: An experimental model of petit mal epilepsy: electrical and behavioral correlates of acute bilateral epileptogenic foci in monkey cerebral cortex. Electroenceph. clin. Neurophysiol. *27*, 735 (1969).

— and C. W. WATSON: Bilateral "epileptogenic" foci in cat cerebral cortex: mechanisms of interaction in the intact, the bilateral cortical callosal and adiencephalic preparation. Electroenceph. clin. Neurophysiol. *17*, 454 (1964).

NAQUET, R., E. BALZANO, and M. PONCET: The light sensitive epilepsy of Papio papio topographic study of cortico subcortical electroencephalographic paroxysmal activity. Electroenceph. clin. Neurophysiol. *24*, 289 (1968).

PENFIELD, W., and H. H. JASPER: Epilepsy and the functional anatomy of human brain. Boston: Little, Brown & Co. 1954.

PETSCHE, H., P. RAPPELSBERGER, and R. TRAPPL: Properties of cortical seizure potential fields. Electroenceph. clin. Neurophysiol. *29*, 567—578 (1970).

POLLEN, D. A., P. PEROT, and K. H. REID: Experimental bilateral wave and spike from thalamic stimulation in relation to level of arousal. Electroenceph. clin. Neurophysiol. *15*, 1017—1028 (1963).

RALSTON, B. L.: Cingulate epilepsy and secondary bilateral synchrony. Electroenceph. clin. Neurophysiol. *13*, 591—598 (1961).

and C. AJMONE MARSAN: Thalamic control of certain normal and abnormal cortical rhythms. Electroenceph. clin. Neurophysiol. *8*, 559—582 (1956).

SAGER, O., V. CHIVU, M. MOISANU et S. BASILESCO: Le complexe pointe-onde obtenu chez les animaux décortiqués bilatéralement pendant l'accès convulsif métrazolique. 1. Int. Congr. Neurol. Sci. Bruxelles, 1957. Excerpta Medica, pp. 75—76.

SASAKI, M.: The importance of subcortical nuclei in epileptic convulsion. Tohoku Med. *51*, 1—13 (1955).

SHIMIZU, K., S. REFSUM, and F. A. GIBBS: Effect on the electrical activity of the brain of intra-arterially and intra-cerebrally injected convulsant and sedative drugs. Electroenceph. clin. Neurophysiol. *4*, 141—146 (1952).

STARZL, T. E., W. T. NIEMER, M. DELL, and P. R. FORGRAVE: Cortical and subcortical activity in experimental seizures induced by metrazol. J. Neuropath. exp. Neurol. *12*, 262—276 (1953).

WEIR, B.: Spike wave from stimulation of reticular core. Arch. Neurol. *11*, 209—218 (1964).

# Discussion

BAROLIN: I am most interested in one problem: the transition of a petit-mal to a grand-mal seizure, is this a question of localization, of aging ore something else?

BANCAUD: I do not know the mechanisms underlying the frequent transformation of a petit-mal absence into a grand-mal seizure. We have usually found that a slight change of the parameters of electric stimulation was sufficient to change completely the clinical seizure pattern. In one case which was published by us, the stimulation of the frontal mesial cortex resulted in a petit-mal absence when a 3/sec pulse rate was applied, and a generalized grand-mal type seizure at 20 pulses per second. Thus it seems as if the way of neuronal recruitment is essential for the type of seizure pattern resulting.

GLOOR: Have you seen cases in which the initial discharges of spikes and waves are found beyond the fissura Rolandi? I am asking this question since we saw a few cases with parietal foci and beautiful secondarily bilateral synchronous spike-wave patterns. By intra-carotid injections of Amytal and Metrazol these foci were clearly lateralized. The therapeutic results obtained by excision of parietal foci were favourable. What is your experience in this matter?

BANCAUD: Sometimes we also found bilateral discharges of spikes and waves which started beyond the fissura Rolandi, particularly in the mesial part. Those discharges, however, were never generalized or accompanied by troubles of consciousness. The clinical semiology observed with such cases was able to be elicited by electrical stimulation of a certain region of the mesial parietal cortex. These types of seizures disappeared after excision of the focus. These cases are rather rare compared with the ones with a focus in the frontal lobe.

# Generalized Spike and Wave Discharges:
## A Consideration of Cortical and Subcortical Mechanisms of Their Genesis and Synchronization

P. GLOOR

Montreal Neurological Institute and the Department of Neurology
and Neurosurgery, McGill University, Montreal, Canada

Generalized spike and wave discharge is one of the most characteristic features of the clinical electroencephalogram. In its classical, regular 3 c/sec form it represents the concomitant of a common form of generalized seizures, the absence attack, which is characterized by a very variable degree of clouding of consciousness and sometimes by minor bilateral motor and autonomic phenomena. No one looking at the electroencephalographic record of such a seizure, be he an expert or a novice, can fail to be impressed by the suddenness with which virtually simultaneously the resting pattern of the electroencephalogram is replaced by high voltage generalized, bilaterally synchronous, rhythmic spike and wave discharge. The cessation of this pattern is equally impressive and so is, often, the symmetry of the waveforms in homologous areas of the hemispheres. Even though the finer analysis of time sequences reveals some lack of perfect synchronization between various areas of the cortex (PETSCHE 1962), one cannot fail to be impressed by the relative synchronicity of the electrical events. This must imply that a system is being activated which is capable of phase-locking within a relatively narrow range of time-dispersion the discharges of a great number of neurons in widely disparate areas of the brain. How to explain this widespread synchronization is the fundamental question which arises when one seeks a satisfactory physiopathogenetic explanation of the mechanism of this discharge.

It is seemingly paradoxical that this electrographic phenomenon, which is perhaps the best defined among all the features visible in the clinical EEG, is the least explained in terms of its underlying pathophysiology. Not all questions, however, are unanswered. If we start with those aspects which are most clearly established, we can say that the spike and wave potentials are cortically generated. They cannot be attributed to purely physical volume conduction to the scalp of some discharge taking place in deep midline structures of the brain. The physical generators of the electric potentials recorded on the scalp lie in the cortex (Hayne et al. 1949). This is confirmed by the detailed electrophysiological studies on the spike and wave discharge produced in experimental animals (Pollen and Sie 1964, Pollen et al. 1964, Weir and Sie 1966). All the features of the spike and wave complex indicate that these potentials are generated by radially oriented neuronal elements of the cortical mantle. Furthermore it can be taken as established that the spike component is an excitatory phenomenon; it is indicative of excitation of cortical neurons brought about by EPSP's generated on their membrane surfaces. The slow wave is an inhibitory phenomenon associated with arrest of action potential discharge of cortical neurons and appears to be generated by prolonged IPSP's in cortical neurons (Pollen 1964). The objection may be raised that these observations were made in animal models of the spike and wave discharge; they may perhaps not be applicable to its human counterpart. However, Weir (1965) has shown, that the experimentally induced spike and wave discharge in animals is morphologically identical to that occurring in the human electroencephalogram. Furthermore, Perot (1963), in a microelectrode recording on human cortex in a patient showing slow spike and wave discharge, has shown that the relationship between excitatory and inhibitory events in the case of human spike and wave activity was the same as that demonstrated by Pollen et al. (1964) and Weir and Sie (1966) for the experimentally induced spike and wave discharge of animals.

Beyond these areas of reasonable certainty, most of the questions concerning the pathophysiology of generalized spike and wave discharge remain unanswered, although conjectures abound. The greatest area of uncertainty, surrounds the mechanism of the widespread synchronization of the discharges. We do not know whether the sudden turning on and turning off of the generalized spike and wave discharge and its relatively precise synchronization over large areas of cortex is dependent upon transcortical synchronizing mechanisms or whether it requires the intervention of subcortical neuronal aggregates in the upper brainstem and/or thalamus. Furthermore the question whether a pacemaker is required for the explanation of this widespread neuronal disturbance, or whether oscillation in neuronal nets without any intervention of a true pacemaker mechanism could be involved in the genesis and maintenance of the pattern still remains a matter of conjecture. The idea of a central, presumably subcortical pacemaker, a kind of neuronal switch mechanism, whereby the generalized pattern can be turned on and off in the dramatic fashion shown in the clinical electroencephalogram remains, of course, an attractive hypothesis for anyone who has ever looked at the abrupt suddenness with which this EEG pattern is initiated and turned off simultaneously over both hemispheres in a widespread manner (JASPER and DROOGLEEVER-FORTUYN 1947, PENFIELD and JASPER 1954). Nevertheless, even in clinical electroencephalography, situations are encountered quite frequently which may cast doubt upon the necessity to postulate a fixed pacemaker mechanism. It is well known that patients with spike and wave discharge in their EEGs may have abortive discharges that involve only one hemisphere or even only focal areas within one hemisphere. These foci are often unstable in localization and may shift from one region to another. This observation by itself of course proves nothing, for it is still conceivable that a hypothetical subcortical pacemaker could, when relatively inactive, project only to limited cortical segments rather than to the entire cortical mantle. It nevertheless raises the possibility that multiple potential pacemakers may exist in the cortex from which, under appropriate conditions, discharges can flare up and spread with great rapidity through cortical association and commissural tracts to involve both hemispheres in a diffuse manner.

There are forms of epilepsy in which stable foci can be demonstrated which give rise to focal spike and wave discharge, from which however variable degrees of unilateral or bilateral spread of the discharge can be observed within one single EEG examination (Fig. 1). It is difficult in a situation like this to assume any other explanation than that of a variable cortical spread of the discharge, depending perhaps

Fig. 1. Three excerpts from the EEG of an epileptic patient showing a variable degree of widespread synchronization of spike and wave discharge. The epileptic discharges were sometimes virtually confined to the left anterior sylvian region (first excerpt), at others they were seen to spread widely over the ipsilateral hemisphere (second excerpt) or even bilaterally (third excerpt). When spreading over large areas of cortex the discharges assume a clearcut spike and wave form

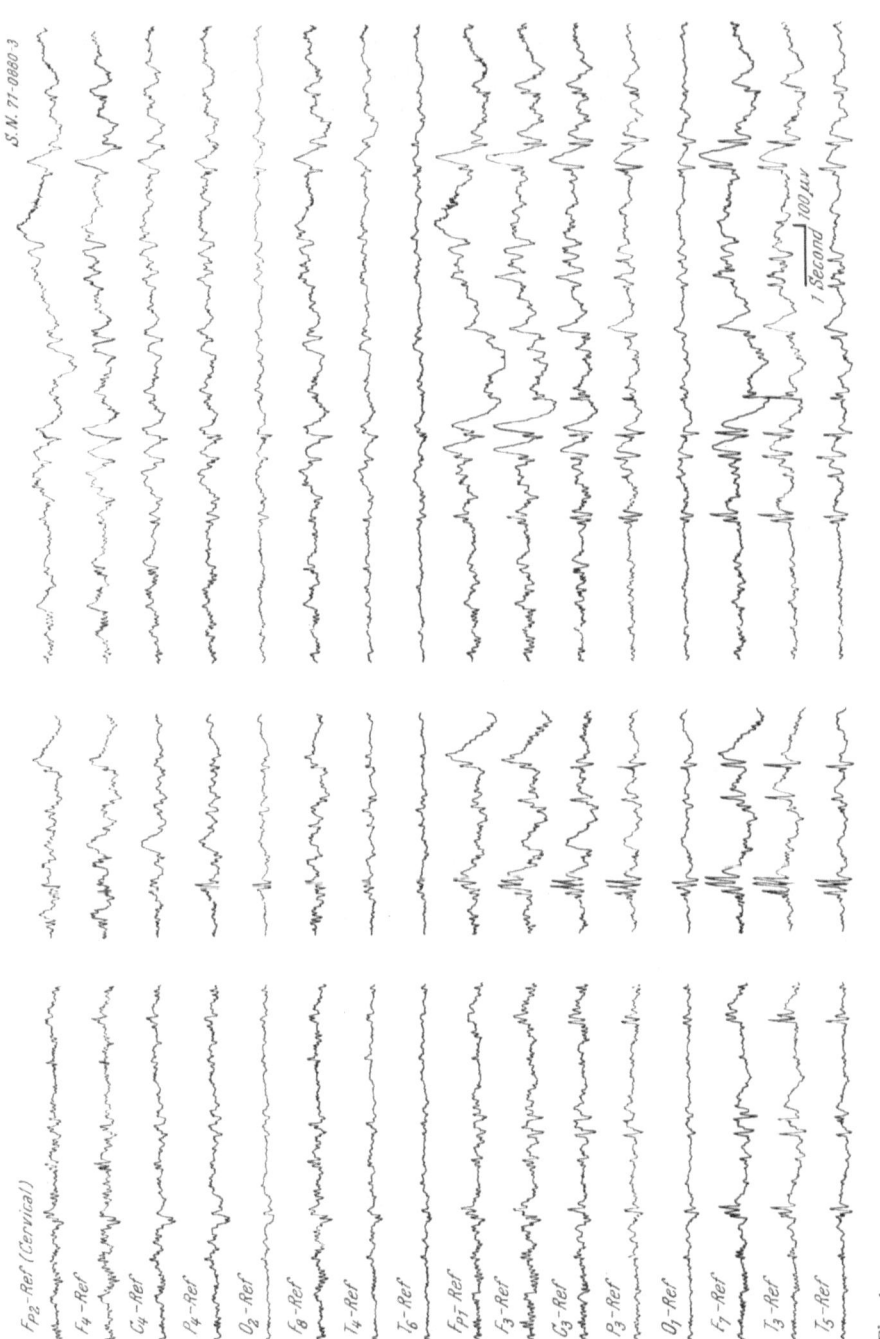

Fig. 1

on the intensity of firing of the initial focus and of the excitatory level of neuronal populations elsewhere in the cortex which receive the impinging afferent inputs from the original focus. It is not always easy to distinguish such cases from those in which the generalized spike and wave mechanism is assumed to depend upon a subcortical pacemaker mechanism.

In spite of all these considerations however, one should not lose sight of the fact that the centrencephalic hypothesis (PENFIELD and JASPER 1954), which postulates a pacemaker mechanism in brainstem and diencephalic structures is buttressed by rather impressive experimental observations. They need not be reviewed in detail in this study, since they are well known to all workers in the field. Suffice it to say that stimulation in the intralaminar thalamic nuclei (JASPER and DROOGLEEVER-FORTUYN 1947), or combined stimulation in the intralaminar thalamus and in the mesencephalic reticular formation (PEROT 1963, POLLEN, PEROT, and REID 1963), can evoke, in experimental animals, bilaterally synchronous generalized spike and wave discharges which are a close facsimile of those occurring spontaneously in human patients with absence attacks. Furthermore, the clinical manifestations elicited by such stimulations resemble human absence attacks (HUNTER and JASPER 1949). Any explanation of the mechanism of bilaterally synchronous spike and wave discharge will have to take these observations into consideration and must be able to account for them in a satisfactory manner.

It is equally well known however that experimental manipulation of cortical mechanisms can reproduce in an equally convincing way bilaterally synchronous spike and wave discharge and clinical concomitants resembling very closely human absence attacks. The work of MARCUS and WATSON (1966, 1968) and MARCUS, WATSON, and SIMON (1968 a, b) has shown this very clearly. They also presented evidence that in their experimental model of bilaterally synchronous spike and wave discharge the synchronization mechanism is dependent upon callosal fibres and does not require intactness of connections from subcortical grey matter to the cerebral cortex.

These two sets of observations stand in apparent contradiction, if one assumes that the role of either the cortex or subcortical grey matter must be that of providing for a pacemaker mechanism.

Theoretically at least, there are ways, however, of reconciling this apparent discrepancy and this paper will attempt to propose some ideas which may resolve this apparent conflict. Before doing so,

I wish to relate briefly some observations made on patients with bilaterally synchronous spike and wave discharge in whom the effects of intracarotid and intravertebral sodium amytal and metrazol injections have been studied (Rovit et al. 1961, Gloor et al. 1964, Garretson et al. 1966). These tests were carried out with a diagnostic and ultimate therapeutic aim in mind. It is known that some patients whose EEGs present bilaterally synchronous spike and wave discharges can be relieved of their seizures by restricted unilateral cortical removals, in spite of the generalized character of the seizure discharge. These patients suffer from a condition which is usually described as "secondary bilateral synchrony" (Penfield and Jasper 1954). The underlying assumption in these conditions, which is vindicated by therapeutic results, is that the generalized bilaterally synchronized discharge is in some way dependent upon a localized cortical mechanism which may act as its pacemaker. Injection of amytal through the internal carotid artery on the side of the pacemaker is capable of stopping the discharges bilaterally, whereas the same injection of amytal given through the contralateral carotid artery fails to do so (Rovit et al. 1961). Conversely the convulsive threshold to fractionized intracarotid injections of metrazol given through the internal carotid artery ipsilateral to the pacemaker focus is significantly lower than that measured with contralateral fractionized intracarotid metrazol injections (Gloor 1969). This presentation will not dwell upon these rather exceptional cases of secondary bilateral synchrony, but relate some observations made on patients which on the basis of these tests were found to suffer from primary bilaterally synchronous generalized spike and wave discharge. These observations shed some light upon possible cortical and subcortical mechanisms in this human form of epilepsy.

To understand the arguments used in the interpretation of these tests, a brief review of the vascular anatomy of the human brain is in order. Blood circulating through the internal carotid artery irrigates the ipsilateral cerebral cortex with the exception of a small posterior territory fed by the terminal branch of the vertebro-basilar system, the posterior cerebral artery. In addition to the cortex, the hemispheral white matter and the corpus striatum are irrigated by the carotid circulation. The diencephalon, with the exception of the anterior hypothalamus and of a thin lateral thalamic shell lie within the vertebro-basilar system and are fed by branches originating from the posterior cerebral artery (Kaplan and Ford 1966). It follows

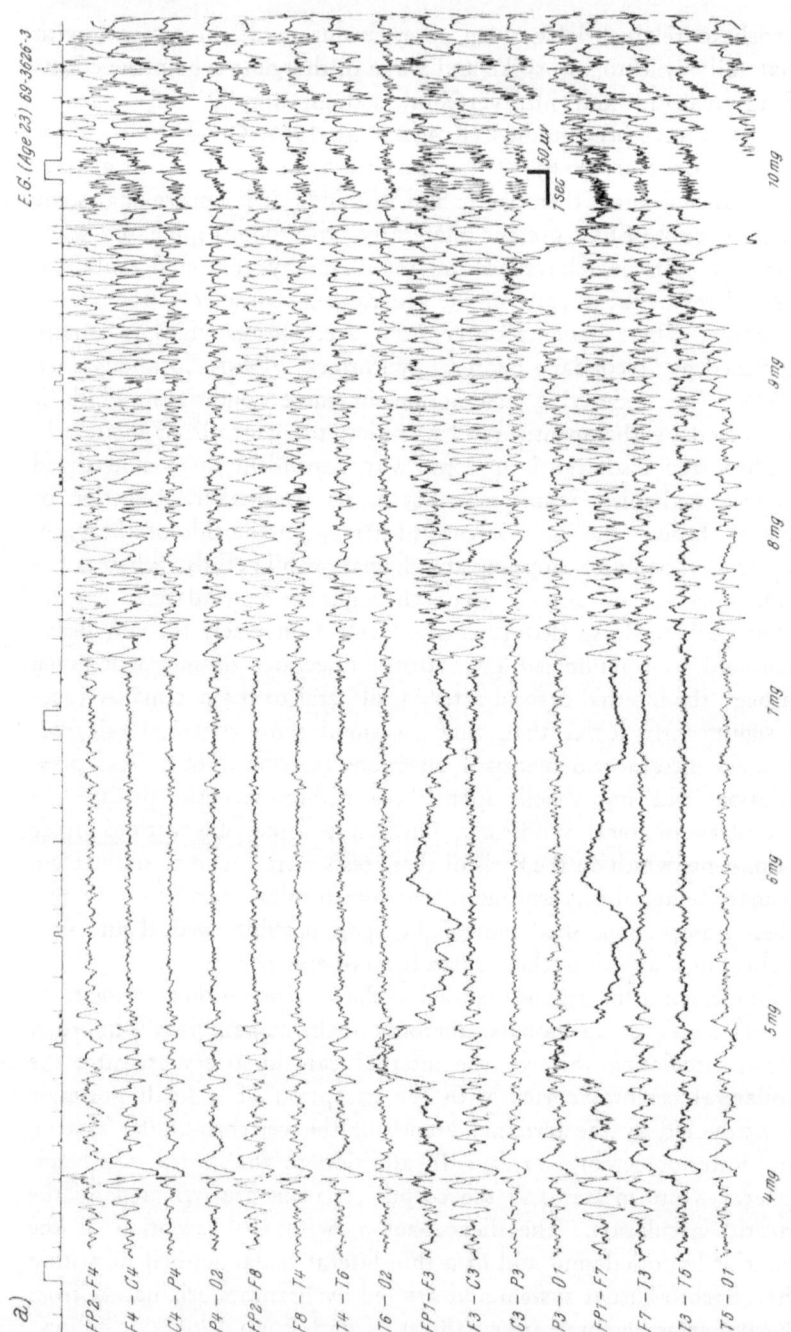

Fig. 2 a

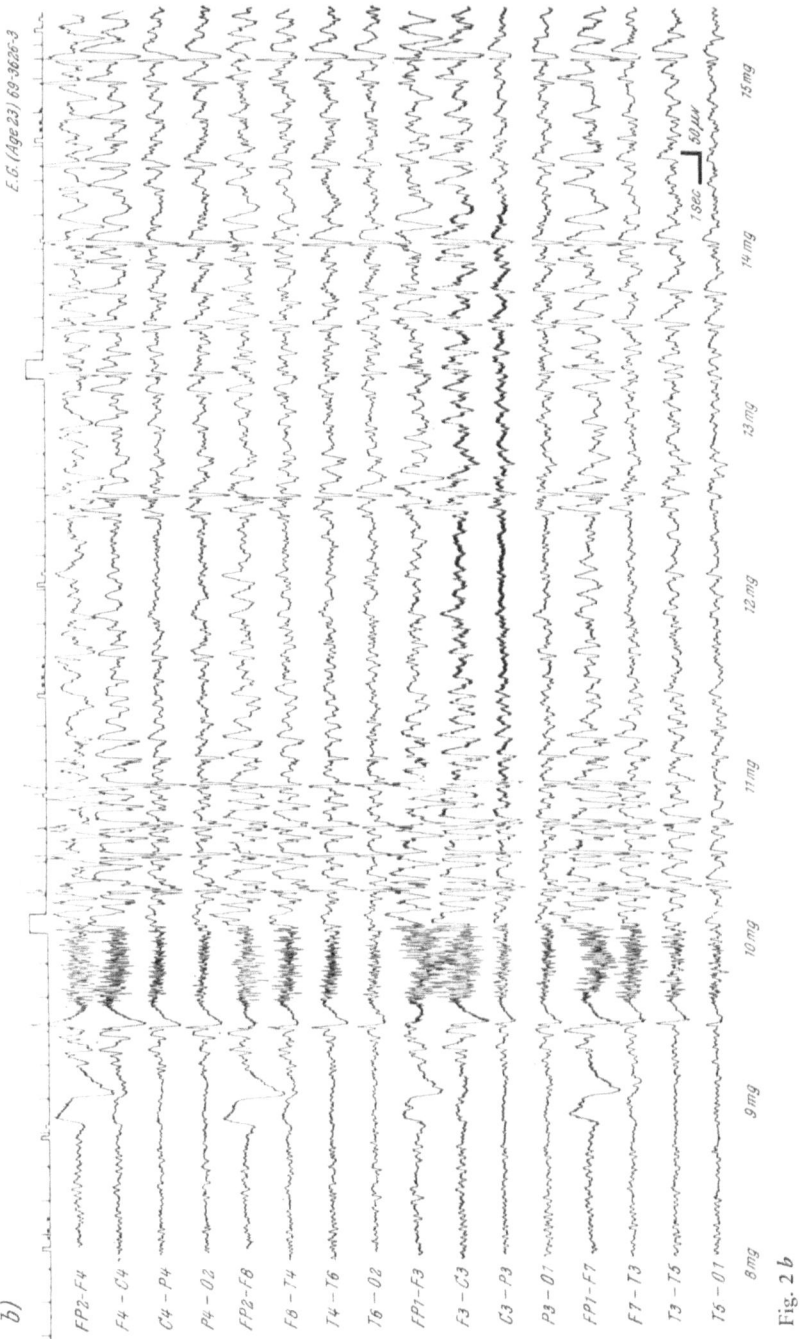

Fig. 2. Effects of left-sided (*a*) and right-sided (*b*) intracarotid metrazol injections. The thresholds for initiation of continuous electrographic seizure activity are nearly identical on the two sides (8 mgs on the left, 10 mgs on the right side). The patient suffered from generalized seizures, absence attacks and akinetic attacks; his resting EEG showed bilaterally synchronous 2/sec second spike and wave activity

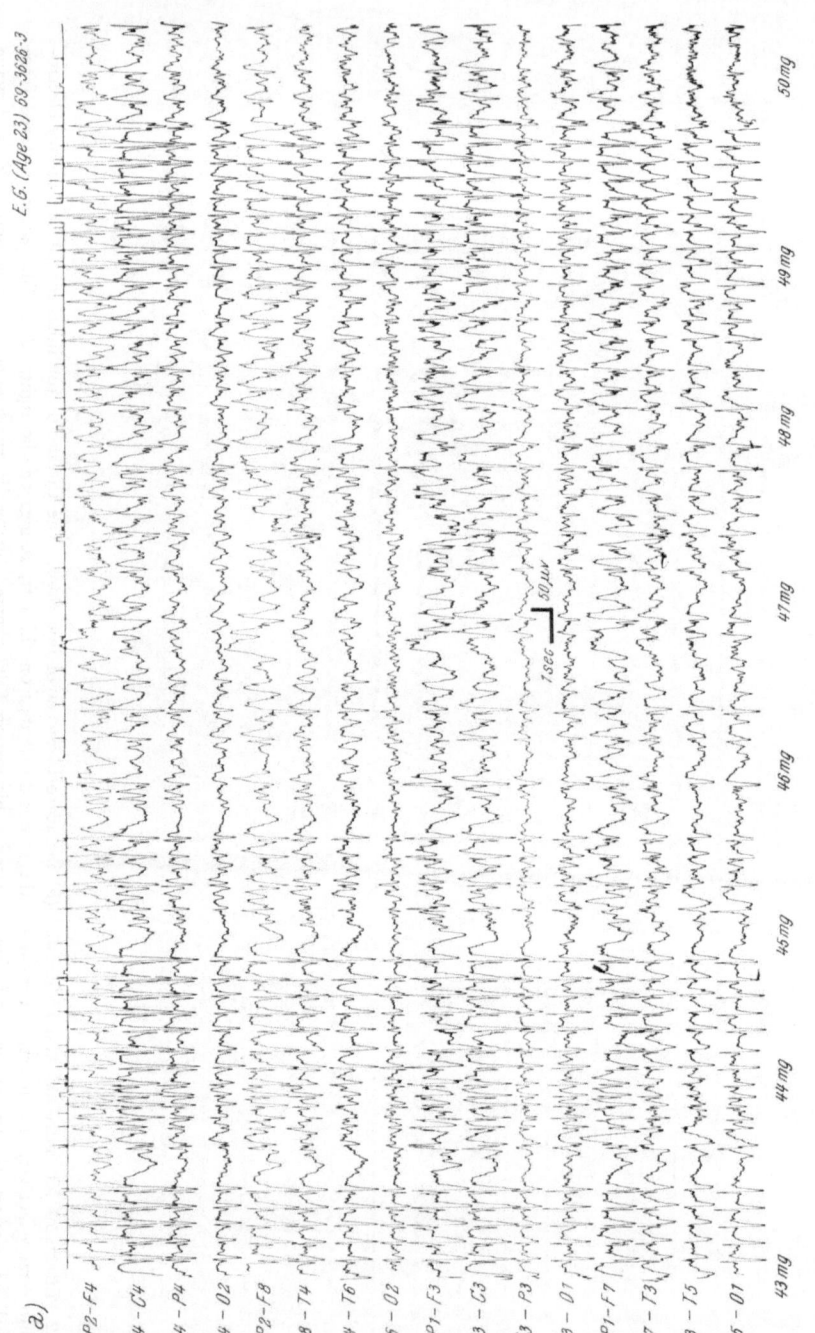

Fig. 3 a

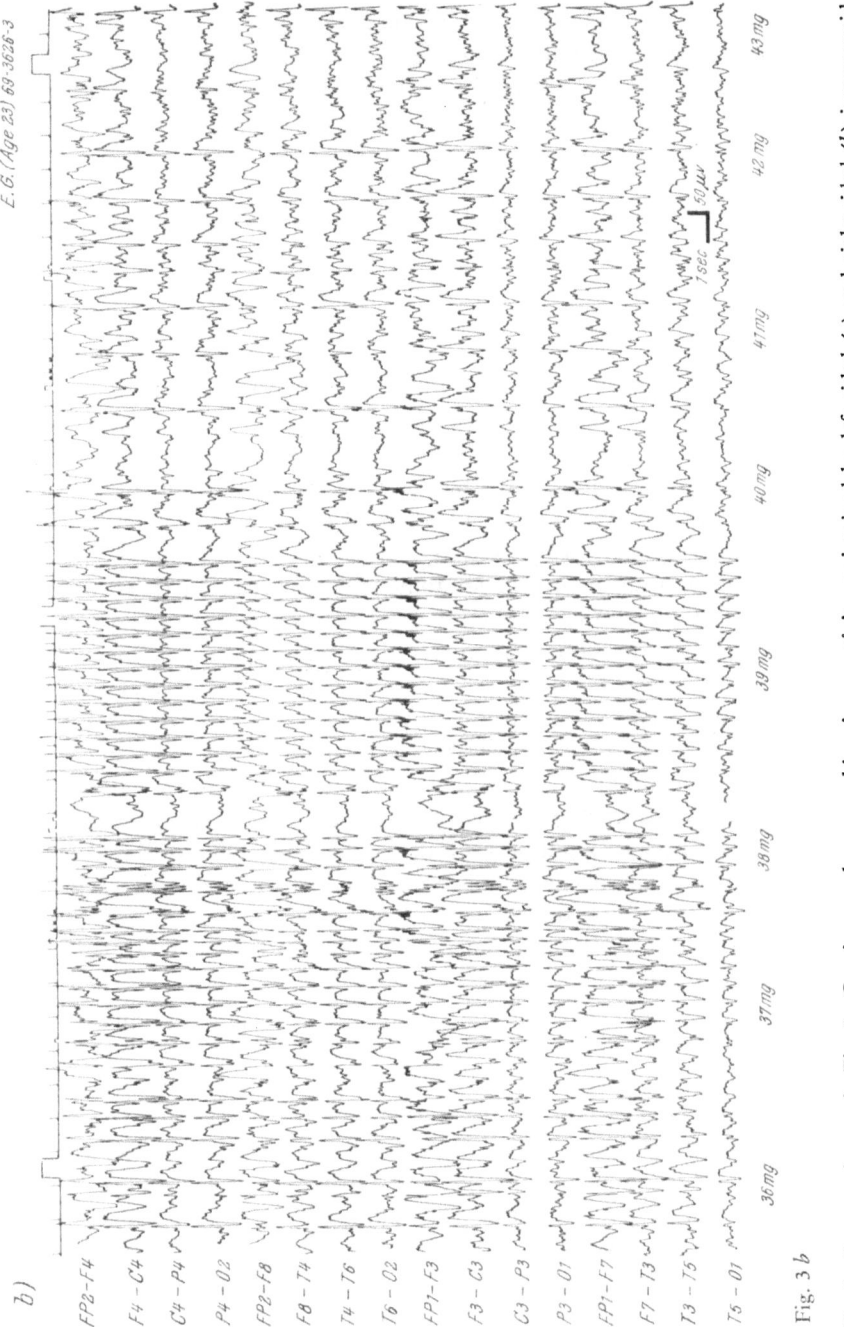

Fig. 3. Same patient as in Fig. 2. Continuous electrographic seizure activity maintained by left-sided (*a*) and right-sided (*b*) intracarotid metrazol injections. The EEG pattern is characterized by bilaterally synchronous spike and wave activity which at times assumes a clock-like regularity. The clinical manifestations at that time were minimal and consisted of fluttering of the eyelids, twitches around the mouth contralateral to the metrazol injection, occasional jerks of the head and slowness in responding to commands

from this that any drug injected through the carotid artery will irrigate cortex and striatum ipsilaterally, but will fail to irrigate any substantial portion of midline cerebral structures, including those of the diencephalon. Conversely any drug that might be injected through the vertebral artery will reach these midline structures, but will only irrigate a small portion of posterior cortex.

For the consideration of the possible pathophysiology of bilaterally synchronous spike and wave discharge, the results obtained with intracarotid and intravertebral metrazol injections are more revealing than those obtained with amytal injections and will therefore be dealt with in greater detail in this paper.

In patients with spontaneous bilaterally synchronous spike and wave discharge, which are not dependent upon a unilateral pacemaker mechanism and thus are not fulfilling the criteria of "secondary bilateral synchrony", unilateral fractionized injection of metrazol through one carotid artery is capable of eliciting ongoing bilaterally synchronous spike and wave activity, often for several minutes, without precipitation of any major tonic-clonic convulsion (Figs. 2 and 3). The onset of this prolonged spike and wave seizure is bilateral in spite of the unilateral injection of the drug and furthermore a repeat test with injection of the drug through the contralateral carotid artery reproduces an identical or nearly identical response (Figs. 2 and 3). Clinically this state of prolonged spike and wave discharge maintained by continued metrazol injection is usually characterized by some clouding of consciousness, mental confusion, slow responsiveness and blinking of the eyes, sometimes associated with minor myoclonic phenomena. In a few patients the pattern has also been observed with full preservation of consciousness or conversely with total loss of it. The important aspect of this observation is that a drug which obviously bypasses the subcortical synchronizing mechanisms of the upper brainstem and thalamus can initiate simultaneously over both hemispheres bilaterally synchronous spike and wave discharge in spite of the fact that only one hemisphere receives the drug. This is a strong argument that a cortical mechanism is initiating the bilaterally synchronous discharge, although a subcortical route of the synchronizing pathways with mediation through intralaminar thalamus or upper brainstem can of course not be eliminated. Equally impressive is the fact that this bilaterally synchronous discharge induced by unilateral carotid metrazol injection can be turned off promptly and simultaneously in both hemispheres by injection of

50 mg of sodium amytal through the same carotid artery through which the metrazol had been given (Fig. 4). It seems therefore that the amytal given through the same carotid artery through which the preceding metrazol injection was administered has counteracted the chemically induced pacemaker mechanism established by the metrazol and that inactivation of this temporary pacemaker is capable of arresting the discharge promptly bilaterally and simultaneously. Again this effect must be exerted at the cortical level since the amytal does not perfuse subcortical synchronizing systems in the thalamus and upper brainstem. Injection of sodium amytal through the contralateral carotid artery also arrests the bilateral discharge but not as promptly. This suggests that to-and-fro bombardment via commissural pathways may be necessary to maintain continued spike and wave discharge.

The results of intracarotid amytal injection upon spontaneously occurring bilaterally synchronous spike and wave discharge in these patients with primary bilateral synchrony are essentially the same whether the drug is injected through the right or the left internal carotid artery. There are however differences in the response pattern between individual patients. Three response patterns can be distinguished.

1. Unilateral depression of voltage of the spike and wave discharge without any effect upon the number of spike and wave complexes per minute.

2. Bilateral suppression or diminution in the number of spike and wave discharges, regardless of the side of the injection.

3. Unilateral or bilateral activation of rhythmic paroxysmal slow wave or more rarely spike and wave discharge localized on the side of the injection, or with voltage predominance on that side. This pattern is often followed by pattern 2.

Pattern 1 could easily be reconciled with a centrencephalic origin of the discharges. The second pattern however suggests that irrigation of the cortical grey matter, even unilaterally, may eliminate the discharges bilaterally. The fact that the response is the same whether injection is carried out on the right or the left side furthermore indicates that neither of the two hemispheres can be considered as the sole obligatory pacemaker for these bilaterally synchronous discharges. The ipsilateral, or ipsilaterally predominant, activation of paroxysmal slow wave or even spike and wave discharge with unilateral injection of sodium amytal is difficult to explain. I have no satisfactory explanation to offer for this observation at the moment.

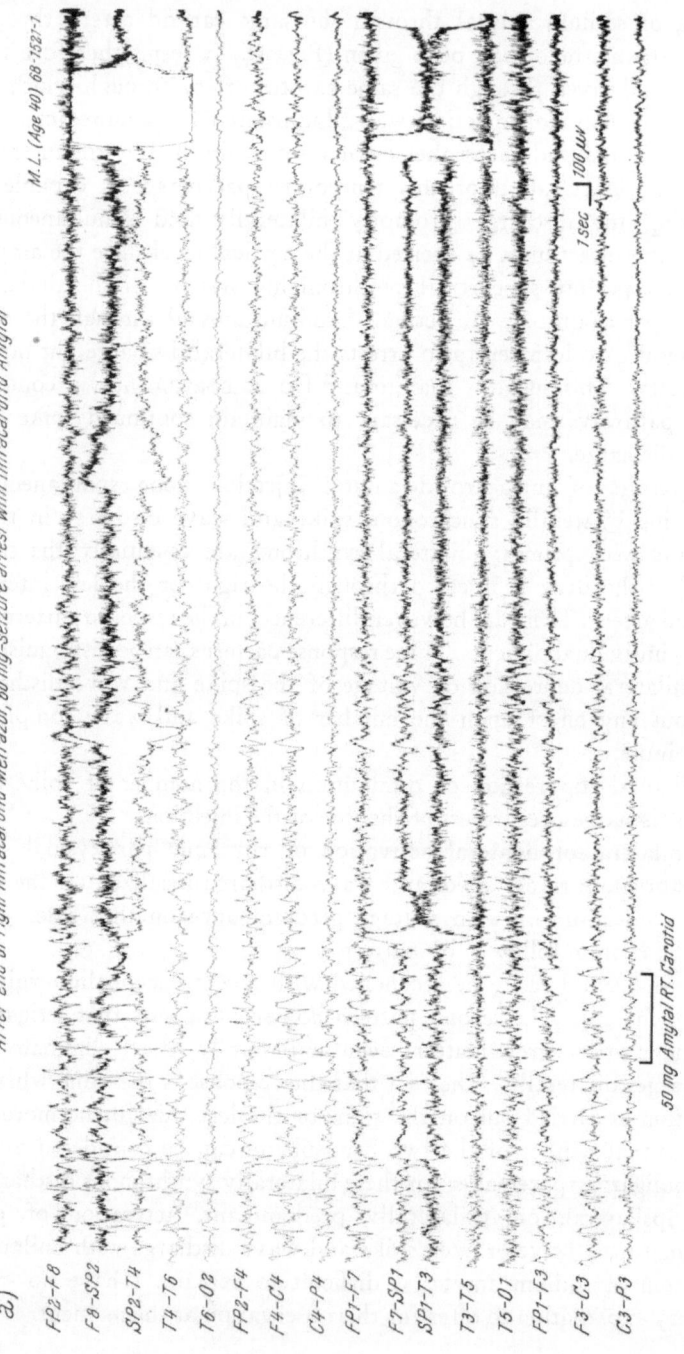

a)

M.L. (Age 40) 68-1521-1

After end of right intracarotid Metrazol, 60 mg–seizure arrest with intracarotid Amytal

FP2–F8
F8–SP2
SP2–T4
T4–T6
T6–O2
FP2–F4
F4–C4
C4–P4
FP1–F7
F7–SP1
SP1–T3
T3–T5
T5–O1
FP1–F3
F3–C3
C3–P3

1 SEC ⌐ 100 µv

50 mg Amytal RT. Carotid

Fig. 4 a

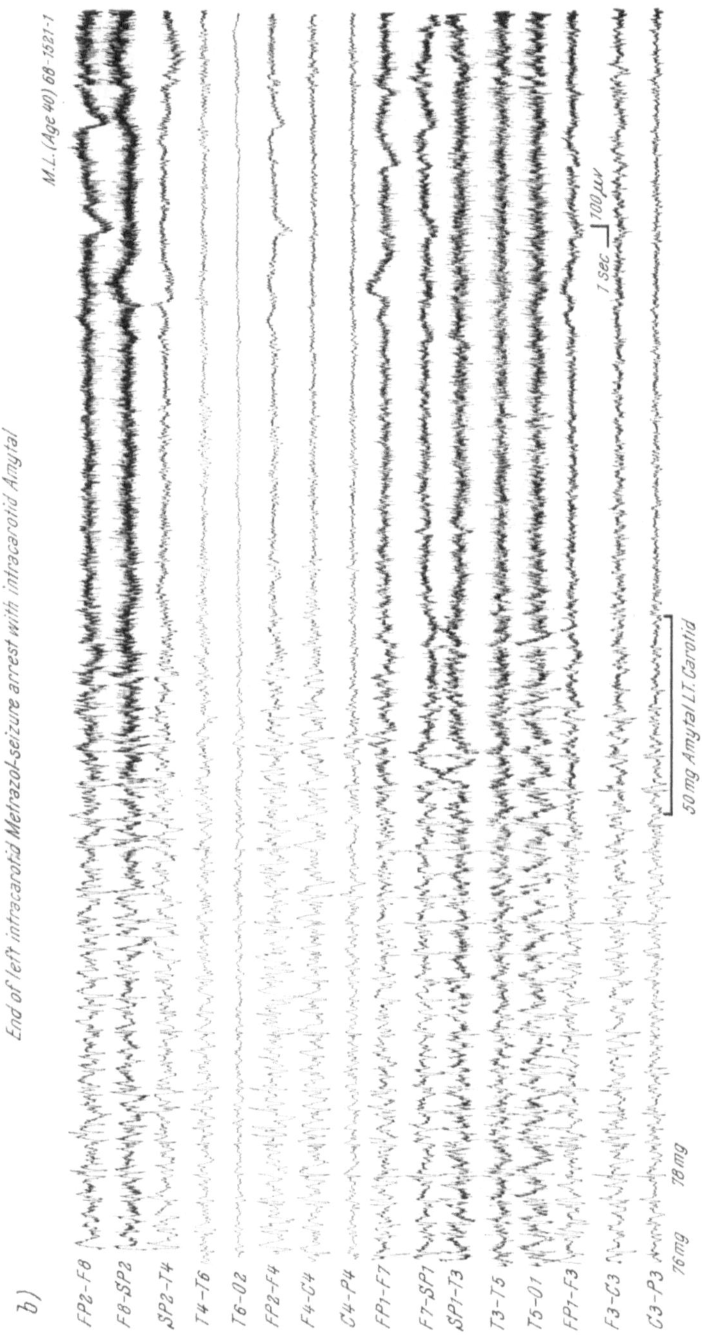

*M.L. (Age 40) 68-1521-1*

*End of left intracarotid Metrazol-seizure arrest with intracarotid Amytal*

*b)*

FP2-F8
F8-SP2
SP2-T4
T4-T6
T6-O2
FP2-F4
F4-C4
C4-P4
FP1-F7
F7-SP1
SP1-T3
T3-T5
T5-O1
FP1-F3
F3-C3
C3-P3

*76mg    78mg*

*50 mg Amytal Lt. Carotid*

*1 sec.*  *100 μv*

Fig. 4 b

Fig. 4. Bilaterally synchronous seizure activity initiated and maintained by intracarotid metrazol injection is arrested by the injection of 50 mgs of sodium amytal through the same carotid through which metrazol had been given previously. The spike and wave activity at the time of the amytal injection had become somewhat irregular and was intermingled with faster rhythmic activity. The patient was unresponsive during the metrazol induced seizure. (*a*) right-sided metrazol followed by right-sided amytal injection; (*b*) left-sided metrazol followed by left-sided amytal injection

In two patients, we had the opportunity to observe the effects of injection of both sodium amytal and metrazol through the vertebral artery. In the first of these patients, the catheter which was to be inserted into the internal carotid artery was accidentally put into the vertebral artery. In the second patient, an intravertebral placement of the catheter was carried out deliberately, because a subcortical stereotaxic procedure for intractable epilepsy associated with bilaterally synchronous spike and wave discharge was planned. It therefore appeared indicated to study the contribution of neuronal structures in the vertebro-basilar territory to the genesis of these discharges.

The surprising observation made in both patients was that fractionized intravertebral injection of metrazol delivered at the same dosage schedule as that used for carotid injections did not produce any activation of the bilaterally synchronous spike and wave discharges, but on the contrary produced a diminution and even in one patient a temporary complete arrest of spike and wave activity (Fig. 5). The clinical manifestations of these injections were of a non-convulsive nature and consisted of nausea. This was the case in spite of the fact that the amount of metrazol injected through the vertebral route was many times larger than that required to activate spike and wave discharge and ultimately precipitate a seizure with either left or right sided intracarotid metrazol injection (Fig. 5).

The effects of intravertebral sodium amytal injections were less dramatic. In the first patient, there was no initial effect upon the spike and wave activity with a delayed diminution of discharges, the delay being sufficiently prolonged to suggest that this may have been due to recirculation of the drug through the general circulation. In the second patient, there was a transient mild activation of bilaterally synchronous spike and wave discharge.

The results of these observations indicate that perfusion of metrazol through brainstem structures, presumably including those postulated as potential pacemakers for the spontaneous bilaterally synchronous spike and wave activity, fails to initiate such discharges, but on the contrary is capable of diminishing or even eliminating those occurring spontaneously. It seems therefore unlikely that the initiating pacemaker at least for the bilaterally synchronous spike and wave activity resides in these deep midline structures. Again, these observations do not eliminate the possibility that these structures might be important for the mediation of bilateral synchrony.

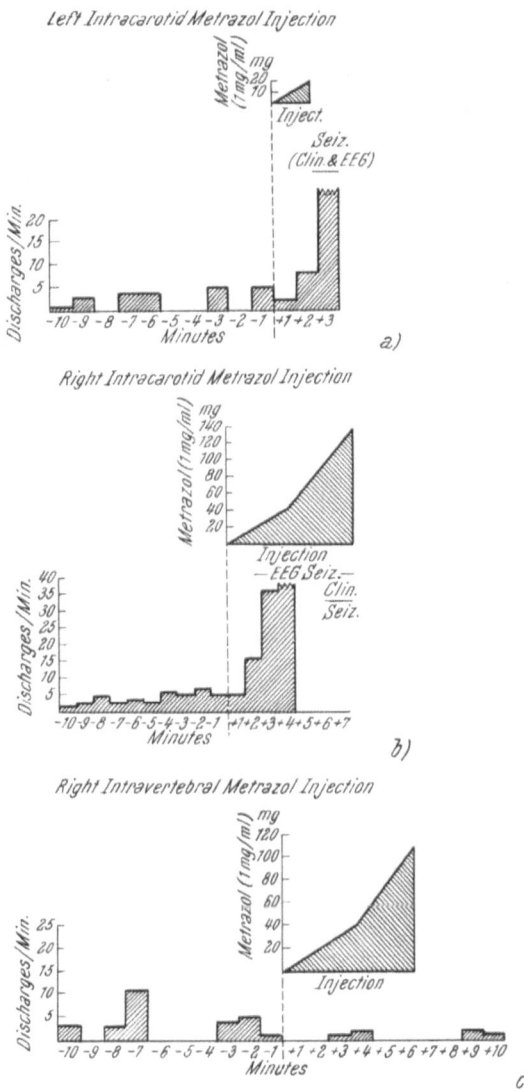

Fig. 5. Responses to left (*a*) and to right-sided (*b*) intracarotid and to right-sided intravertebral (*c*) injections of metrazol. Ordinate of main graphs: number of spike and wave discharges per minute. Abscissa: time; minutes with minus-sign indicate time before, those with plus-sign indicate time after beginning of metrazol injection. The triangle shaped graphs on the top right of (*a*), (*b*) and (*c*) indicate the cumulative dose of metrazol which the patient received. The inflection on the triangle graph in (*c*) indicates that the speed of metrazol injection was doubled after 40 mgs had been given. [From GLOOR, P.: Epilepsia *9*, 249—263 (1968)]

In a recent series of experiments carried out in collaboration with Dr. TESTA we studied an experimental model of bilaterally synchronous spike and wave discharge. Cats given high doses of penicillin intramuscularly develop bilaterally synchronous spike and wave patterns which in many ways are similar to those seen in human absence attacks. Also the animals show minimal clinical manifestations during these discharges, which consist of staring or mild bilateral myoclonic jerks of the face and thus bear some similarity to human absence attacks. This particular experimental model of "centrencephalic epilepsy" was first reported by PRINCE and FARRELL (1969). We have been able to confirm it and we studied the effects of intracarotid and intravertebral drug injections upon the spike and wave discharges occuring in these animals. As in human patients, intravertebral metrazol injection produced arrest of spike and wave activity (Fig. 6), whereas intracarotid injection produced a very prompt activation. The results of Amytal injections were also similar to those obtained in man with either some activation or no effect upon spike and wave activity, when the drug was given through the vertebral route.

It would be easy to conclude from these findings that one need not consider any role of subcortical projection mechanisms in the genesis or maintenance of bilaterally synchronous spike and wave discharge and that therefore this pattern can be accounted for entirely on the basis of purely cortical mechanisms.

This hypothesis which is accepted by some, however, fails to account for a number of observations: it would be difficult, e.g., to conceive how the electrographic pattern and its characteristic clinical concomitants could be reproduced by stimulation of intralaminar thalamic structures or by combined stimulation of these with upper mesencephalic reticular structures. Other observations also suggest some role of subcortical mechanisms in the genesis of this seizure pattern. Bilaterally synchronous paroxysmal activity, although not always of the spike and wave type, is a finding that is invariably

---

Fig. 6. Generalized penicillin epilepsy in the cat induced by 750 000 units of penicillin given intramuscularly. Effect of intravertebral injection of metrazol. Spike and wave discharges are eliminated during the intravertebral metrazol injection except for one brief burst at the end of the injection. The cat was unanesthetized. The injection was carried out through a chronic indwelling catheter. Abbreviations: $RT$ = right; $LT$ = left; $F$ = frontal; $P$ = parietal; $O$ = occipital; $REF$ = reference electrode in frontal bone

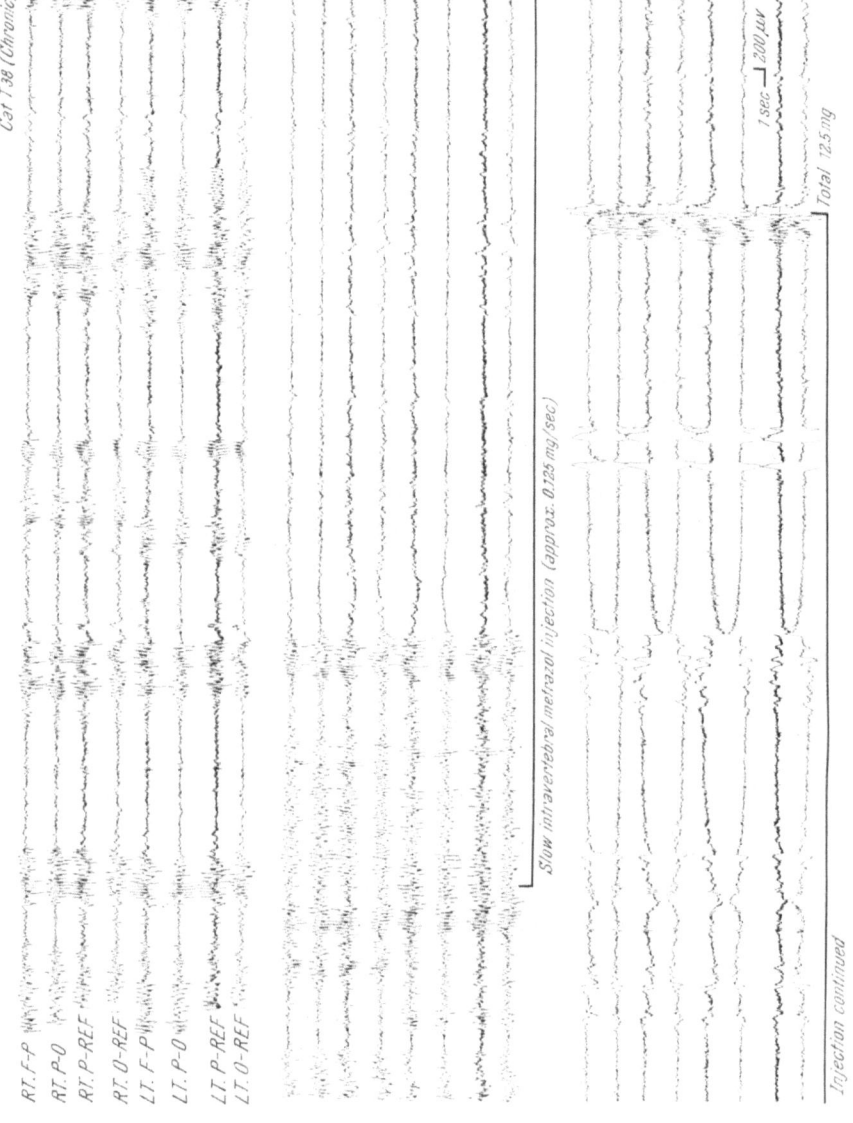

Fig. 6

encountered in diffuse grey matter encephalopathies involving cortical and subcortical structures (GLOOR, KALABAY, and GIARD 1968). In diffuse lipidoses spike and wave discharges are particularly common (GLOOR and ANDERMANN, unpublished observations). Admittedly these are rarely of the classical regular 3/sec spike and wave type, but precise limits are hard to draw between this classical pattern and its variants which are often associated with diffuse neuronal disease. Diffuse cortical lesions without significant involvement of subcortical structures are usually not associated with this degree of EEG synchronization (GLOOR, KALABAY, and GIARD 1968). This suggests that additional damage to subcortical grey matter must greatly enhance the likelihood of the occurrence of widespread cortical synchronization of paroxysmal potentials.

The fact that bilaterally synchronous spike and wave discharge, both in the animal model and in the human patient, are exquisitely sensitive to the level of arousal also suggests that mechanisms located in the upper brainstem are at least capable of greatly influencing the likelihood of occurrence of this pattern (LENNOX et al. 1936, GIBBS and GIBBS 1952, LI et al. 1952, JUNG 1954, POLLEN et al. 1963). High levels of arousal inhibit the occurrence of paroxysmal spike and wave discharge and as the arousal level drops, the probability of its occurrence increases until superficial levels of sleep or anesthesia are reached. During deep sleep and deep anesthesia, again, there is a diminution of the likelihood of the occurrence of this paroxysmal pattern.

These observations suggest that some reduction of drive by the reticular activating system may be an important factor in creating conditions at the cortical level which facilitate the origin and spread of this kind of paroxysmal discharge. Some experimental and clinical observations are in line with this. Lesions in subcortical grey matter, especially in the brainstem, may be associated with the occurrence of paroxysmal activity of the bilaterally synchronous spike and wave type. This has been shown experimentally by HUBEL and NAUTA (1960), GUERRERO-FIGUEROA et al. (1963), and by WEIR (1964). The occurrence of bilaterally synchronous spike and wave activity under these conditions has usually been attributed to some irritative effect of these lesions. Clinical observations, although relatively few, are in line with these findings in animals. Subcortical tumors are sometimes associated with epileptic seizures and with bilaterally synchronous spike and wave discharge. The incidence of both epilepsy and of this EEG pattern in subcortical brain tumors, however, is low (AJMONE MARSAN and LEWIS 1960, STEVENS 1970). Not so low, however, is the incidence of bilaterally synchronous 3/sec rhythmic slow activity with frontal predominance in space occupying lesions of subcortical location, especially those deforming the walls of the third ventricle, as DALY et al. (1953) have shown. This pattern has obvious similarities with the slow wave component of the spike and wave pattern and it is well known that in many patients who exhibit this seizure pattern, the EEG frequently also shows bilaterally synchronous rhythmic slow waves which are hard if not impossible to distinguish from the kind described by DALY et al. (1953).

I would like, therefore, tentatively to put forward the conclusion that the contribution of subcortical structures to the mechanism of bilaterally synchronous spike and wave activity is different from that envisaged in the classical centrencephalic hypothesis. Lesions in systems providing the cortex with a desynchronizing drive associated with arousal, may create in both hemispheres by some release mechanism a peculiar diffuse excitatory state which renders the cortex prone to respond with rapid dissemination of epileptic discharge of the spike and wave type. The fact that injection of a convulsant drug through the vertebral route where it irrigates and presumably stimulates the activating system in the reticular formation is associated with diminution or arrest of spike and wave discharge both in patients and in the animal model would be in line with this interpretation, especially in the light of the observation that unilateral intracarotid injection of the same drug easily activates bilaterally synchronous spike and wave discharge.

What then are the cortical mechanisms of synchronization and maintenance of the spike and wave pattern? The possible importance of callosal fibres in synchronizing the activity between the two hemispheres has already been alluded to. Synchronization within one hemisphere may perhaps depend upon cortico-cortical connections of various lengths, starting with the U-fibre system and ending with the long associational tracts. There are many similarities between these long associational tracts and callosal fibres and, although the former have been little studied from the physiological point of view, the hypothesis is permissible that their functional characteristics may be similar.

Dr. RENAUD has permitted me to quote some of his unpublished observations on the properties of callosal connections and the responses they evoke in cortical neurons. Volleys mediated by callosal fibres seem to have a great propensity in evoking clearcut excitatory-inhibitory sequences in cortical neurons. The inhibitory period is usually followed by excitatory rebound (Fig. 7). Clearly therefore, callosal connections could provide a system capable of producing excitatory-inhibitory discharge cycles which could correspond to the underlying synaptic sequences of the spike and wave complex demonstrated by POLLEN (1964). This callosal system is much more power-

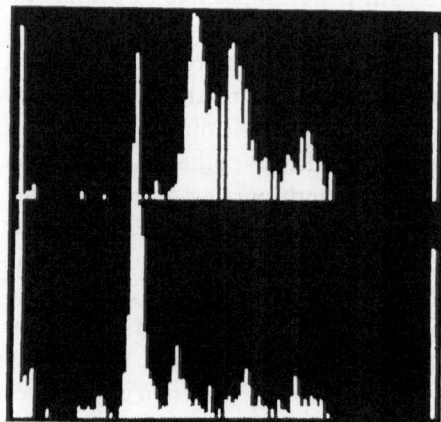

Fig. 7. Poststimulus time histograms of two nonpyramidal tract cells in the anterior sigmoid cortex of the cat. Transcallosal single shock stimuli: 20 volts. Upper graph: cell at a depth of 1.1 mm; lower graph: cell at a depth of 0.7 mm. Sweep: 500 msec. Calibration bar on the right edge of figure: 100 counts. (Courtesy of Dr. L. RENAUD; unpublished data)

ful in eliciting such excitatory-inhibitory sequences than the recurrent inhibitory pathway elicited by antidromic pyramidal stimulation; however similar sequences can be elicited by thalamo-cortical volleys (RENAUD unpublished data). It is thus conceivable that the appropriate conditions for evoking widespread bilaterally synchronous spike and wave discharge are 1. a reduced activatory drive from the reticular formation due to some lesions or a biochemical genetically determined defect and 2. some associated biochemical and/or structural defect in the cortex. Under these conditions callosal excitatory-inhibitory systems may become greatly facilitated and this may lead to a rapid dissemination of spike and wave discharge wherever it originates in the cortex. The hypothetical biochemical factor at the cortical level may be related to a specific transmitter involved in the excitatory-inhibitory sequences of the cortical spike and wave pattern. The self-sustaining character of the discharge may be related to mechanisms similar to the recurrent excitatory-inhibitory drive described for thalamic neurons and other systems of this kind (ANDERSEN and ECCLES 1962). Since thalamic impulses to the cortex also show a tendency to initiate excitatory-inhibitory sequences, it is not entirely unexpected to find that by suitable 3/sec stimulation of the intra-laminar thalamus one may be able to reproduce in animals the sequence of 3/sec spike and wave activity that is so characteristic for the human form of epilepsy characterized by this pattern. Whether in man the callosal or the thalamic mechanism or both may be operative still remains therefore an open question. Our observation with intra-carotid and intravertebral metrazol injections however seem to favor the callosal rather than the thalamic mechanism of spread and syn-chronization of spike and wave discharge.

The tentative explanation given in this paper for the mechanism of bilaterally synchronous spike and wave discharge is, of course, highly hypothetical. Some of the proposed mechanisms however are within the realm of experimental verification. The hypothesis at least does not suffer from the major defects of earlier explanations, namely from the necessity to reject some of the experimentally and clinically available evidence on the mechanism of spike and wave discharge because it could not be reconciled with the tenets of a particular hypothesis. This was true both for the centrencephalic and the cortical hypotheses. In line with the above hypothesis I have pro-posed the term "Generalized Cortico-Reticular Epilepsies" for seizures associated with bilaterally synchronous generalized spike and wave discharge (GLOOR 1968, 1969).

## Summary

The following hypothetical mechanism for the genesis of generalized bilaterally synchronous discharge is proposed:

The spike and wave pattern represents a widely phase-locked alter-nation of excitatory and inhibitory sequences involving cortical neurons.

The anatomical systems responsible for the dissemination and syn-chronization of the paroxysmal discharge are callosal fibres and intra-hemispheric cortical association pathways.

Subcortical structures are important insofar as removal of the activatory drive originating in the ascending reticular formation seems to be an important prerequisite for sensitizing these commissural and associational systems to respond in this rhythmic, widely disseminated manner. This in itself is, however, not sufficient to produce conditions under which this discharge can arise spontaneously. These conditions however are created either by diffuse grey matter lesions which are known to enhance the general tendency to paroxysmal discharge, or in cases in which no demonstrable brain damage exists an as yet unknown biochemical factor may be involved.

# References

AJMONE MARSAN, C., and W. R. LEWIS: Pathological findings in patients with "centrencephalic" electroencephalographic patterns. Neurol. *10*, 922—930 (1960).

ANDERSEN, P., and J. C. ECCLES: Inhibitory phasing of neuronal discharge. Nature (London) *196*, 645—647 (1962).

DALY, D., J. L. WHELAN, R. G. BICKFORD, and C. S. MACCARTY: The electroencephalogram in cases of tumors of the posterior fossa and third ventricle. Electroenceph. clin. Neurophysiol. *5*, 203—216 (1953).

GARRETSON, H., P. GLOOR, and T. RASMUSSEN: Intracarotid amobarbital and metrazol test for the study of epileptiform discharges in man: a note on its technique. Electroenceph. clin. Neurophysiol. *21*, 607—610 (1966).

GIBBS, F. A., and E. L. GIBBS: Atlas of Electroencephalography, Vol. 2. Cambridge, Mass.: Addison-Wesley Press Inc. 1952.

GLOOR, P.: Generalized cortico-reticular epilepsies. Some considerations on the pathophysiology of generalized bilaterally synchronous spike and wave discharge. Epilepsia *9*, 249—263 (1968).

— Neurophysiological bases of generalized seizures termed centrencephalic. In: GASTAUT, H., H. JASPER, J. BANCAUD, and A. WALTREGNY (eds.), The Physiopathogenesis of the Epilepsies, chapter 18, pp. 209—236. Springfield, Ill.: Charles C. Thomas. 1969.

— O. KALABAY, and N. GIARD: The electroencephalogram in diffuse encephalopathies: electroencephalographic correlates of grey and white matter lesions. Brain *91*, 779—802 (1968).

— T. RASMUSSEN, H. GARRETSON, and F. MAROUN: Fractionized intracarotid Metrazol injection. A new diagnostic method in electroencephalography. Electroenceph. clin. Neurophysiol. *17*, 322—327 (1964).

GUERRERO-FIGUEROA, R., A. BARROS, F. DE BALBIAN VERSTER, and R. G. HEATH: Experimental "Petit Mal" in kittens. Arch. Neurol. *9*, 297—306 (1963).

HAYNE, R. A., L. BELINSON, and F. A. GIBBS: Electrical activity of subcortical areas in epilepsy. Electroenceph. clin. Neurophysiol. *1*, 437—445 (1949).

HUNTER, J., and H. H. JASPER: Effects of thalamic stimulation in unanaesthetized animals. Electroenceph. clin. Neurophysiol. *1*, 305—324 (1949).

JASPER, H. H., and J. DROOGLEEVER-FORTUYN: Experimental studies on the functional anatomy of petit mal epilepsy. Res. Publ. Ass. Nerv. ment. Dis. *26*, 272—298 (1947).

Jung, R.: Correlation of bioelectrical and autonomic phenomena with alterations of consciousness and arousal in man. Brain Mechanisms and Consciousness, pp. 310—344. Oxford: Blackwell. 1954.

Kaplan, H., and D. H. Ford: The brain vascular system. Amsterdam: Elsevier. 1966.

Lennox, W. G., E. L. Gibbs, and F. A. Gibbs: Effect on the electroencephalogram of drugs and conditions which influence seizures. Arch. Neurol. Psychiat. 36, 1236—1245 (1936).

Li, C. L., H. Jasper, and L. Henderson: The effect of arousal mechanisms on various forms of abnormality in the electroencephalogram. Electroenceph. clin. Neurophysiol. 4, 512—526 (1952).

Marcus, E. M., and C. W. Watson: Bilateral synchronous spike wave electrographic patterns in the cat. (Interaction of bilateral cortical foci in the intact, the bilateral cortico-callosal and adiencephalic preparation). Arch. Neurol. 14, 601—610 (1966).

— — Symmetrical epileptogenic foci in monkey cerebral cortex. Arch. Neurol. 19, 99—116 (1968).

— — and S. A. Simon: An experimental model of some varieties of petit mal epilepsy. Electrical-behavioral correlations of acute bilateral epileptogenic foci in cerebral cortex. Epilepsia 9, 233—248 (1968 a).

— — — Behavioral correlates of acute bilateral symmetrical epileptogenic foci in monkey cerebral cortex. Brain Res. 9, 370—373 (1968 b).

Penfield, W., and H. Jasper: Epilepsy and the functional anatomy of the human brain. Boston: Little, Brown & Co. 1954.

Perot, P.: Mesencephalic-thalamic relations in wave and spike mechanisms, pp. 220. Montreal: Ph. D. Thesis, McGill University. 1963.

Petsche, H.: Pathophysiologie und Klinik des Petit Mals. Toposkopische Untersuchungen zur Phänomenologie des Spike-Wave-Musters. Wien. Z. Nervenh. Grenzgeb. 19, 345—442 (1962).

Pollen, D. A.: Intracellular studies of cortical neurons during thalamic induced wave and spike. Electroenceph. clin. Neurophysiol. 17, 398—404 (1964).

— P. Perot, and K. H. Reid: Experimental bilateral wave and spike from thalamic stimulation in relation to level of arousal. Electroenceph. clin. Neurophysiol. 15, 1017—1028 (1963).

— K. H. Reid, and P. Perot: Microelectrode studies of experimental 3/sec wave and spike in the cat. Electroenceph. clin. Neurophysiol. 17, 57—67 (1964).

— and P. G. Sie: Analysis of thalamic induced wave and spike by modification in cortical excitability. Electroenceph. clin. Neurophysiol. 17, 154—163 (1964).

Prince, D., and D. Farrell: "Centrencephalic" spike-wave discharges following parenteral penicillin injection in the cat. Neurol. 19, 309—310 (1969). (Abstract.)

Rovit, R., P. Gloor, and T. Rasmussen: Intracarotid amobarbital in epilepsy. Arch. Neurol. 5, 606—626 (1961).

Stevens, J. R.: Focal abnormality in petit mal epilepsy. Neurol. 20, 1069—1076 (1970).

Weir, B.: Spike-wave from stimulation of reticular core. Arch. Neurol. 11, 209—218 (1964).

WEIR, B.: The morphology of the spike-wave complex. Electroenceph. clin. Neurophysiol. *19*, 284—290 (1965).
— and P. G. SIE: Extracellular unit activity in cat cortex during the spike-wave complex. Epilepsia *7*, 30—43 (1966).

# Discussion

VOLANSCHI: I should like to ask Dr. GLOOR two questions:

1. Did you notice any difference in the responsiveness to intracarotid Metrazol between patients with 3 c/s spike-and-wave discharges and those with spike-and-wave-variants? I ask this because the two groups differ from the clinical, therapeutic and prognostic point of view and the difference also suggests diverse pathophysiologic mechanisms.

2. What do you think are the pathophysiologic mechanisms by which petit mal seizures turn into grand-mal ones in some of the patients you studied?

GLOOR: It is difficult to make a very close correlation between the EEG and the clinical aspects in this regard. The limit between "classical" and "variant" spike and wave discharge cannot, in my opinion, be defined very sharply. Epilepsies with spike and wave discharge form a spectrum, the "classical" and the "variant" forms are at the two ends of the spectrum, but there are all sorts of transitional cases which cannot be classified by using rigid criteria. I do not think that they can be differentiated by their reaction to intracarotid Metrazol. The chance of finding somewhat "atypical" cases, not responding rigidly to our criteria of "primary bilateral synchrony" is probably somewhat higher in the "variant" cases, but I would like to emphasize that in our population of patients subjected to the intracarotid drug test, we are dealing with a very selected group of epileptics. This must be obvious, since there is little justification to carry out these tests in patients with absolutely typical absence seizures associated with classical 3/sec spike and wave activity.

As to the second question regarding the mechanisms of transition to grand mal seizures, I can only repeat what I have said, that some patients with primary bilateral synchrony respond with prolonged spike and wave activity in response to intracarotid metrazol; they seem somewhat resistant to the development of grand mal attacks. This certainly sets this group somewhat apart from the others, but beyond this we have no precise information with regard to the matter you raised in your question.

PETSCHE: Much trouble has come from the fact that two quite different concepts, namely the concept of spike-and-wave pattern and the concept of absence, have been confused. These two concepts have to be clearly separated. One cannot foresee, from the clinical point of view and without any knowledge of the EEG, whether a patient with absences will present typical spike-and-wave patterns or, for instance, a temporal focus. Spike-and-wave patterns seem to be quite primitive reactions of the cortex to some kind of stimuli. Otherwise it wouldn't be possible for them to occur in such primitive brains as those of frogs and lizards, as SERVIT has observed, brains the morphology of which is totally different from mammalian brains. This primitive pattern may be, under certain and still unknown circumstances, connected to the clinical syndrome of absence.

GLOOR: I am in partial agreement with you on this. The correlation between absence attacks and the spike and wave patterns is far from perfect, but on the other hand let's not overlook the obvious fact that the chances are extremely high that a patient who presents with spike and wave discharge in his EEG also presents absence attacks, although he may in addition have other seizures. Some difficulties arise from the definition of what constitutes an absence attack. If the definition is sufficiently rigid the number of typical absence seizures shrinks, but conversely the correlation with spike and wave discharge seems to get better.

The place where the discharges start is probably in the cortex. They may become rapidly synchronized. Some particular cortical excitatory state is probably necessary to make the generalization possible and this may well depend upon brainstem mechanisms.

SPECKMANN: My question concerns the two different types of application of metrazol. Did you see any differences in the reactions of blood pressure, cerebral blood flow or $CO_2$?

GLOOR: In patients, opportunities to accurately measure all these values are of course very limited. No significant changes in respiratory rate and cardiac rate were observed with intracarotid Metrazol and Amytal injections.

BRAZIER: We electroencephalographers have to shake off this 35-year old habit of regarding the EEG in petit mal epilepsy as a phenomenon. It is only an epiphenomenon. Your clinical observation of the *"absence"* may well be a stronger correlate of the real phenomenon than spike-and-waves in the EEG. The latter pattern is determined by the physical properties of nerve cells and, as clinical electroencephalographers know, cells can be provoked to behave in this way in disorders other than petit mal epilepsy.

GLOOR: I can only repeat what I said in my answer to Dr. PETSCHE. The correlation between spike and wave pattern and absence attacks is not perfect, but neither is it insignificant. I would not go back all the way to 1936.

# Interactions of Deep Structures during Seizures in Man

MARY A. B. BRAZIER [1]

Brain Research Institute, School of Medicine,
University of California, Los Angeles, California

The question raised at this symposium relates to the mechanisms by which electrical seizure discharge spreads within the brain and to the question whether any of this spread merits the name of "synchronization" in the absolute meaning of the word. Exact synchrony would demand either that the underlying neuronal mechanism be by volume conduction or that a single deeply situated trigger zone could activate many regions of the cortex, even in the two hemispheres, through fibre tracts and synaptic relays having precisely the same rate of conduction.

The studies reported here were not designed to examine volume conduction in brain tissue, although, as will be seen, the results have some bearing on this problem; they were designed to seek data as to whether the apparent synchrony of discharge in epileptic seizures was the result of excessive activity in known anatomical connections or whether there were synaptic breakthroughs in pathways not normally used.

This research paradigm imposes recognition of the established anatomical connections within the human brain and the establishment of a means of detecting the functional use of these connections in the interictal state for comparison with the electrical seizure.

To establish the normal usage it seemed important to avoid the extremely unphysiological technique of stimulating electrically in one site and looking for a response or an afterdischarge in another, for

1 The work of this investigator is supported by Career Award #5-K-6 NS 18,608 and Grant NS 09774 from the National Institutes of Health of the U.S. Public Health Service.

electrical stimulation has the effect of synchronizing all discharge in the stimulus site, thus impinging a severely abnormally concentrated barrage of impulses at the next synapses, a barrage condensed in temporal terms and probably excessive in threshold. Such a barrage may open trans-synaptic connections not normally traversed, and give false evidence for the relationships operating in the interictal state.

To this end the least unphysiological technique available has been adopted, namely that of following the on-going wave-activity from one locus in the brain to another as recorded by long-indwelling electrodes in the unanaesthetized brain. This is a method very similar in concept, though different in technique, to that which PETSCHE (1960, 1962, 1965) originated in his studies of lower animals. PETSCHE's initial observation (1960) was that electrodes implanted in a number of regions of the rabbit's brain all showed the same monorhythmic theta activity though not exactly synchronously. There was always some detectable phase-difference when the wave-trains were compared. Later he and his colleagues were able to establish time-differences, and found that the phase of the septal waves always led the thalamus, hippocampus and hypothalamus; therefore the septum appeared to be a pacemaker for the theta activity, a conclusion for which a great deal of additional evidence was collected.

The work reported here undertook to explore whether similar occurrences of wave-trains exist in deep structures of man's brain and whether they exhibit the locked phase-relations found in the rabbit. Such analyses of necessity require the help of a computer and, for this purpose, use has been made of a programme [2] for determining coherence and phase.

The calculation of coherence detects the presence of frequencies common to wave-trains in any two recording sites and, for each frequency band chosen, calculates the percentage activity common to the two sites. The coherence function is not dependent on the amplitude of the wave-trains, only their frequency. The programme also gives, again for each frequency band, the phase-lag existing between them and, from this phase-displacement, can be derived the time-displacement in the pathway in milliseconds.

2 Program X92 available from the Health Sciences Computing Facility in the School of Medicine at the University of California Los Angeles. This Facility is supported by Grant FR-3 from the National Institutes of Health of the U.S. Public Health Service.

*Subjects*

All the results reported here have been obtained from human subjects. These are mostly patients with temporal lobe epilepsy who are being studied under a joint clinical and neurophysiological research programme [3] at the Brain Research Institute of the University of California Los Angeles, in which Dr. PAUL CRANDALL of the Division of Neurosurgery and Dr. RICHARD WALTER of the Department of Neurology are the leading figures.

The patients in this series, who now number over 50, are cases of epilepsy whose seizures have proved uncontrollable by medication and whose electroencephalograms, as recorded from the scalp or from sphenoidal leads, give insufficiently clear lateralizing signs, either when they are awake, or asleep, or with any of the usual activating techniques used in clinical electroencephalography. These patients are, therefore, candidates for therapeutic surgery by Dr. CRANDALL. In the cases of temporal lobe epilepsy, who form the majority of cases studied so far, lateralization of the more impaired hemisphere is of importance, bilateral resection being ruled out owing to the severe impairment of memory that ensues (SCOVILLE and MILNER 1957).

The series also includes some cases of epilepsia partialis continua and three cases of non-epileptic chronic psychosis in whom indwelling electrodes had been placed in the course of a research project in the Department of Psychiatry. These, as non-epileptic brains, form the only available "controls" that can be offered for the major series of epileptic patients.

*Electrode Placements*

The patients, by their own consent and by that of their next of kin, have electrodes inserted through twist-drill holes into various structures as indicated by the clinical signs and the need for refinement of the diagnostic information. The electrodes are implanted according to stereotactic coordinates determined from internal landmarks viewed by X-ray using contrast media (CRANDALL 1963).

Electrode position is checked by X-ray for both hemispheres and later by histology in the removed lobe. The close agreement of the histo-

---

3 This programme is supported by Grant NS 02808 from the National Institutes of Health of the U.S. Public Health Service.

logical check with the previous X-ray has led to considerable confidence in the placements in the unoperated hemisphere and in those cases which do not come to operation including the psychotic "controls". Choice of sites varies according to the patient's symptoms. In the largest series, the temporal lobe epileptics, in whom lateralization and possible localization is sought, the usual placements include 3 bipolar electrodes inserted into the hippocampus of each hemisphere; three in each hippocampal gyrus; and at least one pair in each amygdala (usually in the basal nucleus). In those patients whose symptoms include some mid-line thalamic or brain stem signs, electrodes have also been placed in the centre médian.

Some patients, whose clinical signs so indicate, have had electrodes placed in the dorsal medial nucleus of the thalamus and others in the anterior thalamic nucleus and the cingulate gyrus. In every instance the placements have been determined by the clinical problem presented by the case. In addition to the deep electrodes, all patients have a minimum of 12 cortical electrodes inserted through the skull at operation so that they just reach the dura.

The important feature of these deep implantations is that the electrodes are left in place for 4 to 6 weeks so that recordings may be made during many varying states of the patient's behaviour: waking, sleeping, actively engaged in various situations including interviews, and during different forms of sensory stimulation: photic, auditory or somato-sensory, as well as under the influence of various drugs, and during their spontaneous seizures. It is, of course, the comparison of this last with the waking state that gives us the information sought in the context of the present report.

## Results

Because of the nature of the clinical material available (as described above) the majority of the data concerns the relationships of limbic system structures. The following are some of the relationships found.

### Amygdalo-Thalamic Relationships

Fibre tracts running between the amygdaloid complex and the magnocellular part of the dorsal medial nucleus of the thalamus have been known in monkey since the work of Fox in 1949 and have been followed in detail by NAUTA and VALENSTEIN (1958). But more important for this work in man are the anatomical studies of the human brain by KLINGLER and GLOOR (1960) who identified a fibre tract passing from the amygdala to the medial thalamus.

The coherence has therefore been examined, in the few available

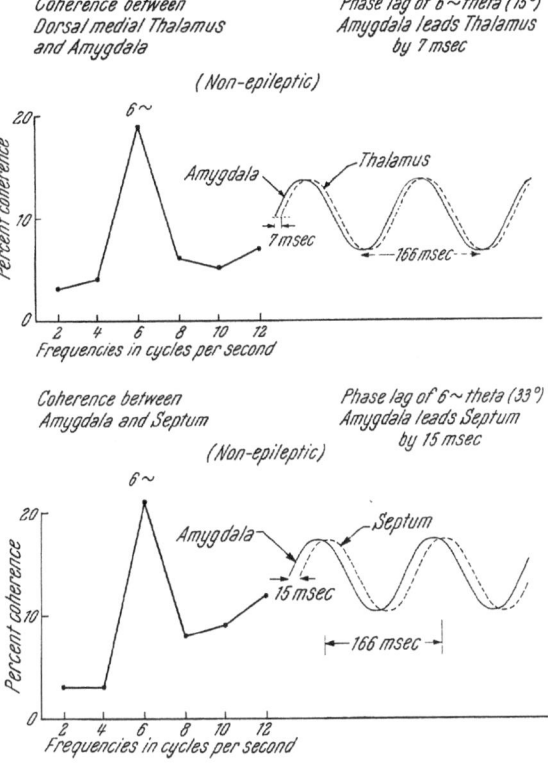

Fig. 1. *Upper Section*: Coherences averaged over fifteen 30-second epochs for each of the frequency bands from 2 to 12 cps recorded simultaneously from the amygdala and dorsal medial nucleus in a patient fully awake. The schematic diagram on the right shows a sinusoidal model that would provide similar phase characteristics. *Lower Section*: Coherences between amygdala and septum illustrated in the same way as above

cases, of the apparently simultaneous on-going activity in the amygdala and dorsal medial nucleus of man. The upper section of Fig. 1 illustrates a result from one of these non-epileptic brains. The only frequency band in which a prominent coherence was found was that centering on 6 per second which was the dominant frequency of the amygdala in this patient. (Indeed, in all patients in this series the dominant frequency of the amygdala was either 4 or 6 cps.) At this frequency there was a consistent phase lead by the amygdala averaging 15° which, for a wave 166 msec in duration, is equivalent to a time-difference between trains of 7 msec.

## Amygdalo-Septal Relationships

Another strong anatomical connection that has been identified in man is a fibre tract running from the amygdala to the septum through the diagonal band of Broca and the stria terminalis.

When speaking of the "septum" one must realize that some workers use this term for a very large region in the human brain, but the intensive studies of ANDY and STEPHAN (1968) indicate that the "septum verum", i.e., the true septum in man (as distinguished from the septum pellucidum) is restricted to a small zone. The septum pellucidum in man is proposed by KLINGLER and GLOOR (1960) as the human homologue of the lateral septal nuclei in lower animals. ANDY and STEPHAN restrict this term to the dorsal part which contains fibre tracts and glia, but no cell bodies.

Opportunity for obtaining septal recordings in human cases has come only from the non-epileptic psychotic patients and as the number of these observations is necessarily few, the results must be considered as preliminary. In the patient whose results are shown in the lower section of Fig. 1, the dominant frequency in both amygdala and septum was found to be 6 cycles per second, and at this frequency (and at this frequency only) there was a consistent coherence in all fifteen 30-second epochs analyzed, representing 7½ minutes of recording in real time. In every 30 second sample of record, the amygdaloid waves led the septal waves. The average phase angle was 33° which, for a 6 cps wave with duration of 166 msec is equivalent to a lead-time of 15 milliseconds.

From this extremely consistent result it seems reasonable to infer that a pacemaking influence for this theta rhythm may be exerted on the septal activity from the amygdala. The known anatomical connections running in this direction are presumed to be the structural substrate of this electrophysiological phenomenon in the living human brain.

This result was, however, in some ways (as is so often the case in work on man) not entirely satisfying, for these non-epileptic patients do not come to operation and therefore exact histological localization of the electrode sites is lacking and reliance has to be placed on the X-ray.

## Amygdalo-Hippocampal Relationships

In terms of temporal lobe epilepsy, the relationships of the amygdala with the hippocampus are of obvious interest. One has to be, however, careful in making comparisons with the findings in lower mammals for those have largely been derived from the dorsal hippocampus which has so scant an analogous structure in the human brain —just the thin strip adjacent to the corpus callosum, a hippocampal rudiment, the induseum griseum.

For hippocampal connections running to the amygdala there is little anatomical evidence but several categories of electrophysiological observations imply such connections e.g.: responses evoked in the amygdala by stimulation of the hippocampus (GREEN and ADEY 1956) in cats; BRAZIER (1964) in man and afterdischarges evoked in the same way (ANDY and KOSHINO 1967). Both these techniques, however, use the artificiality of the electrical stimulus and may by forcing activity through normally unused pathways.

In the opposite direction there is more anatomical evidence for amygdalo-hippocampal connections though some controversy still exists. Any connection that does exist must surely be polysynaptic in nature, possibly via the pyriform lobe or the long circuit through the septum. With the goal of looking for such evidence by the technique of coherence this form of analysis has been made in the patients in this series.

The coherences between some regions of the hippocampus (but not all) and the ipsilateral amygdala are marked in almost all patients, but the phase-displacement of the waves is far less consistent and very large. In previously published work (BRAZIER 1970), the contrasts in electrogenesis of the different regions of the human hippocampus have been described. Zones along the length of the hippocampus have clearly different characteristics in terms of their electrical properties, and it is therefore not surprising that only in a restricted region does the wave-activity cohere with that of the amygdala.

Fig. 2 illustrates the point that the high coherence with the amygdala is with certain sections of this patient's hippocampus but not all.

416    MARY A. B. BRAZIER

Only the linkage $R_4 R_5$ (as labelled in the sketch) gave wave-activity which cohered with that present in simultaneous recordings from the basal nucleus of the ipsilateral amygdala. The average phase-displacement in 10 samples was 28°, equivalent to 13 msec for the 6 ~ waves, the amygdala leading the hippocampus.

*Septal-Hippocampal Relations*

Here we drew a negative result in terms of finding any suggestive pacemaking influence. Coherences were high but no consistently

Coherence between amygdala and various sites in the hippocampus in man (alert) (averages of 10 samples)

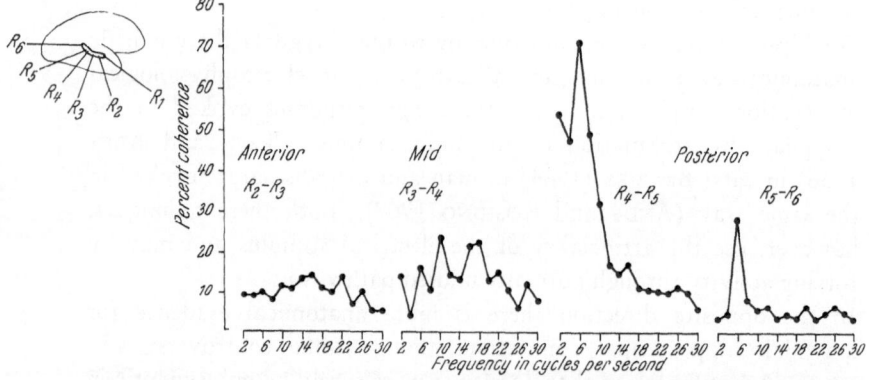

Fig. 2. Six unipolar indwelling electrodes along the length of the hippocampus (as shown in the sketch of the brain) were recorded from in bipolar pairs simultaneously with recordings of the on-going activity in the ipsilateral basal nucleus of the amygdala. Significant coherences with the amygdala were found only in the theta frequencies and only from the electrode pair $R_4$, $R_5$

constant phase-relation was found, such as is present in lower animals. Probably the reason for this is the marked species difference in the anatomy of the hippocampus (as already mentioned) so that the electrodes were not in tissue analogous to the dorsal hippocampus of rabbit or cat.

*Hippocampus—Anterior Thalamic Nucleus—Cingulate Gyrus*

The anatomical pathways between these structures are well-known. Fibres originating in the hippocampus pass in the fornix to terminate in the mammillary bodies and synapse there with the neurones of the mammillo-thalamic tract projecting to the anterior nucleus of the thalamus which has two-way connections with the anterior cingulate gyrus on the mesial surface of the brain. In lower animals, other fibres from neurones of CA 1 and CA 2 of the hippocampus pass directly to the anterior nucleus (RAISMAN 1966). However, this latter connection has not been demonstrated in man.

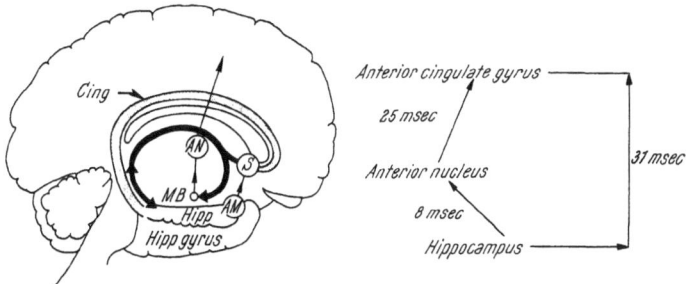

Fig. 3. In this patient coherences were found between the mid hippocampus, the anterior nucleus of the thalamus and the anterior cingulate gyrus. The direction of leading phase is shown by the arrows in the diagram together with the equivalents in terms of time-displacement

There has been no justification for placing electrodes in the mammillary bodies of these patients but, in a few, they have been implanted in the other stations of this chain. Fig. 3 is an example from a patient in whom simultaneous recordings were made from two regions of the hippocampus, from the anterior nucleus of the thalamus and two sites (anterior and posterior) in the cingulate gyrus, all in the same hemisphere. This patient had a dominant theta rhythm of 4 cps in these structures. Significant coherences in this frequency band were found between the following pairs of recordings: mid hippocampus and anterior nucleus; anterior nucleus and anterior cingulate gyrus; mid hippocampus and anterior cingulate gyrus. No significant coherence was found between the anterior hippocampus and either the anterior thalamic nucleus or the cingulate gyrus; the posterior cingulate gyrus had no significant correlation with any other site from which activity was simultaneously recorded (including the anterior cingulate gyrus).

The average phase displacements are entered in Fig. 3 in terms of milliseconds between wave-trains. The fact that the time-lag of the activity in the anterior cingulate gyrus behind that of the hippocampus equals the sum of the two intermediate steps gives an indirect

Coherence between hippocampus and gyrus in waking and sleeping states in man

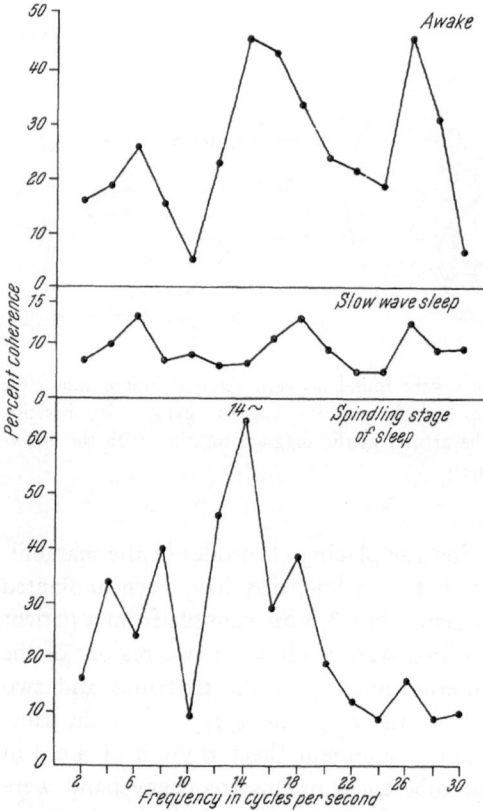

Fig. 4. Graphs of coherence in recordings from a patient when at three levels of awareness illustrating the great differences found in limbic system relationships in these contrasting behavioural states

confirmation of these figures but gives no suggestion of a more direct route from the hippocampus. There is also no evidence from this technique of traffic flow in the opposite direction.

## Hippocampal Gyrus-Hippocampus

Some of the strongest coherences so far found in the human brain are between the hippocampal gyrus and the ipsilateral hippocampus (BRAZIER 1968 a, b). This is hardly surprising for the anatomists have identified (in lower animals) the inflow into the hippocampus from its gyrus (LORENTE DE NÓ 1934, BLACKSTAD 1958). According to the test used in this research, some (but not all) regions of the hippocampal gyrus were found to cohere well with certain zones in the ipsilateral hippocampus, the relationships being stronger posteriorly.

Fig. 4 introduces another characteristic of the coherences between limbic structures that has been published previously (BRAZIER 1967 a, 1968 c) namely the change in these relationships when the subject falls asleep. Coherences fall into insignificance as though all relationships are lost. There is a striking exception to this, however in the spindling stage of sleep when strong coherences develop in the frequency bands of which the spindles consist. In the particular patient whose graphs are shown in Fig. 4, the average phase-displacement between the bands centering on 14 cycles was $24°$, an approximately 5 msec time-difference with the gyrus leading.

## Thalamo-Cortical Relationships

In this work on man we have no data to report on the relationships of specific thalamic nuclei with their specific projection cortices, such as the lateral geniculate and the striate cortex, for there has been no clinical justification for such placements in these patients. A few have, however, had electrodes placed in the dorsal medial nucleus. The dominant frequency, unlike that in the limbic structures is usually 10 cps both in the thalamus and in leads from the fronto-central cortex and, at this frequency a 60 per cent coherence has been found with a fairly consistent phase-difference of the cortex lagging behind the thalamus by $48°$ on the average. This represents a time-difference of 13 msec between wave-trains. These coherences are lost as soon as the subject falls asleep. Only in the spindling stage are coherences once again high and then only in the "spindle" frequencies of 12–16 cps.

## Anaesthesia

Returning to the original goal of using this kind of test to explore relationships between brain structures in various states of the organism, coherences have been examined in anaesthesia for comparison with natural sleep. In contrast to the tradition of regarding loss of consciousness in natural sleep as having similarities with loss of consciousness in barbiturate anaesthesia, evidence for markedly different mechanisms emerges. Coherences in barbiturate anaesthesia increase not only during the prenarcotic stage when fast activity is

Contrasting effects of slow wave sleep and thiopental anesthesia on coherence between hippocampus and hippocampal gyrus in man

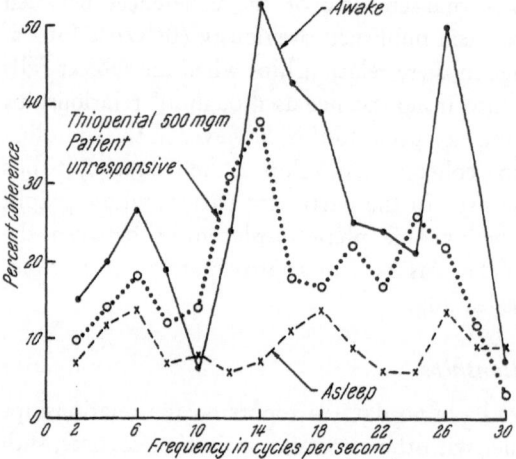

Fig. 5. Curves of coherences at frequencies from 2 to 30 cps found in ipsilateral limbic structures in three contrasting states in the same patient: full awareness, natural sleep and anaesthesia

present in the EEG, but remain high even when the patient loses conscious awareness and becomes unresponsive (see Fig. 5). Some reports, on this phenomenon have been published previously (BRAZIER 1967 c, 1969) so the subject will not be elaborated further here though in the context of this symposium, it should be remarked that barbiturates are great synchronizers, a property that should be remembered in experiments under this anaesthetic.

*Epilepsy*

As is well-known to electroencephalographers who have recorded from deep structures in patients with epilepsy, severe electrical seizure

activity may occur in depth without overt clinical signs and without
the discharges reaching the cortex. Hence the disturbance goes un-
observed if only scalp recordings are available.

During seizures strikingly high coherences develop which are not
present in the interictal state. They develop between deep structures

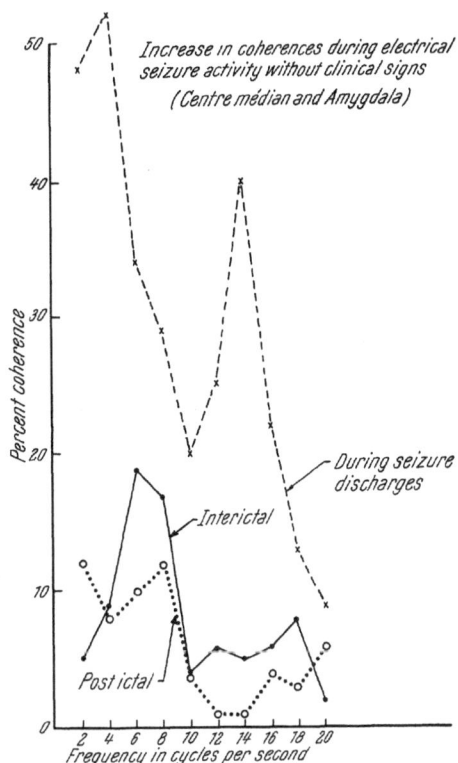

Fig. 6. Averaged values of coherences in frequency bands
from 2 to 20 cps during interictal, ictal and postictal states
in a patient with temporal lobe
epilepsy. The involvement of
the mid-line thalamus by the
seizure discharges initiated in
the limbic system led to a
spread of this activity to the
frontal cortex, sparing the
more posterior surface leads

not normally related and between limbic structures and scalp
(Brazier 1969); also transhemispheric coherences may emerge that
are not normally present.

An example of a case of abnormal spread is shown in Fig. 6 which
illustrates this change in the case of the relationship, during a spontaneous seizure, between the on-going activity of the amygdala and
the centre médian. Other coherences that showed this increase during
this same episode were those of the centre médian with all ipsilateral
limbic sites including multiple sites in the hippocampus and hippo-

campal gyrus. During this seizure activity all relationships within the limbic system also developed abnormally high coherences (amygdala, hippocampus, hippocampal gyrus) but the only involvement with the cortex was in the most frontal leads. The temporal and occipital cortices were not affected.

The suggestion is made that unless neurones of the thalamus are drawn into play by excessive discharge originating in the limbic structures, there will be no spread to the neocortex of the convexity and therefore no possibility of detecting the ictal event by scalp electroencephalography in cases of epilepsy arising in the depths of the temporal lobe.

In this admittedly small series in which data have been gained during seizures, the thalamic sites usually involved have been in the centre médian and hence the scalp signs have appeared in the surface leads on the anterior part of the head.

## Discussion

If the observations of coherences reported here for the normal waking state are to be considered as representing the influence of one neuronal aggregate on another, the neurophysiologist has to suggest the mechanism. For example, does the waxing and waning of activity in the amygdala pace the emission of impulses in the amygdaloseptal tract and do they, on arrival at the septum, initiate excitatory post-synaptic potentials that follow their timing and sum to produce, in their turn, the slow wave activity that gross electrodes record? One can hardly go further towards answering this question by work in man, but the results in lower animals, and especially those of Petsche's group, gives credence to this explanation.

In the context of the questions raised in this symposium one has to consider, in the case of the epileptic seizure discharge, three possible mechanisms for the abnormal spread:

1. that it consists of excessive traffic over normal but little used connections;

2. that there are synaptic break-throughs to pathways not used normally including possibly even antidromic conduction;

3. that there is an element of volume conduction that influences the threshold for discharge of neurones in the neighbourhood of a focus.

It is hoped that further work along the lines reported here may be able to distinguish in some sense between these three possibilities.

The criterion of a phase-difference is certainly useful in ruling out volume conduction (when no phase-lag is present), though all three mechanisms may, of course be operating at the same time. The results of the present studies have produced no evidence for massive spread of seizure activity by volume conduction.

## Summary

Studies have been made of the on-going wave-activity of various deep structures of the human brain in cases of temporal lobe epilepsy. A computer programme for analysing these wave-trains into their component frequencies, detecting those that are common to any pair of sites, and determining the phase-difference of their occurrence, has been used.

A comparison of the interictal results with those recorded during electrical seizure activity points to spread via normally unused pathways as well as excessive discharge in normal connections. No evidence for massive intracerebral spread by volume conduction was found. For epileptic discharges from the limbic system to reach the cortex involvement of the thalamus with its cortical projections seems mandatory. Only when the cortex is invaded can scalp electrodes (by volume conduction through the skull) give witness to these discharges.

## References

ANDY, O. J., and K. KOSHINO: Duration and frequency patterns of the after-discharge from septum and amygdala. Electroenceph. clin. Neurophysiol. 22, 167—173 (1967).
— and H. STEPHAN: The septum in the human brain. J. comp. Neurol. 133, 383—410 (1968).
BLACKSTAD, T.: On the termination of some afferents to the hippocampus and fascia dentata. Acta anat. (Basel) 35, 202—214 (1958).
BRAZIER, M. A. B.: Evoked responses recorded from the depths of the human brain. Ann. N. Y. Acad. Sci. 112, 33—59 (1964).
— Absence of dreaming or failure to recall? In: CLEMENTE, C. D. (ed.), Physiological Correlates of Dreaming. Exper. Neurol. Suppl. 4, 91—98 (1967 a).
— Thiopental effects on subcortical mechanisms in temporal lobe epilepsy. Anesthesiology 28, 192—200 (1967 b).
— Studies of the EEG activity of limbic structures in man. Electroenceph. clin. Neurophysiol. 25, 309—318 (1968 a).
— Étude électrophysiologique de l'hippocampe et du thalamus chez l'homme. Actualités Neurophysiologiques, Vol. 8, pp. 149—160. Paris: Masson. 1968 b.

Brazier, M. A. B.: Analysis of sleep activity as revealed by deep recording in man. In: Gastaut, H., E. Lugaresi, G. Berti Ceroni, and G. Coccagna (eds.), The Abnormalities of Sleep in Man. Proc. of the XVth European Meeting on Electroencephalography, Bologna 1968, pp. 35—43. Bologna: Aulo Gaggi. 1968 c.

— Prenarcotic doses of barbiturates as an aid in localizing diseased brain tissue. Anesthesiology 31, 78—83 (1969).

— Regional activities within the human hippocampus and hippocampal gyrus. Exper. Neurol. 26, 354—368 (1970).

Crandall, P. H.: Clinical applications of studies on stereotacticly implanted electrodes in temporal lobe epilepsy. J. Neurosurg. 20, 827—840 (1963).

Fox, C.: Amygdalo-thalamic connections in Macaca Mulatta. Anat. Rec. 103, 537 (1949).

Green, J. D., and W. R. Adey: Electrophysiological studies of hippocampal connections and excitability. Electroenceph. clin. Neurophysiol. 8, 245—262 (1956).

Klingler, J., and P. Gloor: The connections of the amygdala and of the anterior temporal cortex in the human brain. J. Comp. Neurol. 115, 333—355 (1960).

Lorente de Nó, R.: Studies on the structure of the cerebral cortex. II. Continuation of the study of the ammonic system. J. Psychol. Neurol. (Lpz) 46, 113—167 (1934).

Nauta, W. J. H., and E. S. Valenstein: Some projections of the amygdaloid complex in the monkey. Anat. Rec. 130, 346 (1958).

Petsche, H.: The quantitative analysis of EEG data. In: Schadé, J. P., and J. Smith (eds.), Computers and Brains. Progress in Brain Research, Vol. 33, pp. 63—86. Amsterdam: Elsevier. 1970.

— and Ch. Stumpf: Topographic and toposcopic study of origin and spread of the regular synchronized arousal pattern in the rabbit. Electroenceph. clin. Neurophysiol. 12, 589—600 (1960).

— — and G. Gogolák: Significance of the rabbit's septum as a relay station between the midbrain and the hippocampus. Electroenceph. clin. Neurophysiol. 14, 202—211 (1962).

— G. Gogolák, and P. A. van Zwieten: Rhythmicity of septal cell discharges at various levels of reticular excitation. Electroenceph. clin. Neurophysiol. 19, 25—33 (1965).

Raisman, G.: The connexions of the septum. Brain 89, 317—348 (1966).

Scoville, W. B., and B. Milner: Loss of recent memory after bilateral hippocampal lesions. J. Neurol. Neurosurg. Psychiat. 20, 11—21 (1957).

# Discussion

Ganglberger: The spreading of limbic seizure discharges is a great problem for us. Your findings may explain why stereotaxic fornicotomy combined with anterior commissurotomy does not result in a higher percentage of success.

While applying rather high frequency stimulation to amygdala we were able to produce a distant extinction phenomenon in the amygdaloid-hippocampal complex (explains slide). This distant extinction is probably elicited via prefrontal cortex (area 9, 10, 46), the projection of the dorsomedial nucleus.

In accordance with your observation of the thalamic breakthrough this can perhaps be explained by conduction along pathways from amygdala to the dorsomedial nucleus, which have been described for animals but have not yet been found in man.

GLOOR: What are the mechanisms which one could expect to produce this abnormal break-through at synapses? One may assume that at some synapses, incoming impulse traffic only produces subthreshold excitation which may modify the excitability level of the post-synaptic element. When the discharge rate of the afferent fibre system increases this may lead at some point to a break-through at the synapses. The other possibility would be to consider a highly hypo-thetical situation of "abnormal permeability" at a synapse with no increase in impulse traffic, a situation which I think is never identifi-able as such in an epileptic condition. Even if such an abnormal mechanism at a synapse would exist, I think one could never hope to demonstrate this in an epileptic discharge, because one always has exaggerated impulse traffic anyway in this case. I think it is very difficult to differentiate between two mechanisms. In my early studies on the projection of the amygdala, it was quite obvious that amygdaloid afterdischarges spread much farther in subcortical struc-tures than responses to single shock or to 10/sec amygdaloid stimu-lation. In the latter case, the projection remained localized to the basal diencephalon, hypothalamus and to the septal region. Amyg-daloid afterdischarge, however, spread to the entire diencephalon (in-cluding the thalamus). I could restrict it only in very deep anaesthe-sia. Obviously increased impulse traffic during the afterdischarge leads to "abnormal" break-through at many diencephalic synapses, but the two mechanisms, in this instance, were inseparably linked together.

LUX: Dr. PRINCE showed me intracellular records of thalamic nerve cells which are activated antidromically during cortical seizure activ-ity starting from a penicillin focus. With an elegant technique to study the occlusive effects of direct cortex stimuli which were trig-gered by the action potential of the thalamic neuron. GUTNICK and PRINCE (in preparation) were able to demonstrate that the thalamic reaction was not orthodromic. This study presents a first direct evidence of synaptic back-firing.

BRAZIER: This seems to be a question really for Dr. PETSCHE. It was he who showed, in lower animals, that although many regions in the brain had wave-trains of the same theta frequency, these were not exactly synchronous. The septum acted as a pacemaker and led the activity on the other loci. It was his work that suggested to me the exploration of coherence between wave-trains in deep structures of the human brain and an examination of their phase-lags.

PETSCHE: This leads us to ask how such coherent wave patterns may be propagated if there is only white matter between the two recording points. Because of the short time constant of nerve fibres, a direct propagation is unthinkable. Probably some transformation into a pattern of nerve pulses takes place. When we were studying (together with STUMPF and GOGOLÁK) the septum-hippocampus relationships, we found that the total nerve pulse output of the upper part of Broca's diagonal band nucleus in the septum has an envelope identical to the hippocampus theta wave but preceding it by some 20 msec. Our conclusion was that the theta waves may be composed of synaptic events elicited by pulses produced by a group of cells in the septum. At that time we considered this to be an extremely special case in nature. According to Dr. Brazier's findings, however, this pulse-to-wave-pattern transformation seems to be a rather common phenomenon.

VERZEANO: Did you study these processes with microelectrodes in the hippocampus itself?

PETSCHE: Yes, but we never found any evidence of theta waves being composed of hippocampal unit discharges.

SZENTÁGOTHAI: When you speak about "abnormal break-through", do you mean that excitation might pass from one dendrite to another through adjacent membranes? Tight-gap-junctions—or what has been earlier called membrane fusions—have not so far been observed in the normal neocortex. In some human biopsy material taken from epileptic foci the impression was gained that the normal distance between adjacent dendrite or dendrite-glial membranes has decreased (HÁMORI and PÀSZTOR 1970), but the possibility of artefacts cannot be ruled out completely.

BRAZIER: We have no direct evidence from the human brain, but I would certainly consider all the types of synapses that you mention as possible routes for spread of the discharge.

GERIN: Did you have any opportunity to compare coherence between the same points, the same electrode situations, in epileptic patients on the one hand and in non-epileptic ones on the other? If so, was there a significant difference between them as far as the inter-seizure rhythms are concerned?

BRAZIER: Yes indeed. As I mentioned in my paper, we have had the opportunity to record from 3 non-epileptic patients and find that their data agree well with that from the interictal periods of our epileptic cases.

# General Discussion

PETSCHE: As to the phenomenon we have been discussing these last two days, we don't know what its basic mechanisms are nor what we exactly understand by it, nor what term we should choose to characterize it in the best possible way. Let me, therefore, first outline a few theoretical considerations that may perhaps help us to see a few key questions.

Scientific thinking operates with *concepts*. For scientific research it is useful to consider each concept as being composed of two parts, a *core concept* and a *zone of vagueness* around it. The core concept contains those attributes characteristic of the concept and indispensible to its unambiguous definition. The zone of vagueness around it is formed by diffuse and equivocal attributes which may prejudice the precision of the concept. The proper aim of science is to classify the phenomena discovered by means of concepts and to restrict the conceptually uncertain zones of vagueness around these concepts.

Thus, our first task should be an attempt at elaborating the core concept of the phenomenon "synchronization". This is most difficult, as the very criterion to which this concept owes its denomination, namely synchrony, is not correct, as has been shown by most of the papers presented at this Symposium. The high degree of vagueness of the concept of synchronization manifests itself in the pleonastic word "hypersynchrony" that is still very common among electro-encephalographers.

Before attempting to look for a proper definition, one has to define the kind of phenomena denoted as synchronized. The term is not used unequivocally by the electroencephalographers. One part of them understands by synchronized activities only phenomena occurring on the whole skull with uniform shape, such as the classical spike-and-wave pattern. Another part also considers uniform EEG patterns on a limited zone as being synchronized. Nobody has ever tried to define the size of the region concerned and a question such as how large may be the area taken by a local spike in order that the event can be considered as synchronized cannot be answered.

Any attempt to define the concept in question by an approach from this direction will fail. It reminds one of the joke about how many bristles are needed for an instrument to be called a brush.

Another approach seems to be more profitable, however. Synchronization may be defined by postulating elementary units of the EEG. It was COOPER and *collaborators* who were among the first to have thought along these lines (1965). From simultaneous recordings from the human cortex and the skull, these authors came to the conclusion that each potential difference as seen in the EEG may be due to a large number of elementary events being active at roughly the same time. We follow this hypothesis.

The phenomenon to be described has to have several properties. One of these is its graded character. If an EEG potential is composed of a large amount of more or less well timed unitary events instead of being a homogeneous individual phenomenon—there is no doubt that the degree of synchronization of a normal trace should be lower than that of an EEG in an epileptic seizure. In the first case, the EEG potentials are composed of a great number of different elementary activities in different regions, whereas, in a seizure, large regions of the brain surface display almost uniform electrical discharges.

Whether or not the *voltage* of EEG waves, as in seizures, may be a proper parameter for estimating the degree of synchronization, as several electroencephalographers believe, must remain undecided.

It is true that potentials of high voltage may generally be considered as being built up by an assembly of unitary events being rather exactly timed—this was shown by us to be the case with penicillin spikes. However, the number of active elements may also be a basic factor for an EEG potential to be large or small. Therefore, as a possible determinative parameter of the synchronization phenomenon, the degree of potential difference seems to be out of the question.

The same holds true for the *wave duration*. It is true that most electroencephalographers confer a higher degree of synchronization to a generalized delta pattern than, for instance, to a beta EEG. But I do not think this criterion is of any value for better determining the concept of synchronization; its entire uselessness for this purpose becomes immediately obvious if one puts the question: which of the two events have a higher degree of synchronization, the "spike" or the "wave"?

However, one point most electroencephalographers agree on is that the *uniformity of a pattern* at different recording points may be used

as an indicator for synchronization. I would like to propose, therefore, to define synchronization, for practical electroencephalography,
as the degree of *"synmorphism"* of the EEG pattern found at two
recording sites. No spatial parameter is involved in this definition
so that "synmorphism" can extend over areas of different size.

For exact scientific use, however, we need a more accurate definition
for the degree of synmorphism, a dimension that may be expressed
by numbers. Let us, for the moment, call this parameter "S" and
consider the possibilities of "S" taking on different values between
0 and 1.

If "S" = 0, then, three different situations may be thought of as
causal:

a) no EEG is present at either recording point: the EEG generators,
if active at all, may be working at random;

b) an EEG is present only at one recording point, the other being
inactive;

c) an EEG is indeed present at both recording points, but these two
activities bear no relation whatsoever to each other. The two EEG
patterns are completely different.

0 < "S" < 1: an EEG is present in both recording sites; there is
also some similarity between the two records.

"S" = 1: the EEG is identical in shape (*i.e.*, it has the same frequency
spectrum) in the two recording points.

Such a definition makes possible an exact quantitative expression of
the degree of similarity between the two recording points. Mathematically, this concept is well defined by the coherence function.

The theoretical and even clinical usefulness of this concept has been
well demonstrated by Dr. Brazier's work. It seems to me that the
proper use of the coherence function may result, one day, in a very
profitable tool for determining the functional connections between
different regions of the brain. It may be that the study of coherence
will eventually bring us more knowledge about the instantaneous
functional interrelationships within the brain and may be able to
represent a physiological and very effective "neuronography" which
Dusser de Barenne's method with locally applied strychnine failed
to be.

I should like you to discuss this proposal.

Shaw: The relationships between two signals may be measured by
means of the correlation function. If the bandwidth of the signals
is reduced by narrow band filters we can eventually get the cor-

relation function for one frequency component. If we do this for all frequency components, we get correlation as a function of frequency, which is the coherence function. Also, as the bandwidth is made smaller, the correlation functions approach cosine waves so that one also gets phase as a function of frequency, which is the phase spectrum.

BRAITENBERG: Dr. SHAW, could you explain in a few words, what is the advantage of using the coherence function compared with cross-correlation?

SHAW: It depends on what questions one wants to ask. Sometimes the correlation function is sufficient. Usually the individual signals will have components over a wide spectrum. It may be that for a pair of signals, there is little correlation at one part of the spectrum whilst at some other band of frequencies the correlation is quite large. The coherence function will separate these; in other words it will indicate the degree of correlation as a function of the frequency components.

BRAZIER: Dr. SHAW has shown us that the choice of method we should use (cross-correlation or coherence) depends on the question we are asking of the data. Cross-correlation examines the time-series as a whole. Coherence examines each frequency band for the activity common to that band, the width of which you can choose when you instruct the computer. For the slides shown here, I have used a bandwidth of 2 cps throughout a whole spectrum of frequencies from 2 to 30 cps. This search for coherences of individual frequency bands is useful when the spectra of the two activities under examination are quite different. An example was the data I showed on the coherence between the amygdala and the dorsal medial nucleus. Their full spectra and their dominant frequencies were quite different but there was one frequency band in which their wave-trains cohered well— that centering at 6 cps.

VERZEANO: One of the basic characteristics of what we call synchronization is, precisely, that neurons and groups of neurons don't do the same thing at the same time: They fire in a specific order, within the pathway of circulating activity.

SCHERRER: It would perhaps be useful to have different words to describe different types of synchronization. Simultaneity of two spikes is rather different from simultaneity of two waves and simultaneity of two stages of an epileptic seizure.